ATLAS *of* MATERIAL WORLDS

Atlas of Material Worlds is a highly designed narrative atlas illustrating the agency of nonliving materials with unique, ubiquitous, and often hidden influence on our daily lives.

Employing new materialism as a jumping-off point, it examines the increasingly blurry lines between the organic and inorganic, engaging the following questions: What roles do nonliving materials play? Might a closer examination of those roles reveal an undeniable agency we have long overlooked or disregarded? If so, does this material agency change our understanding of the social structures, ecologies, economies, cosmologies, technologies, and landscapes that surround us? And, perhaps most importantly, why does material agency matter? This is the story of the world's driest nonpolar desert, pink flamingos, and cerulean blue lithium ponds; industrial shipping logistics, pudding-like jiggling substrates, and monuments of mud; galactic bodies, radioactive sheep, and the yellowcake of uranium.

Put simply, this book dares readers to see the world anew, from material up. *Atlas of Material Worlds* offers this new relationship to our host environment in a time of mounting crises—accelerating climate change, ballooning socioeconomic inequality, and rising toxic nationalism—uniquely telling materialist stories for practitioners and students in landscape, architecture, and other built environment disciplines.

Matthew Seibert is Assistant Professor of Landscape Architecture at the University of Virginia and former co-founder of Landscape Metrics, a visualization studio that specialized in data and design. Beyond his present studies in the agency of nonliving materials, his work employs representation as interrogative and speculative tools, from the employment of game engines as new model systems to study the experience of place, to the intervention within historical trajectories by crafting rich parafictions as both critique and potential future.

"This beautiful and insightful collection stretches the idea of the atlas to offer a meditation on the material elements with which we are historically entangled. Just listen to the names of the chapters—uranium, lithium, clay, crude, sand, mud, metabolite—to hear the resonance of industrial worlds in motion. Through a fantastic array of images, maps, and words, the atlas offers stories that need to be told."

- Anna Tsing,
co-editor, *Feral Atlas: The More-than-Human Anthropocene*

"*Atlas of Material Worlds* delves into the earth's lithosphere, presenting a series of mineral narratives that animate the so-called inanimate world. Matthew Seibert's expertly edited and illustrated volume challenges the capitalist extraction enterprise by mapping the very agency of elemental minerals, moving seamlessly across scales from the microscopic to the cosmic. Much like Alexander von Humboldt's 1845 *Kosmos*, the atlas seeks to radically redefine relationships between the biosphere and the geosphere, while asserting that we humans are inseparably, fluidly entangled with the vibrant matter of our planet. This timely volume repositions elemental materials as dynamic agents of power, and calls for new materialist assemblages to address our crises of climate, health, and inequity."

- Catherine Seavitt Nordenson,
City College of New York

"Matthew Seibert's *Atlas of Material Worlds* reorients us by asking us to consider the earth from the perspective of seven materials—uranium, lithium, clay, crude oil, sand, mud, and metabolite—seven nonhuman protagonists whose fascinating stories take us far from home and deep into our own bodies. Through radical cartography, image, and text, Seibert and his fellow landscape architects map out alternative, non-utilitarian, non-anthropocentric ways of thinking and being in our world, that, if we take this new materialist sensibility seriously, may just lead us away from the brink of climate catastrophe."

- Susan Barba,
author of *geode*

Mapping the Agency of Matter

for

a New Landscape Practice

ATLAS

of

MATERIAL
WORLDS

Edited

by

Matthew Seibert

Routledge
Taylor & Francis Group

LONDON AND NEW YORK

First published 2021
by Routledge
2 Park Square, Milton Park, Abingdon, Oxon OX14 4RN

and by Routledge
52 Vanderbilt Avenue, New York, NY 10017

Routledge is an imprint of the Taylor & Francis Group, an informa business

British Library Cataloguing-in-Publication Data
A catalogue record for this book is available from the British Library

Library of Congress Cataloging-in-Publication Data
A catalog record has been requested for this book

ISBN: 9780367624163 (hbk)
ISBN: 9780367624156 (pbk)
ISBN: 9781003109358 (ebk)

Typeset in Avenir Book
by Matthew Seibert

Publisher's Note
This book has been prepared from camera-ready copy provided by the editor.

To my students, who have taught me as much I them;

To my teachers, who have endured my lasting obstinacy;

To my parents, Bill and Janet, who gifted me opportunity;

To Erica, my love and champion,

who won't let me plant the basil where I want;

And to the world's landscapes, composed of living and nonliving lives;

may we finally look upon you as kin, equal and expressive in voice.

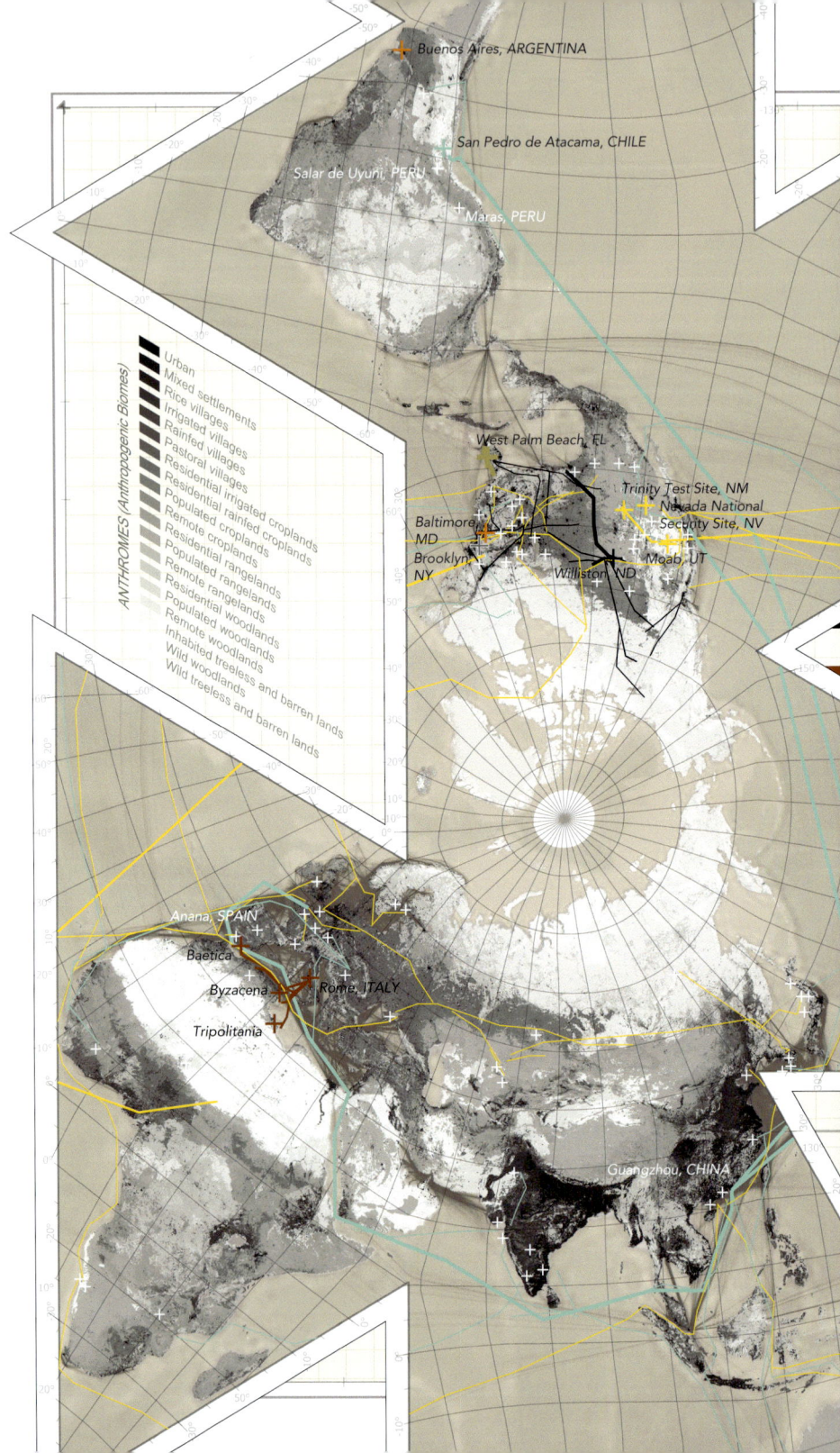

ANTHROMES (Anthropogenic Biomes)

Urban
Mixed settlements
Rice villages
Irrigated villages
Rainfed villages
Pastoral villages
Residential irrigated croplands
Residential rainfed croplands
Populated croplands
Remote croplands
Residential rangelands
Populated rangelands
Remote rangelands
Residential woodlands
Populated woodlands
Remote woodlands
Inhabited treeless and barren lands
Wild woodlands
Wild treeless and barren lands

Buenos Aires, ARGENTINA
San Pedro de Atacama, CHILE
Salar de Uyuni, PERU
Maras, PERU

West Palm Beach, FL
Trinity Test Site, NM
Nevada National
Security Site, NV
Baltimore, MD
Brooklyn, NY
Williston, ND
Moab, UT

Anana, SPAIN
Baetica
Byzacena
Rome, ITALY
Tripolitania

Guangzhou, CHINA

COVER + TABLE OF CONTENTS DATA SOURCES: Map Layer: Ellis, E. C., K. Klein Goldewijk, S. Siebert, D. Lightman, and N. Ramankutty. 2010. Anthropogenic transformation of the biomes, 1700 to 2000. Global Ecology and Biogeography 19(5): 589-606; Benjamin Halpern, et al. Raw Stressor Data: A Global Map of Human Impact on Marine Ecosystems, 2008. Knowledge Network for Biocomplexity. doi:10.5063/ F1JW8C4R; References: North America Pipeline & Oil and Gas Infrastructure Proposals. January 2014. Mapped by FracTracker Alliance on FracTracker.org; Fractracker Webapp. Mapped by FracTracker Alliance on FracTracker.org. Retreived October 31, 2020; World Nuclear Association, "Uranium Mining Overview" (September 2020). https://www.world-nuclear.org/ information-library/nuclear-fuel-cycle/mining-of-uranium/uranium-mining-overview.aspx; World Nuclear Association, "Storage and Disposal of Radioactive Waste" (March 2020). https://www.world-nuclear.org/ information-library/nuclear-fuel-cycle/nuclear-waste/ storage-and-disposal-of-radioactive-waste.aspx; Mining Technology, "The 10 biggest uranium mines in the world" (November 3, 2013). https://www.mining-technology. com/features/feature-the-10-biggest-uranium-mines-in-the-world/; David Coffin and Jeff Horowitz, "The Supply Chain forElectric Vehicle Batteries" (December 2018). Journal of International Commerce and Economics, United States International Trade Commission; Mining Technology, "Top ten biggest lithium mines in the world" (August 30, 2019). https:// www.mining-technology.com/features/top-ten-biggest-lithium-mines/
Research Assistant: Theodore Teichman

I set out on this project something like five years ago now, likely six by the time this book makes it to digital shopping carts and those few physical shelves still enduring in an era increasingly immaterial. Back then it felt less daunting to see it as an edited volume, drawing and leaning on the bodies of knowledge being actively created by colleagues, former bosses, and prior teachers. I can't feign it was always easy to corral these peers ricocheting across personal and professional lives, especially as the COVID-19 pandemic set in and any semblance of routine was upended, but I wouldn't have had it any other way and am indebted to their perseverance, rigor, and creativity in making this collection of works real. Clearly this atlas would not be what it is without their unique and powerful voices giving form, texture, and movement to a suite of nonliving materials. In a field often driven by ego and competition, I am grateful to be able to call the contributors of this collection friends. Relationships I hope to continue far into the future.

I would be remiss to not acknowledge my foreword and afterword authors independently. Asked to respond to the developing manuscript and accompanying graphics as the submission deadline barreled ever closer, they performed magnanimously. Stacy Alaimo, even under the ballooning demands of teaching during a pandemic and an apocalyptic West Coast fire season, crafted a most thoughtful and insightful entrance to the collection. I am sincerely appreciative for the monumental obstacles she overcame to make such a meaningful contribution. Thomas Woltz, forever on the road, sensitively preaching the landscape gospel to an ever-growing audience of admirers, tied many of the loose ends of the project together in the afterword through the important lens of landscape architectural practice. Truly a maker and builder of place. Thank you.

I am also thankful to the Dean of the University of Virginia's (UVA) School of Architecture, Ila Berman, for her multiple years of professional and financial support for this project. Without her assistance I would not have been able to travel to numerous sites across three continents for meaningful research of place. Blistered feet, sunburnt skin, bleeding hand from an Andean fox bite, this is the difference between a mere description and a story with the power to catalyze change.

Without Dean Berman and Bradley Cantrell, Chair of Landscape Architecture at UVA, I wouldn't have had the opportunity to work with those who this project is perhaps most indebted to: my graduate research assistants over the last two and a half years, when the vast majority of graphic production and design took place. Xiaonian (Vida) Shen, my first research assistant upon arriving at UVA, was instrumental in the early conceptualization and representation of the Crude chapter with her rapid productivity and fearlessness. Gaelle Gourmelon contributed immensely to the Clay chapter, with her meticulous attention to detail and admirable organization. Tian Wang has perhaps been part of the project the longest, contributing to most all of the book's graphics. Her efficiency and great skill in visualization has proved without parallel. Leah Kahler led the early efforts with the Mud chapter, moving across GIS and particle simulation with great fluidity and passion. Theodore Teichman, a polymath of incredible breadth, dug deep for Crude and applied his bottomless creativity and scientific rigor to Metabolite. Chloe Nagraj came in the final summer to finalize the Sand chapter and led the effort to pull all the various materials together in an actual book form. Aleksander De Mott, to which I am in deep gratitude for coming in during the final months to shoulder the monotony of citations, but also for flexing his creative muscles on several lingering drawings that proved particularly challenging. And lastly, Fanke Su, who aided the compilation of the index with grit and grace. I learned as much, if not more, from my students than I taught them. And I will be forever appreciative of their tolerance of my often unrealistic specifications and growing stress during the final weeks of production. Thank you.

I would also like to thank those who I met during and following my travels: David Pacaur, kind-hearted salinero at the Salt Pans of Maras who taught me what can only be learned by those who've lived on their land for generations and who, in turn, requests the lyrics of American pop songs from me to this day; Ramón Morales Balcázar, Chilean activist, organizer, and co-founder of Fundación Tantí, fighting for a better future; Jorge Muñoz Coca, Atacameño Leader of the Solcor Community in Antofogasta and member of Observatorio Plurinacional de Salares Andinos, understanding life as a generational project.

My gratitude also goes out to those at Routledge, an imprint of Taylor & Francis Group, who helped shepherd this to press: Grace Harrison, Sean Spears, Kathryn Schell, and Adam Guppy. And to independent copyeditors Kirsty Kay and Emily Sekine, who read over the introduction and lithium chapters.

And a final thanks to Erica Hellen, who—with the heavy panting of Mr. Fox ever near in our one-room studio—meticulously copyedited pieces of the volume and put up with my neologisms and affection for sentence fragments. Love you two.

Thank you to all who have played a role in this atlas, big and small. Rarely is a book ever the product of a singular individual, especially one that calls for the valuing of all things. This atlas is authored as much by the sands, soils, and elements of the Earth as it is by us walking, talking assemblages of its matter.

by Stacy Alaimo

THE ATLAS AT HAND:
Mapping as New Materialist Practice, Politics, + Design

Few things thrill me a much as an atlas, sparking wanderlust and promising adventures while keeping complete disorientation at bay. Despite the appeal, it is difficult to relate the flat pages of the road atlas we hold in our hands to the brawny Greek figure of Atlas who spans the oceanic depths, earth, and sky. Atlas, often depicted as struggling under the weight of the world, was originally condemned to lift the heavens. He is a "fearful" figure in Emily Wilson's translation of *The Odyssey*, "who holds up the pillars of the sea, and knows its depths—/those pillars keep the heaven and earth apart."[1] When considering what an atlas could do or how it should proceed in the anthropocene, both of these figures may be all too human, anthropomorphically and heroically figured on the one hand, flat, distant, and fueled by touristic petro-mobilities on the other. Moreover, Western epistemologies, which tend, like Atlas, to hold things apart, cannot adequately convey the interacting agencies and the multispecies perspectives necessary for environmentally oriented mappings that defamiliarize our anthropocentrism. Wallace Stevens begins his famously enigmatic modernist poem, "Thirteen Ways of Looking at a Blackbird," with an eerie scene: "Among twenty snowy mountains,/The only moving thing/Was the Eye of the blackbird."[2] Stevens stages human practices of ordering the world, using the blackbird as a geographical and conceptual coordinate for thirteen different mappings. Yet the blackbird's own potential for perception initiates the poem, keeping the human surveys and ruminations that follow in check. Craig Santos Perez's recent poem "Thirteen Ways of Looking at a Glacier," echoes Stevens, with a more explicitly environmentalist message about climate change, concluding with this uncanny, intimate moment, in which even our bodily and affective warmth becomes culpable, as it makes massive glaciers—ecosystems, habitats, polar ice caps—vanish: "The last glacier fits/in our warm hands."[3] In the face of climate change, industrialized peoples become Atlas, bearing the burden of the earth, not the heavens, in our hands.

New materialist theories, which scramble the Western dualisms of nature and culture, and insist on the significance and agency of things and substances that are often undetected, ignored, or placed off stage, pose particular challenges for mapping. Where does one begin when all sorts of perspectives, beings, places, elements, xenobiotic chemicals, and climates could be acting within or altering the frame? The entangled, interconnected environmental, economic, and political systems of the anthropocene demand that, even in our everyday lives, we attempt to make sense of what is at hand, through mapping transcorporeal[4] connections across our own bodily inhabitations, as they intersect with social, economic, and planetary systems. Such new materialist mappings cross disciplines, shift scales, and grapple with scientific and political modes of capture, mediation, and transfer. We cannot externalize a world to be mapped, as our own anthropocene bodies are imbued with substances and material agencies that interconnect us with wider forces, enmeshing the knower with what they seek to know. Moreover, for Western settler-colonialists, our sense of mapping is always already suspect, as global visions are complicit in colonialist violence and capitalist routes of plunder. David A. Chang, reminds us that "all geographical knowledge is perspectival and relative," as he analyzes how J. H. Kanepu'u proposed in a Hawaiian newspaper in 1877, "a distinctly Hawaiian and ocean-centered perspective on the world, a perspective that countered the Western, colonialist, and land-centered perspective."[5] Karin Amimoto Ingersoll, *in Waves of Knowing*, a Kanaka (native Hawaiian) epistemology, argues that the "power of a seascape epistemology lies in its inability to be mapped absolutely, and its required interaction with the intangible sea."[6] While indigenous epistemologies do not need new materialism, which developed as a critique of Western philosophies, new materialisms can respectfully echo indigenous thought, in this case, the sense of mapping as a process that is neither removed nor absolute, but palpably in contact with material worlds and mobile forces.

As I hold this atlas in my hands, albeit at this point in a merely digital form, I relish the gorgeous design, which reminds us that while the eloquent essays may draw upon scientific and other captures of materiality, this is no transparent window onto the world, but instead, a making, a poesis. The salty aperitif, in Matthew Seibert's introduction, a sensual, experiential, and historical account, entices readers into "ontological attunement(s)." The aim is not epicurean, however, but rather to ask, Seibert writes, "abnormal questions about very normal materials," in order to trace the "complex network of relations across varied scales of space and time," adequate for new materialist political ecologies (ibid.) While some modes of new materialism have been critiqued as not being properly political, the essays in this volume demonstrate how inquiry into unforeseen interconnections can reveal matters of political concern. As discussing Diné miners and uranium, Denise Hoffman Brandt argues that without maps that refuse the division between nature and culture, "we are doomed to repeat errors and injustice." Seibert's conclusion, brilliantly traces ethical practices and scales of intervention across vital political movements, including Buen Vivir, indigenous movements, and the rights of nature, underscoring their resonance with new materialist practices that embrace not one world but the pluriverse.

In consumerist capitalism, it is not easy to maintain our focus on matters that are not readily apparent. Elizabeth Hénaff's scintillating "Coda" warns that it's "far too easy to avoid the history of toxicity and extraction when gossiping over a boulevardier," for example. She shifts our gaze back to many of the materials in the volume, magnifying them in order to reveal how microbial assemblages are at play in many of the landscapes, sites, and materials we may routinely overlook, while urging us to develop and design more "cross-species collaborations." In this era of the Sixth Great Extinction, every built environment, every human practice, and every design need be propelled by the possibility of making space for a multitude of species and forms of life. A new materialist atlas is already and always at hand, while forever recalibrating and expanding to map new beings, assemblages, and interrelations.

ENDNOTES

1   Homer, *The Odyssey*. Translated by Emily Wilson. New York: Norton, 2018: 107.
2   Wallace Stevens, "Thirteen Ways of Looking at a Blackbird," Poetry Foundation, https://www.poetryfoundation.org/poems/45236/thirteen-ways-of-looking-at-a-blackbird: lines 1-3.
3   Craig Santos Perez, "Thirteen Ways of Looking at a Glacier," in *Habitat Threshold*. Oakland: Omnidawn Publishing, 2020: 20.
4   See Stacy Alaimo, *Bodily Natures: Science, Environment, and the Material Self*. Bloomington: Indiana UP, 2010.
5   David A Chang, *The World and All the Things Upon It: Native Hawaiian Geographies of Exploration*. Minneapolis: University of Minnesota Press, 2016: 22, 101.
6   Karin Amimoto Ingersoll, *Waves of Knowing: A Seascape Epistemology*. Durham: Duke UP, 2016: 6.

Matter feels, converses, suffers, desires, yearns and remembers.

- Karen Barad[1]

Introd

by Matthew Seibert

It matters what thoughts think thoughts. It matters what knowledges know knowledges. It matters what relations relate relations. It matters what worlds world worlds. It matters what stories tell stories.

- Donna Haraway[2]

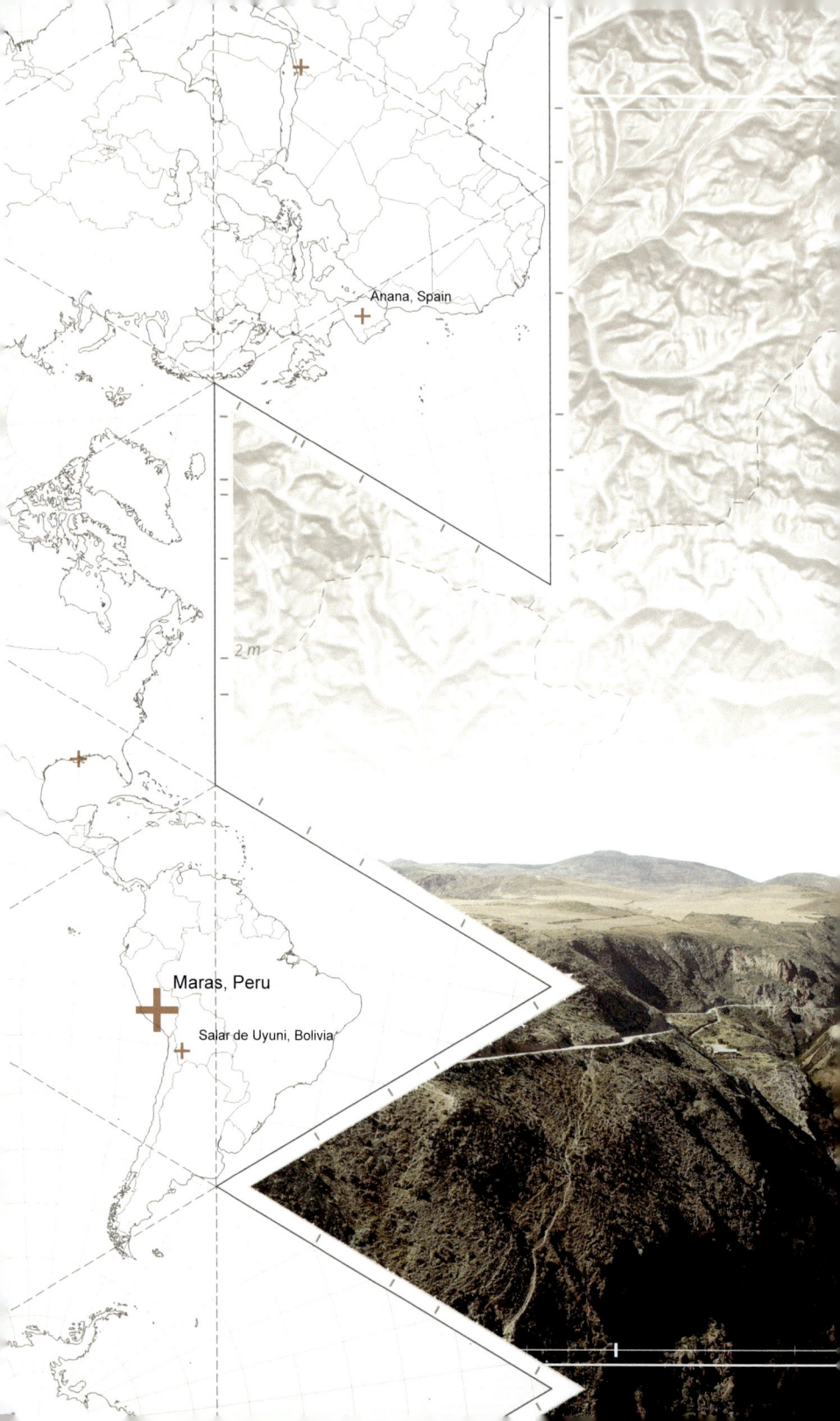

Añana, Spain

2 m

Maras, Peru

Salar de Uyuni, Bolivia

Cusco

Urubamba

Maras

Basemap sources: Esri, Airbus DS, USGS, NGA, NASA, CGIAR, N Robinson, NCEAS, NLS, OS, NMA, Geodatastyrelsen, Rijkswaterstaat, GSA, Geoland, FEMA, Intermap and the GIS user community

Earth + Clay, the Berms of Maras

alt – THE ONLY ROCK WE EAT

Smooth.

A trace of the vegetal.

An undeniable, mineral, salty percolation too.

With a translucent pearl hue and a sprig of thyme gracing its rim, I savor the creation. It's not as ornamental as my companion's across the table, held in a bespoke ceramic vessel and topped with a tuft of jet black seaweed, but its visual minimalism clothes layers of sensory revelation, realized best through tongue and nose. When sipped, the weight belies its ghostly appearance. A strange blend of lithic heaviness and rarefied thinness intimates its high-altitude origins. Perhaps a synesthetic projection, yes, but its creator likely had just this in mind.

Peruvian Chef Virgilio Martínez Véliz is known for guiding his patrons through space—from beneath the crashing waves of the Pacific Ocean, up and over the great continental divide of the Andes, and down again into the humid tropics of the Amazon jungle—never leaving their seat at the dining table. His plates are built around ecosystems, drawing on site-specific species of plants and animals, often incorporating indigenous cultural practices, to create gastronomical portals into other worlds. His cocktails are no different. In fact, the drink I am so enamored by was built around a mineral, "from the only family of rocks eaten by humans": salt.[3] But not just any salt. This was salt from Las Salineras de Maras, or the Salt Pans of Maras, in Peru's Sacred Valley, between Cusco and Machu Picchu. A sip from this coupe transports one over 300 miles and up about 10,000 feet in elevation.

Dating back to pre-Incan times, the Salt Pans of Maras are like a vast community garden carved into a valley's steep shoulders, but instead of squash and sunflowers, families and individuals farm sodium chloride in designated plots. It's quite the striking landscape to encounter. Most visitors hail a car from the nearby town, entering the valley from above. My pilgrimage, however—whistle having been wettened days earlier in Lima—had me hike in on foot from below. A brisk 30 minute ascent from the Urubamba River, snaking upwards between arid mountain folds, eventually reveals a stark white mass rising above the trail's horizon as the veritable goat path bends back against the cliff. Several minutes further, the gleaming mass grows and unfolds into an intricately sculpted painter's palette. Around 5,000 shallow ponds caked in white crystals bleed with ochres, umbers, roses, and rusts, beautifully terraced along the valley's contours.

Each step gives a satisfying crunch as one finally steps into the network of paths and pans.[4] Runnels and channels, some carved into the soil, others fashioned with overlapping ceramic shingles, burble pass and direct salty brine from a nearby spring into the sepia-hued wells. The fruiting font of a process millions of years in the making, the spring eventually emerged as tectonic plates shifted, first impounding part of an ocean into a lake, then burying the lake's salt deposit deep beneath a mountain now called Qaqawiñay, only to be slowly dissolved and brought to the surface by a subterranean stream. When a pool fills with this primeval solution, its earthen walls and floor muscled into shape by hand yet still abiding by the valley's form, the runnel is plugged, its flow continuing down to other pools or draining into the valley below. The brine, up to 2 inches in depth, then rests under the sun's heat. After a few days of evaporation, the sodium and chlorine ions—having been dissolved by the water's polarization—bond and crystallize into flakes, the flakes into geometric shards, and the shards eventually into a sedimented crust precipitated along the bottom of the pool. The process is repeated until 3 to 4 inches of salt accrues. Salineros, or salt miners, scrape this into mounds with flat rakes, shoveling them with baskets into greater mounds in adjacent dry plots to drain further, then port their harvest to storage sheds, and eventually funnel the glistening grains—a spectrum from white to pink to brown based on the fluctuating minerality of the brine and pools—into bags to be loaded on trucks and sold throughout Peru and neighboring regions (and online through Amazon, of course).

Each earthen pool, no more than a foot deep and roughly the size of a single-car garage floor, produces an average of about 400 lbs of salt per month during the dry season (March to November).[5] Possibly continuing a communal governance similar to the Incans, the pools are designated by the people of neighboring towns Maras and Pichingoto to local families for production, the number of pools and location determined by family size and seniority. Pools are inherited by children from their parents, with many working their wells from an early age into late life, just as their parents, grandparents, and Incan ancestors did before them. In fact, legend has it this all began with Ayar Chachi, one of the four siblings believed to have founded the Incan empire: after being trapped by his brothers and sisters in a cave, jealous and fearful of his power, Chachi wept salty tears—tears that continue to emerge from the ground to feed the salt pans.[6]

I can't help but think I consumed a few tears in my cocktail back in Lima, an unsubtle metaphor for modern tourism's infiltration of Indigenous Peru and a multi-century legacy of Western exploitation.

This convergence—of colonization, governance, mythology, economy, transportation, landform-making, familial enterprise, solar evaporation, salt crystallization, tectonics, material aesthetics, tourism, all the way to Michelin-starred gastronomy hundreds of miles away with cocktail in hand—revolves around a rock, is *animated* by it. This seemingly innocuous inanimate material, a simple one-to-one mineral compound of sodium and chlorine, quite prevalent around the world, dissolved in oceans and buried in mountains,

Productive Topographies

catalyzes a vibrant web of dynamics and relationships across all areas of human life. And beyond. Salt is more than just sodium, chlorine, and its useful marriage, but also cultures, economies, and ecologies.

But we seldom think of this liveliness—salt as an actor, an agent, an expression in interaction. With such influence it seems inadequate to relegate it to the impotent terms that are traditionally paired with such a material: inanimate, inert, lifeless. Salt has played a leading role in everything from the Union winning the American Civil War ("the salt famine"), to serving as insurance against energy crises (the US Strategic Petroleum Reserve, storing over 600 million barrels of crude oil in subterranean salt domes along the Gulf Coast), to housing the largest lithium reserve in the world (Bolivia's Salar de Uyuni), to winter road de-icing across the globe, to a plethora of industrial uses, to various culturally significant salt harvesting operations similar to the Salt Pans of Maras (the elevated structures of the Añana Salt Valley in Spain, the camel-led commutes to the Danakil depression in Ethiopia, etc). From there, salt wages its own subtle battles against other materials, systems, and human constructs, from salt corrosion of vehicles in northern climates, to the salinization of agricultural fields in the Aral Sea basin of Central Asia, to saltwater intrusion and the loss of land in South Louisiana. Labeling a material like salt inanimate paints a misleading picture.

Might we be missing something in how we conceive of and live with the nonliving?

THE MISSION

The story of salt in Peru, with its far-reaching connections and constructions, is an aperitif for what follows, for the kinds of questions this book attempts to energize: What roles do nonliving materials have in our lives? Might a closer examination of those roles reveal an undeniable agency we have long overlooked or disregarded? If so, does this material agency change our understanding of the social structures, ecologies, economies, cosmologies, technologies, and landscapes that surround us? Does an altered understanding change how we intersect and entangle across human, nonhuman, and nonliving systems?[7]

And, perhaps most importantly, why does this matter? How might this knowledge and outlook empower us?

By asking abnormal questions about very normal materials, we enter into complex networks of relations across varied scales of time and space. Composed as an atlas, the chapters of this book thus accustom the reader to the agency of nonliving material by assisting in the navigation of uncommonly charted territories and times. It's an ontological attunement of sorts. The agenda is a metaphysical repositioning of the physical—particularly the inanimate matter of elements and elemental materials— as equal agents of power. The purpose is to catalyze a more productive and ultimately richer way of approaching the world than current neoliberalism, or even noble environmentalism, as we accelerate into the global climate crisis. This is a project where material is understood as a type of kin that runs through us, composes us, directs us, as much as we exert ourselves upon it and its source landscapes. This kinship is achieved by rendering the agency of nonliving materials through an unusual leveling of subjects, objects, and environments into their respective elemental materiality. A new sensitivity and engagement—*a new practice*—with a world increasingly in social, ecological, and cognitive crisis is thus set into motion. Put simply, this book dares readers to see the world anew: from material up.[8]

The subtle yet paradigmatic recalibration of political ecology, from a new materialist lens, is achieved with an aesthetic caress and a tickle of entertainment through a proposed genre of *onto-cartographic* stories.[9] Fusing ontology (the study of being), by calling on a metaphysics of the relational nature behind all things, and cartography (the science or practice of map-making), by calling on the world-building and wayfinding powers of mapping, with the narrative arc and emotive power of the story, this nascent genre of onto-cartography attempts to render an ideologically flat view of the world. A flatness where humans, fungi, minerals, and things perform upon an equal playing field, where "all forces and flows (materialities) are or can become lively, affective, and signaling," often as grouped networks, assemblages, or ecologies of actors.[10] Onto-cartographic protagonists are often not human, often not even living, but rather mineral, chemical, elemental. Their stories are almost always telescopic, scalar. The characters, or materials in this case, require understanding as both the irreducible matter

Material-Cultural Entanglements

from which they come, to the larger, planetary, multi-actor assemblages and ecologies within which they live. This understanding of the nature of being, or way of seeing the world—a weaving of new materialism, actor-network theory, natureculture entanglement, and transcorporeality, to employ the argot of academia explored in subsequent pages—with an underscored focus on nonliving material is at the heart of the practical sensibility attempting to be cultivated within these pages.

It's not intuitive. It's not effortless for the reader, or writer, to approach the world in such a way, as we are up against a few millennia of human-centered thought, of a human-commanded hierarchy of agency. Moreover, how can a story, written by and in the language of humans, successfully achieve such a position of worldly flatness, of material equality? It appears futile by definition. There are a few strategies in response to this, such as illustrating the alien hybridity of human bodies with respect to the plethora of micro-organismic communities supporting their health and physiology, that "insofar as anything 'acts' at all, it has already entered an agentic assemblage" with a multitude of bacterial consortia.[11] But, in truth, a story is a medium of communication, and thus necessarily requires species-specific symbols and syntax. Still, the onto-cartographic story is a unique undertaking, attempting to capture something inherently resistant to representation—a flat ontology, or worldview, of all matter—and for that reason this book embraces a collagic technique of word and image, synthesizing quantitative analysis, qualitative representation, and radical cartography.

The chapters in this book weave a story read as much through language as through its cartographic mappings, co-constructing the narrative. They also require a loosening, a dislodging of readers' normative perceptions of space, time, and agency. This is both preparatory advice for the reader and desired result. As illustrated in Maras, Peru, salt is multitudes more than a tasty rock supporting gastronomical inventions. Salt is elemental, sodium and chlorine, whose atoms, packed tightly together in a cube to form crystals of halite (salt), were sedimented and buried by geological forces at the scale of tectonic plates and millions of years, well before humans were humans. Salt figures into the genesis of a people, bridging man and land, a way of life, an identity. Salt expresses itself as an undeniable force, challenging human-centered and human-designed worlds. Salt makes worlds too, in lively participation with other living and nonliving actors. Vast scales of time, space, and agency such as this will be navigated in the following chapters, challenging the reader's projective flexibility and imaginary prowess. Spaces from the subatomic to the galactic, periods of time from the instant to the astrogeological will be traversed. It will be exciting. A material guide will accompany, part human voice, part materialist expression, part wayfinding cartography, entangled in composition.

Although the authors of this volume mostly come from the practice and research of landscape architecture, when I pose the question "what might this knowledge and outlook afford us?" this *us* is equally the designer reading a site and its context, or speccing source materials for a project, as it is the quotidian consumer shopping for the latest iPhone, or driving through the fracking fields of North Dakota on their way to the county fair. Landscape architecture is a particularly versatile lens through which to study such multi-scalar and complicated dynamics as it is inherently interdisciplinary, navigating the social justice issues of public space, the socioecological issues of ecosystem performance and interface, the space-making and engineering of architecture, the horticulture of the garden, the economics and logistics of infrastructure, all the way to the demystification of those far off landscapes of extraction, seldom seen and rarely considered. Thus the book holds a certain irreverence for binary categorizations (designer/quotidian citizen, architect/landscape architect, garden/infrastructure, living/nonliving) in pursuit of empowering a readership of all designers of the built environment, including those untrained in the design professions.[12] As discussed in more detail momentarily, the selected materials composing the atlas guide one through a diverse collection of nature-culture entanglements, from the recreation across Florida's beaches to the lithium mining in Chile fueling the world's "green" energy transition. Investigating these domains of life—human and nonhuman—through onto-cartographic stories affords new ways of seeing, knowing, and thus acting in a complex world. Consider it the proffering of a new power.

A UNIQUE TIME ACROSS DISCIPLINES

New powers are needed when one is living through such a particularly dark time in history. It is the summer of 2020 as I write these words. A viciously divided America, fueled by a potent mix of white supremacy,[13] willful ignorance,[14] and extreme profiteering,[15] leads an increasingly nationalistic world[16] into an era of mass extinction,[17] extreme political polarization,[18] and grave climate change.[19] Great Britain voted to leave the European Union largely due to xenophobia; Alternative for Germany entered the country's national parliament, the first far-right party to do so in decades; Indian Prime Minister Narendra Modi forges on with his Hindu-nationalist agenda and anti-Muslim Citizenship Amendment Act. In his first year in office, far-right Brazilian President Jair Bolsonaro oversaw the highest one-year loss of rainforest in a decade, "a swath of jungle nearly the size of Lebanon," having opened it to industry, scaled back protections, and refused outside help in fighting out-of-control fires.[20] Adding to the volatility, the COVID-19 pandemic continues to tally hundreds of thousands of deaths, with a likely resurgence of fatalities in the fall.[21] At the risk of dating the book, this temporal acknowledgment is intentional. Though most thoughts and narratives within these pages have been forming for some time now, materializing in a more stable and rational political period, how they shape and are shaped by current events of the late 2010s is of particular relevance—even revelation. The racist and self-serving worldview of the American government during this time,[22] a festering sore mirroring similar strongman governments around the world, rings new truth in the book's chapters. And brings new implications as a result. Myopic living—short-sighted and self-centered—largely defines our present time, and in various ways is a particular target of this book's call to action: a new materialist sensibility predicated on a relational living for future prosperity of all things. It's also important to note that this period in time is not populated purely by destructive voices. It is concurrently a time of unique and powerful thinkers giving voice to progress, exhilarating paradigm shifts in science, and promising new philosophies. This book is positioned within this new landscape, both materially and conceptually. So though the book is framed by present politics' growing need for relationality—thinking across human and nonhuman agents and their interrelationships— the landscape of political ecologies it renders extends well beyond the last four years of the 2010s. Material agency has been around since material itself after all. The timelessness and omnipresence of material action is a principal point. Current events are often just punctuated proof or further

Maras Salineros

example of the everywhere, the all the time, the always of material agency. However, like a metalsmith's steel, the call for a new materialist practice is undeniably tempered in the stark inequity, injustice, and environmental destruction of today.

To better contextualize the ultimate intentions of this book, its allies and precedents, I'd like to take a moment to give form to the present landscape upon which this collection of stories stands. These contemporary events and foundational ideas, some theoretical, some practical, give specific background to the book's provocation of new worldviews. The diverse landscape thus formed is best understood through three lenses: that of politics, science, and philosophy.

Politics

In a world struggling under multiple and intersecting crises, several new, powerful, and just voices have emerged in politics. They come bearing the start of a map, braiding policy and design, with a better future as destination. None are quite so resonant these days as the youngest woman to serve in the United States Congress: Alexandria Ocasio-Cortez. Sharp, compelling, and resolute, she has given voice to a new politics propelled by her rhetorical command. Her politics aside, Ocasio-Cortez is masterfully skilled in telling a story with targeted purpose. Her unapologetic worldviews, informed by her blue-collar upbringing in one of the nation's most diverse districts, have both invigorated progressives and affronted conservatives. For example, the call for a Green New Deal (House Resolution 109,[23] co-sponsored with Senator Ed Markey) was met with palpable derision upon its voicing in early 2019. Described as a request for green job-creating proposals, or as a position that the American government should commit the country to the climate crisis at the same scale and scope as "President Franklin D. Roosevelt's New Deal" committed "the federal government to dig the United States out of the Great Depression in the 1930s,"[24] the Green New Deal has since gained considerable traction within the public discourse. This occurred slowly, in fits and starts, until widespread media began to take note. Almost all the leading presidential candidates for the 2020 Democratic Primary eventually supported such expansive, intrepid, and committed action.

Billy Fleming, Director for the Ian L. McHarg Center at the University of Pennsylvania School of Design, has taken up the torch as well:

> Of course, a request for proposals is not a plan. But FDR's New Deal was not a plan either; it was an improvisational series of programs, some of which were successful and some not, and all of which evolved over time. Similarly, the Green New Deal is a generational investment in planning and design that will radically transform the social and physical landscape of the United States. It is the biggest design idea in a century.[25]

The Green New Deal is a bold umbrella framework for specific policies and programs. It's a statement that design plays a critical role in materializing the future we want at this precise moment in history. It's als o an idea. And every good idea needs a story in order to manifest, to affect change. Or to put it in the words of Tyrian Lannister following the ruination of King's Landing, "There is nothing in the world more powerful than a good story. Nothing can stop it. No enemy can defeat it."[26] Ocasio-Cortez understands this of course, and, in collaboration with writer and theorist Naomi Klein, released the short animated film A Message from the Future in April 2019.[27] A brilliant and clever painting of the Green New Deal with a narrative reflecting back from a future date, it's strategically and conceptually grounded in the saying "You can't be what you can't see." In giving voice, image, and imaginary to a challenging but vibrant future, the temporal framing of the short film illustrates what we can be, what the American people, the world's people,[28] can achieve following brave imagination: "We can be whatever we have the courage to see."[29] A well told story holds the power of great change.[30]

Ocasio-Cortez, Fleming, and the Green New Deal are not the only bold figures and proposals calling on design for future projection. They are representative of significant, progressive, extensive calls for political action through the crafting of new imaginaries, new ways of seeing the world meant to provoke new ways of living in it. This is not new, but rather newly urgent in today's realities. The original New Deal, for instance, also visionary at the time, focused significant attention and resources on the material world by building parks, planting trees, constructing dams and massive infrastructure, and electrifying large parts of rural America. Roosevelt's New Deal "was literally stamped on the American landscape."[31] We must learn from the successes and failures[32] of past projects and dedicate ourselves to fashioning the world we want to see. Powerful voices and powerful stories are needed.

Science

Tracking alongside emergent voices and movements in politics is an exciting time in science. Recent advancements in the natural sciences and bioengineering have increasingly called into question the once firm line separating the organic and inorganic, and thus life and mere material. In addition to the highly interdisciplinary fields of systems theory, complexity theory, and chaos theory that emerged in the 1960s and revolutionized thinking across many spheres of thought—promoting concepts such as feedback loops, phase transitions, self-organization, and emergent behavior—the fields of *synthetic biology* and *mineral evolution* might best illustrate the dissolving line between life and matter.

Synthetic biology, or bioengineering, became possible with paralleling improvements in nanotechnology and molecular biology. Once one could tinker at the molecular and eventually atomic level, new techniques opened entirely new methods for medicine and the editing of living cells and organisms. This was a brave new frontier of genome editing, individually tailored medicine, and synthetic tissue cultivation, accompanied by a myriad of legal and ethical controversies. On one side of the field is the goal of "recreat[ing] in unnatural chemical systems the emergent properties of living systems, including inheritance, genetics, and evolution."[33] Meanwhile, the other side strives to "extract from living systems interchangeable parts that might be tested, validated as construction units, and reassembled to create devices that might (or might not) have analogues in living systems."[34] The former effectively results in inorganic components functioning as organic, while the latter results in organic components functioning as inorganic. Living and nonliving are thus not quite so black and white. These categories, once conceived as timeless, are now, at best, placed along a continuum, if not altogether thrown out as small-minded and short-sighted.

If synthetic biology created new knowledge through making and doing in the lab, mineral evolution shifted thought through observation and analysis. "Life arose from minerals; then minerals arose from life," states Robert Hazen.[35] Stirring controversy with its mere name when introduced in 2008 by Hazen and his team, as the word "evolution" suggests competing species, mutation, and the passing of genetic information through generations, mineral evolution is meant to communicate that "the mineralogy of terrestrial planets evolves as a consequence of a range of physical, chemical, and *biological* processes [emphasis added]."[36] In other words, through an irreversible sequence of events increasingly complex and diverse, assemblages of minerals develop. (It is also sometimes referred to as mineral ecology, or, less frequently, geobiology.) There are three processes that drive this diversification of mineral identity: 1) separation and concentration of elements as a solar system begins to amass, and as a result of planetary accretion, 2) greater ranges of temperature and pressure paired with the action of volatiles like water, carbon dioxide, and oxygen, and 3) new chemical pathways and conditions provided by living organisms.[37]

Mineral evolution can also be expressed in the words of a chronology, a veritable cosmology, or what is described as the three eras of mineral evolution pertaining to Earth. Following the Big Bang, pre-stellar molecular clouds wafted through space as microscopic dust particles performed a gauzy choreography in a rapidly expanding universe. Gravity began clumping these drifts into stars, planets, and asteroids. Hydrogen, helium, and a speck of lithium were the first and only elements, but with the heat of stars and their explosive supernova came the synthesis of minerals made possible with newly formed heavier elements. This first era increased a primordial 12 minerals into a respectable 250. Earth, newly accreted into existence, then commenced its initial mineral evolution through a varied palette of geochemical and petrologic processes within its crust. This included "volcanism and degassing, fractional crystallization, crystal settling, assimilation reactions, regional and contact metamorphism, plate tectonics, and associated large-scale fluid-rock interactions."[38] And with the first continents that resulted came an estimated 1,500 different mineral "species." This was the second era. The third and last era is when life began to self-organize within this necessarily diverse palette of minerals, elements, and chemistry. Biological processes began, multicellular life emerged, and skeletal biomineralization "irreversibly transformed Earth's surface mineralogy."[39] With the final number of minerals being a touch over 5,400, more than two-thirds are attributable to mineral-biological *coevolution.* "Turns out, establishing where geochemistry ends and life begins isn't always so easy," states geobiologist Lisa M. Pratt of Indiana University.[40] Or as Hazen puts it, "The geosphere and biosphere have become complexly intertwined, with numerous feedback loops driving myriad critical natural processes in ways that are only now coming into focus."[41] Organic and inorganic, life and matter, the line between the two is anything but stark.

Philosophy

As the once clean boundaries of life and matter are increasingly muddied by mineral evolution and synthetic engineering, and bold new political imaginaries are conjured for a climate crisis-fueled future, significant philosophical work has opened new portals into "the arts of living on a damaged planet"[42] with nonhuman and, specifically, *nonliving* brethren. Political theorist Jane Bennett's seminal work *Vibrant Matter: A Political Ecology of Things,* published in 2010, ushered in an exciting era of *new* (or *vital) materialism,* or the belief in an agency or capacity of all things—plant, animal, mineral, chemical, event—"not only to impede or block the will and designs of humans but also to act as quasi agents or forces with trajectories, propensities, or tendencies of their own."[43] This thing-power, or material vitality, as she often puts it, is a considerable expansion of the traditional definition of agency. Though a slippery word to begin with, even in a purely anthropocentric context, if one begins to look more deeply into this loaded concept of agency, clarity and consensus is tellingly absent:

> No one really knows what human agency is, or what humans are doing when they are said to perform as agents. In the face of every analysis, human agency remains something of a mystery. If we do not know just how it is that human agency operates, how can we be so sure that the processes through which nonhumans make their mark are qualitatively different?[44]

Bennett is not alone in her philosophy of looking upon nonhumans, animate and inanimate, as imbued with a

Salar de Uyuni, Bolivia
The largest lithium deposit in the world

certain liveliness and agential significance. Like conifers cross-pollinating over the wind or bacteria sharing DNA through horizontal gene transfer,[45] there are a bevy of cross-fertilizing figures preaching the new materialist gospel.

Theorist Bruno Latour set the stage for this in 1991 with *We Have Never Been Modern*, powerfully challenging the common nature/culture, subject/object, human/nonhuman dualities with lasting consequence:

> *the human, as we now understand, cannot be grasped and saved unless that other part of itself, the share of things, is restored to it. So long as humanism is constructed through contrast with the object... neither the human nor the nonhuman can be understood.*[46]

This deconstruction of "modern" dichotomies and subsequent dethroning of human exceptionalism lifted the nonhuman in stature. Power, or agency, was then explicitly bestowed to the nonhuman by Latour and his collaborators' actor-network theory, where everything exists through their ever-changing networks of relationships, be they material (relations between things), semiotic (between concepts), or simultaneously both. Additional actor-network theorists like Isabelle Stengers, Michel Serres, and Donna Haraway have further enriched this dialogue with often provocative, complicating, and entangling effect, describing these networks as rather

> *the worlds in which the axes of the technical, organic, mythic, political, economic, and textual intersect in optically and gravitationally dense nodes... function[ing] like wormholes to cast us into the turbulent and barely charted territories of technoscience.*[47]

Feminist new materialists like Jane Bennett, Karen Barad, and Stacy Alaimo have further rendered the agency of the nonhuman through often corporeal attentiveness:

> *thinking across bodies may catalyze the recognition that the environment, which is too often imagined as inert, empty space or as a resource for human use, is, in fact, a world of fleshly beings with their own needs, claims, and actions.*[48]

Last of the new materialism gospel choir, we have the speculative realists, the philosophers granting objects existence in reality independently of human conceptions, perceptions, and language. These practitioners notably include Graham Harman, Reza Negarestani, and Levi Bryant, among others. But of these, it is to Bryant—following Bennett, of course—that this book is perhaps most indebted to philosophically.

Decisively positioned that the world is composed entirely of matter—and "by 'matter,' all I mean is 'stuff' and 'things'"[49]—Bryant concludes that our reality, our world, is elementally material, and not constructed solely through signifying or discursive practices. This vision obviously includes stuff and things like "trees, rocks, planets, stars, wombats, and automobiles" but also sees ideas and concepts as possessing materiality, that these semiotic objects "only exist in brains, on paper, and in computer data banks, and that ideas can only be transmitted through physical media such as fiber optic cables, smoke signals, oxygen-rich atmospheres, and so on."[50] The world then is a shifting, almost fractal composition of material entities at a variety of scales and nested levels of structure. Bryant calls these material entities *machines* to "emphasize the manner in which entities dynamically *operate* on inputs producing outputs."[51] The Maras salt flats are thus a material entity composed of earth, brine, and human entity inputs. The output of salt it produces is an entity composed of sodium and chloride crystallizing under the infrared rays (heat) of that ball of gas in the sky—the sun. And these two elements could be broken even further into their compositional entities, their atomic particles.[52] These material entities—atoms, elements, compounds, organic compositions and organisms, etc—subsequently afford, delimit, and catalyze semiotic entities: those of ideas, concepts, and beliefs. From the indigenous cosmology of the salt ponds' creation to the regional economics of salt sales, the material matter of the salt pans of Maras expresses itself in language and culture, carried in vibrations through an all-too-material air, written in words on dried vegetal pulp, or magnetized in bits as 1's and 0's. That the seemingly intangible is "always embedded in a particular space or place, …communication takes time to travel through space and requires media to travel, and … geographical features of the material world play an important role"[53] cannot be understated when cultivating a materialist sensibility to a world grounded in matter and defined by its material-semiotic networks, feedback loops, and reciprocities.

As both a student and professor of landscape (I research and teach within the discipline of landscape architecture), this idea—belief, rather—of material relationality is consequential. Composed of biological, mineral, chemical, cultural, and processual *things* (i.e. events and processes such as storms and hydrology), landscapes not only resist definition in their apparent infinite variety, but pose great difficulty in comprehension, let alone their engagement, through their emergent, indeterminate, web-like complexities. Landscapes are not just background settings in which we act or live. Nor are they simply cultural constructs formed by our social conceptions. Rather, they're continuously pushing and pulling against and with us.[54] Not only are the plants and animals that inhabit a landscape seen as possessing formative, designerly abilities—a relatively new revelation in the sciences—but so too are the minerals of a salt flat, the sediment of a river, the dunes of a beach. The nonliving are just as elemental and creatively influential as those two-legged egos that claim ultimate supremacy. In fact, if you begin to pull back and stretch your sense of time, elongating it along the remote lines of geology and evolution, elements and their composite minerals begin to reveal themselves as the true drivers of life, the real agents of change, leaving human bodies and minds as mere product (or object). Or viewed through the eyes of the Russian-Soviet mineralogist Vladimir Vernadsky, widely considered one of the founders of biogeochemistry:

> What struck me most was that the material of Earth's crust has been packaged into myriad moving beings whose reproduction and growth build and break down matter on a global scale. People, for example, redistribute and concentrate oxygen, hydrogen, nitrogen, carbon, sulfur, phosphorous, and other elements of Earth's crust into two-legged, upright forms. …We are walking, talking minerals [emphasis added].[55]

Inorganic minerals are the life-giving force, the driving material, that bestowed the history of all human civilization.

But again, why should we care? Yes, we are complex fusions composed of the resulting matter of the universe's explosive genesis; everything is. The hope is that by seeing the world in a new way, one might carry themselves differently in this new and *familial* world. That is certainly what the environmentalists intended in the 1960s and 1970s. But new materialism, cultivating an "aesthetic-affective openness" to the agency in all things,[56] brings renewed nuance to human and nonhuman relationships, suggesting a way of life beyond environmentalism. The environment is not simply an idyllic background upon and against which we live, to be conserved and preserved, but rather part and player of a flattened view of the world where relationships and entanglements between humans and all other things exist without hierarchy. Additionally, nature is not a harmonious ideal, acting with purpose toward an end state of equilibrium, nor is it blindly chugging along like a machine following deterministic laws. It bifurcates, shifting dramatically in response to indeterminate actors (hurricanes, ice ages, arrivals of new species). There is nonlinear emergence in our worlds. Environmentalism necessarily fails in identifying what, and how, to protect under such conditions.

Lastly, we are not human. At least a majority 57% of us is not if you count human cells versus the cells of the many microorganisms that live in and on our bodies, many of which we could not survive without. Building on this, Stacy Alaimo, professor of English and materialist feminist, argues for a *trans-corporeality*, or an understanding "in which the human is always intermeshed with the more-than-human world, underlin[ing] the extent to which the substance of the human is ultimately inseparable from 'the environment.'"[57] A new, vital materialist view of the world better accommodates this truth than that of an environmentalist. Powerful ethical and political implications germinate as a result. The environment literally lives within us. Our bodies are fundamentally alien—compromised as ideal, sovereign, superior stewards as environmentalists might like to conceive of them:

> If environmentalists are selves who live on earth, vital materialists are selves who live as earth... of the various materials that they are. If environmentalism leads to the call for the protection and wise management of an ecosystem that surrounds us, a vital materialism suggests that the task is to engage more strategically with a trenchant materiality that is us as it vies with us in agentic assemblages.[58]

Environmentalism may be virtuous, honorable, productive, but it is also insufficient in a world nearing a human population of eight billion, with a warming climate that has snowballed past its point of no return, and host to a species able to engineer DNA, the building block of life.[59]

RE-LEARNING THE WORLD THROUGH ONTO-CARTOGRAPHIES, or: *Parables for the New Materialist Wayfarer*

So where does this leave us as the climate crisis envelops the globe? It's certainly tempting to "treat our collective ruin as inevitable," to resign ourselves to a "fatalistic passivity," to watch Louisiana—and other coastal regions—wash into the sea, as journalist David Wallace-Wells exemplifies in his bestselling *The Uninhabitable Earth*, and as much of the media ecosystem's climate change coverage suggests in default and inexperience.[60] Alternatives exist, however. Alternative futures, alternative philosophies, alternative stories. It's in our own self-interest. As Bennett argues, "there will be no greening of the economy, no redistribution of wealth, no enforcement or extension of rights without human dispositions, moods, and cultural ensembles hospitable to these effects."[61] This is where the story comes in—in the case of this book, the *onto-cartographic story.* Though the term "parable" might be less of a mouthful, let's first unpack the meaning and intention that the term *onto-cartography* affords and around which this book finds its form.

The term is meant to capture and evoke this idea of graphic and textual hybridity, with the act of map-making as a historical form of the story. For both maps and stories can be understood as guides in traveling across space and time.[62] The book draws on cartography's important co-constitutive powers, employing the role of the map itself in defining space through the viewer's reading as much as the map-maker's authoring of it. Or as landscape architect and theorist James Corner suggests, "Mapping is a fantastic cultural project, creating and building the world as much as measuring and describing it."[63] Additionally, building on our speculative realist guide of prior, Levi Bryant frames his philosophical work under the very title of Onto-Cartography, crafting a practice of "geophilosophy." He defines onto-cartography as the work of mapping the "relations or interactions between machines or entities and how they structure the movements and becomings of one another" in order to "expand our possibilities for intervening in the world to produce change."[64] A new form of political philosophy is thus produced, or in his words, "a work of meta-politics and meta-ethics."[65] This book holds similar intentions. Text and carto-graphics—sometimes in parallel, sometimes in divergence—equally map new ways of seeing the world and chart new ways of intervening in its assemblies of relation. Instrumentalizing a cartography both ontologically and geographically in such a way involves several key objectives: identifying and describing 1) key drivers—frequently overlooked or unseen—of spaces and/or processes; 2) the relationships and linkages between things—be they animate/animate, organic/inorganic, etc; 3) the sources of energy these things draw upon—and thus their dependencies; and 4) the outputs, products, or waste that result from these interacting materials, things, and beings.[66] Bryant's onto-cartography is not that different from how the discipline of landscape architecture—an intervener in systems—analyzes a site to inform the shaping of a design proposal. Seen from the lens of narrative craft, it's also a modern-day parable for the new materialist wayfarer. In this book, an onto-cartographic story thus renders a nuanced portrait of a material's many-tentacled entanglements, defamiliarizes one's view by lifting nonhumans and the nonliving to the level of its reader, and thereby suggests moral implications and directions in carrying oneself forward in the world.

The following stories are guided travels through unusual perspectives, times, and agencies, employing a singular material as gateway into new worlds illustrative of new morals of matter. They are intended to cultivate a universal material monism, worlding worlds[67] where patience and "sensory attentiveness" develop with respect to "nonhuman forces operating outside and inside the human body."[68] This universal material monism might be better understood as a practice. Like an ethics, this spatial-temporal-material practice aspires to foster new actionable thought. It aspires to foment an enduring reflection of action meant to ferment and effloresce over time. It is not an instantaneous switch. It will take time to reconfigure and dismantle humankind from an ontological center to merely a player—a particularly powerful player, yes—within a larger and messier milieu of being.[69] The design of the book is thus an expanded,

Salt Valley of Añana, Spain

nuanced conception of our shared world, promoting action and behavior not only respectful, but admiring, embodying, of all things, living and inanimate, unshackled by venerable yet limiting beliefs like environmentalism, and empowering a paradigmatic shift—in no uncertain terms—of current capitalistic modes of operation, production, consumption, and thinking. Let us recalibrate self-interest to one valuing all of which we are made. It is a question of survival, after all.

ATLAS STRUCTURE: The Geographic Scales and Nature-Culture Entanglements of Seven Materials

The nonhuman protagonists of this atlas of onto-cartographic stories—the materials—were identified for their socioecological entanglements woven and enmeshed across human cultures and environmental systems. Such entanglements are often called *naturecultures* in the humanities, employing Donna Haraway's lexicon,[70] while they are frequently referred to as *anthromes*, or anthropogenic biomes, in the sciences.[71] Both terms acknowledge that nature and culture are not separate, but rather deeply engaged in an evolving web of relational dynamics. As such, each chapter's material touches on one or more domain of human life, all unquestionably entangled with the nonhuman and nonliving systems in which they operate: from recreation to warfare, from the driest desert on Earth to the aqueous slurries of dredged shipping routes, from the promise of green energy to the fracking of crude oil, from Roman landfills to the microscopic landscapes of the human microbiome. They not only challenge both reader and writer in emphasizing nonhuman agency, but map the structures of entanglement and thus chart potential points of intervention.

The authors were not only selected for their compelling and distinctive voices, but as most hail from the discipline of landscape architecture, they have attuned themselves to unique landscape sensitivities over years of practice and research, and can illuminate these points of intervention with particular skill and to great effect. And as each material selection tracks a mobilization of matter and action across space, the sequence of chapters are organized along a spectrum of geographic scales, reaching the cosmographic at times even. A type of spatial fingerprint of manipulation is thus revealed, from the scale of site, to the region, all the way to the scale of the planet and cosmos. The journey starts with a bang, the Big Bang, when time and space themselves were born, before accumulating and cooling to the planetary scale. Chapters then incrementally dial down to the site over the next five voyages. Finally, as a type of coda, we'll oscillate between scales in a reflective refrain. Each author thus employs a single element or elemental material from which to craft an onto-cartographic story and render a distinctive, new materialist view of the world.

Beginning with an element named after a planet, Denise Hoffman Brandt engages with the explosive worlds of uranium in the first chapter: tracing the entanglement of war, law, health, and indigenous rights. As a pervasive lens of radioactive dust to be discovered by future archaeologists, waste buried in receptacles of unknown lifespan, animator of weapons abstract in scale, and part of star-stuff like us, uranium is everywhere. This is a tale of galactic proportion punctuated with site-specific parables of quantum mechanics and radioactive sheep.

Moving from the deserts of the US to the Martian vistas of Chile's Atacama Desert, I follow the "green" energy future of lithium to the technicolor evaporation ponds of the Salar de Atacama. The economic might of the lightest metal on the periodic table reveals a many-tentacled, scalar dynamic not without consequence. Lithium is not just a component of the batteries in our phones, or the promise of a clean electric future, it is also the material stabilizing our psyches, an element embedded in larger landscapes, and existentially tied to the health of unique ecosystems across the globe. The supply chains of consumer goods and the promise of a renewable energy future reveal deep connections to environmental destruction.

Back in the land of Big Macs, Ian Quate and Colleen Tuite had been hearing about it: a massive migration to a new post-peak-oil frontier, recession-proof, $20/hr McDonald's jobs, and gas flares that can be seen from space. Soon enough they were watching YouTube videos about how to get jobs in the oil patch, how to live in a man camp, how to live out of your truck in a Wal-Mart parking lot until you could afford to score a spot in a man camp. The endless landscape of the Bakken Formation in North Dakota provides the lens through which to consider the culture and ecology of fracking for crude, both the lifeblood and curse of modern civilization.

Salt Crystallization at Añana, Spain

Traveling back in time, Kristi Cheramie follows the tangled histories of clay across the Mediterranean. Built from the byproducts of Imperial Rome's flourishing market-driven economy and a physical record of Mediterranean trade, taxation, and the reach of an empire, the city's ancient landfill, Monte Testacccio, condenses the material record of the productive landscapes that served Ancient Rome to one object and one material: clay. Through the clay amphora vessel that facilitated the Roman Empire's transportation of olive oil, the socioecological domain of food is partially unraveled for study.

From the coasts of the Mediterranean we move to the sunny shores of Palm Beach, Florida with Rob Holmes leading us through a world constructed atop sand. As a fundamental substrate for modern life, the story of sand is intimately interwoven in the dynamics of the Gulf Coast. It is a vital ingredient in the concrete we build cities and infrastructures with; it becomes the glass that composes everything from car windshields to smartphone screens; and it is often literally substrate, the ground beneath our feet. Beaches, as Florida's park proxies and the foundation for recreation, reveal their power in identity formation and the lengths by which we attempt to control them.

Bridging across the Americas, Brian Davis plays in the mud. As the raw material for many aqueous communities, this chapter takes a low country trip through a few muddy landscapes of the Chesapeake Bay, tying mud narratives, properties, and consequences across the globe to specific tales in and around Buenos Aires, Argentina's Reserva Ecologica. Port towns, with their maritime history and high value access to the coast, are surprisingly enmeshed with the muds beneath their conduit waters.

Finally, computational biologist Elizabeth Hénaff leads us in a reflective refrain, asking the reader to see evidence of material agency through the physical history of a relationship. With microbes and products of their biological processes—metabolites—we are welcomed into a microscopic world where microorganisms act as a revealing bridge between matter and human. Brooklyn's hippest Superfund, the Gowanus Canal, enables a new lens through which to understand material agency and look back on the authors' previous materials anew. This also offers the reader a new landscape to make sense of the built environment's material role in communal health in the time of the global COVID-19 pandemic.

The book's conclusion then fleshes out in more detail what this all means not only for the design of cities and landscapes, but the design of daily lives and what a new materialist practice might mean for a world besought by crises from all sides: health, climate, and socio-economic inequity.

More than twenty years ago Bruno Latour, French philosopher and pioneer in the field of science and technology studies, asked "Are you ready, and at the price of what sacrifice, to live the good life together?"[72] He then reflected "that this highest of moral and political questions could have been raised, for so many centuries, by so many bright minds, *for human only* without the nonhumans that make them up, will soon appear, I have no doubt, as extravagant as when the Founding Fathers denied slaves and women the vote."[73]

Perhaps we might finally embark upon this long sought good life together: an embraced web of human, nonhuman, and nonliving, sacrificing domination for conviviality, autopoiesis for sympoiesis,[74] hierarchy for a leveled vibrancy. Let us cast off into the radical, novel worlds of material agency that lay the foundational insight for this life. And find new power in a new materialist practice.

ENDNOTES

1 Rick Dolphijn and Iris van der Tuin, *New Materialism: Interviews & Cartographies* (London: Open Humanities Press, an imprint of Michigan Publishing, 2012).

2 Donna Haraway, *Staying with the Trouble* (Durham: Duke University Press, 2016), 35. All rights reserved. Republished by permission of the copyright holder. www.dukeupress.edu.

3 Mark Kurlansky, *Salt: A World History* (New York: Penguin Books, 2003), 6.

4 The salt pans have since been closed to visitors entering the operational site, directed to the perimeter for sanitary purposes.

5 "Salt Mines of Maras," UNESCO World Heritage Center, August 5, 2019, https://whc.unesco.org/en/tentativelists/6412/.

6 David Paucar (Maras salinero), in discussion with the author, July 20, 2020; "Salt Mines of Maras," UNESCO World Heritage Center.

7 Jane Hutton, *Reciprocal Landscapes* (London: Routledge, 2019). Hutton's book expertly investigates some of these questions in relation

to projects in New York City, but my questions long for something more elemental, an insight into the nature of being itself.

8   This builds upon many predecessors to which I am indebted and who will be brought into discussion in subsequent paragraphs—writers, theorists, and practitioners like Donna Haraway, Karen Barad, Levi Bryant, and Jane Bennett, for instance.

9   Jane Bennett, "In Parliament with Things," in *Radical Democracy: Politics between Abundance and Lack*, ed. Lars Tønder and Lasse Thomassen (New York: Manchester University Press, 2005); Levi R. Bryant, *Onto-Cartography: An Ontology of Machines and Media*, Speculative Realism (Edinburgh: Edinburgh University Press, 2014). Bennett speaks of onto-stories while Bryant speaks of onto-cartography, or geophilosophy.

10 Jane Bennett, *Vibrant Matter: A Political Ecology of Things* (Durham: Duke University Press, 2010), 117.

11 Bennett, *Vibrant Matter*, 120. Lynn Margulis, *Symbiotic Planet: A New Look at Evolution* (New York: Basic Books, 1998). Ed Yong, *I Contain Multitudes: The Microbes within Us and a Grander View of Life* (New York: HarperCollins Publishers, 2016).

12 Ezio Manzini, *Design, When Everybody Designs* (Cambridge: The MIT Press, 2015). This will be elaborated upon in the conclusion, but for now can be read as an agreement with Ezio Manzini's *Design, When Everybody Designs* and the belief that everyone is a designer, regardless of their training, due to their actions, beliefs, and choices in the world.

13 Jamil Smith. "Violent White Nationalism is All-American," Rolling Stone, August 5, 2019, https://www.rollingstone.com/politics/ political-commentary/white-nationalism-el-paso-dayton-shootings-867603/. The Southern Poverty Law Center defines a hate group as a group of people "that have beliefs or practices that attack or malign an entire class of people, typically for their immutable characteristics; The Southern Poverty Law Center, "Hate and Extremism," Southern Poverty Law Center, accessed September 29, 2020, https:// www.splcenter.org/issues/hate-and-extremism. "In 2019, there were 940 hate groups operating in the US. There has been a 55% increase in white nationalist hate groups since 2017."

14 The Editorial Board, "America's Willful Ignorance about Black Lives," *The Boston Globe*, May 31, 2020, https://www. bostonglobe.com/2020/05/31/opinion/americas-willful-ignorance-about-black-lives/.

15 Martin Levine, "Billionaires' Fortunes Rise by $434 Billion amid the Pandemic," *Nonprofit Quarterly*, June 8, 2020, https://nonprofitquarterly.org/billionaires-fortunes-rise-by-434-billion-amid-the-pandemic/.

16 Yasmeen Serhan, "Populism is Morphing in Insidious Ways," *The Atlantic*, January 6, 2020, https://www.theatlantic.com/international/ archive/2020/01/future-populism-2020s/604393/.

17 Gerardo Ceballos, Paul R. Ehrlich, Peter H. Raven, "Vertebrates on the brink as indicators of biological annihilation and the sixth mass extinction," *Proceedings of the National Academy of Sciences* 117, no. 24 (2020), accessed July 18, 2020, DOI: 10.1073/pnas.1922686117.

18 Gordon Heltzel and Kristin Laurin, "Polarization in America: two possible futures," *Current Opinion in Behavioral Sciences 34* (May 6, 2020): 179-184, DOI: 10.1016/j.cobeha.2020.03.008.

19 Sebastian Sippel et al., "Climate change now detectable from any single day of weather at global scale," *Nature Climate Change* 10, no. 1 (2020): 35–41, accessed September 2, 2020, https://doi.org/10.1038/s41558-019-0666-7; Trisos, C.H., Merow, C. & Pigot, A.L. "The projected timing of abrupt ecological disruption from climate change," *Nature* 580 (2020): 496–501. https://doi.org/10.1038/s41586-020-2189-9. "We project that future disruption of ecological assemblages as a result of climate change will be abrupt, because within any given ecological assemblage the exposure of most species to climate conditions beyond their realized niche limits occurs almost simultaneously. Under a high-emissions scenario (representative concentration pathway (RCP) 8.5), such abrupt exposure events begin before 2030 in tropical oceans and spread to tropical forests and higher latitudes by 2050."

20 Matt Sandy, "'The Amazon is Completely Lawless': The Rainforest after Bolsonaro's First Year," *The New York Times*, December 5, 2019, https://www.nytimes.com/2019/12/05/world/americas/amazon-fires-bolsonaro-photos.html.

21 Len Strazewski, "Harvard Epidemiologist: Beware COVID-19's Second Wave This Fall," American Medical Association, May 8, 2020, https://www.ama-assn.org/delivering-care/public-health/harvard-epidemiologist-beware-covid-19-s-second-wave-fall; Lena H. Sun, "CDC Director Warns Second Wave of Coronavirus Likely to Be Even More Devastating," *Washington Post*, April 21, 2020, https://www.washingtonpost.com/health/2020/04/21/coronavirus-second-wave-cdcdirector/.

22 Adam Serwer, "The Cruelty is the Point," *The Atlantic*, October 3, 2018, https://www.theatlantic.com/ideas/archive/2018/10/the-cruelty-is-the-point/572104/?gclid=CjwKCAjw2dD7BRASEiwAWCtCb9eGFO-sdnyXlA2HHZZyaPeU5Ys8F5ToizomGmdE0WLdvvhw9ei8NxoCBV0QAvD_BwE.

23 US Congress, House, *Recognizing the duty of the Federal Government to create a Green New Deal*, H. Res. 109, 116th Cong., 1st sess., introduced in House February 12, 2019, https://www.congress.gov/bill/116th-congress/house-resolution/109/text.

24 Andrew Chatzky, "Envisioning a Green New Deal: A Global Comparison," *Council on Foreign Relations* (October, 21, 2020), https://www.cfr.org/backgrounder/envisioning-green-new-deal-global-comparison.

25 Billy Fleming, "Design and the Green New Deal," *Places Journal* (2019), accessed August 6, 2020, https://placesjournal.org/article/design-and-the-green-new-deal/.

26 *Game of Thrones*, "The Iron Throne," Episode number 6, Season 8, Directed and Written by David Benioff, D.B. Weiss, HBO, May 19, 2019.

27 Jimm Batt and Kim Boekbinder, "A Message from the Future with Alexandria Ocasio-Cortez," *The Intercept* video, 7:35, April 17, 2019, https://theintercept.com/2019/04/17/green-new-deal-short-film-alexandria-ocasio-cortez/

28 Kate Aronoff et al., *A Planet to Win: Why We Need a Green New Deal* (New York: Verso Books, 2019). This book expands upon a Global Green New Deal as well.

29 Batt and Boekbinder, "A Message from the Future with Alexandria Ocasio-Cortez."

30   Shannon Mattern, "Maintenance and Care," *Places Journal* (2018), accessed September 29, 2020, https://placesjournal.org/article/maintenance-and-care/. It's notable that these are stories that take into account the unsexy concepts of maintenance, repair, and care in a time of deterioration, ruination, and despair, "the sudden collapse of dams and bridges; the slow deterioration of power grids and sewer systems; the hacked data, broken treaties, rigged elections."; Green Stimulus Proposal, "A Green Stimulus to Rebuild Our Economy," *Medium*, March 22, 2020, https://medium.com/@green_stimulus_now/a-green-stimulus-to-rebuild-our-economy-1e7030a1d9ee. In light of an impending economic crash across the globe triggered by the novel coronavirus, The Green New Deal lead to calls for a Green Stimulus. The authors of the call to rebuild the economy identify three converging crises (the COVID-19 pandemic driving an economic recession, the climate crisis, and extreme inequality) and implore an "immediate and sustained intervention" for recovery. Billy Fleming and Alexandra Lillehei, "To Rebuild Our Towns and Cities, We Need to Design a Green Stimulus," *Jacobin Magazine*, April 18, 2020, https://jacobinmag.com/2020/04/green-stimulus-new-deal-infrastructure-buildout-coronavirus. It is argued that this stimulus should be directed at sites of maintenance and care: the retrofit and repair of schools, parks, libraries, sidewalks, and sustainable transportation networks, to name a few (Fleming, 2020).

31   Time-Life Books, *The United States* (New York: Time-Life Books, 1989).

32   It's also important to note significant flaws and failures of the New Deal: its damaging practice of draining wetland ecosystems, its racism and segregation of working groups, and its sexist recruitment and granting of opportunities to men.

33   Steven A. Benner, A. Michael Sismour, "Synthetic biology," *Nature Reviews Genetics* 6, no. 7 (2005): 533, https://doi.org/10.1038/nrg1637.

34   Ibid.

35   Robert Hazen, "Mineral Fodder: We May Think We Are the First Organisms to Remake the Planet, but Life Has Been Transforming the Earth for Aeons," *Aeon*, June 24, 2014, https://aeon.co/essays/how-life-made-the-earth-into-a-cosmic-marvel.

36   Robert M. Hazen et al., "Mineral evolution," *American Mineralogist* 93, no. 11–12 (2008): 1693, https://doi.org/10.2138/am.2008.2955.

37   Ibid., 1693–1720.

38   Ibid., 1693.

39   Ibid.

40   Colin Nickerson, "The Blurry Line between Life, Nonlife," *The Boston Globe*, January 12, 2009, http://archive.boston.com/news/science/articles/2009/01/12/the_blurry_line_between_life_nonlife/.

41   Hazen, "Mineral Fodder."

42   Anna Tsing, Heather Swanson, Elaine Gan, and Nils Bubandt, eds., *Arts of Living on a Damaged Plant: Ghosts and Monsters of the Anthropocene* (Minneapolis: University of Minnesota Press, 2017).

43   Bennett, *Vibrant Matter*, viii.

44   Ibid., 43.

45   Horizontal gene transfer is the sharing of genetic material between organisms laterally, in contrast to vertically from parent to offspring. This lateral movement can occur through multiple mechanisms: transformation, transduction, bacterial conjugation, and gene trans-

fer agents (virus-like elements). It is important in many organism's evolution.

46 Bruno Latour, *We Have Never Been Modern*, trans. Catherine Porter (Cambridge: Harvard University Press, 1993), 136.

47 Donna Haraway, "A Game of Cat's Cradle: Science Studies, Feminist Theory, Cultural Studies," *Configurations* 2, no. 1 (Winter 1994): 64, https://muse-jhu-edu.proxy01.its.virginia.edu/article/8029.

48 Stacy Alaimo, *Bodily Natures: Science, Environment, and the Material Self* (Bloomington, IN: Indiana University Press, 2010), 2.

49 Bryant, *Onto-Cartography*, 6.

50 Ibid.

51 Ibid.

52 Whether these atomic particles end up being composed of even smaller entities, be they energy, strings, or something else, is beyond the scope of this text.

53 Ibid, 7.

54 John Wylie, "Depths and Folds: On Landscape and the Gazing Subject," *Society and Space* (August 1, 2006), https://doi.org/10.1068/d380t

55 Russian scientist Vladimir Ivanovich Vernadsky, quoted in Lynn Margulis and Dorion Sagan, *What is Life?* (Berkeley: University of California Press, 1995), 49.

56 Bennett, *Vibrant Matter*, x.

57 Alaimo, *Bodily Natures*, 2.

58 Bennett, *Vibrant Matter*, 111.

59 "Greenhouse Gas Emissions: Global Greenhouse Gas Emission Data," United States Environmental Protection Agency, accessed September 30, 2020, https://www.epa.gov/ghgemissions/global-greenhouse-gas-emissions-data#:~:text=Since%201970%2C%20CO2%20emissions,increase%20from%201970%20to%202011. "Since 1970 [the first Earth Day], CO_2 emissions have increased by about 90%, with emissions from fossil fuel combustion and industrial processes contributing about 78% of the total greenhouse gas emissions increase from 1970 to 2011."

60 Fleming, "Design and the Green New Deal." To use the words of Billy Fleming imploring designers, specifically landscape architects, to re-politicize the profession in line with Frederick Law Olmsted and Ian McHarg and re-value the public sector in tackling the climate crisis at the scale and scope that is required.

61 Bennett, *Vibrant Matter*, xii.

62 Denis Cosgrove, ed., *Mappings* (London: Reaktion Books, 2011); Daniel Rosenberg and Anthony Grafton, *Cartographies of Time: A History of the Timeline* (New York: Princeton Architectural Press, 2012). See Cosgrove for a larger discussion of socio-spatio-temporal dynamics captured in the practice of cartography, or for more temporally-specific examples see Rosenberg.

63 James Corner, "The Agency of Mapping: Speculation, Critique and Invention," in *The Map Reader: Theories of Mapping Practice and Cartographic Representation*, ed. Martin Dodge, Rob Kitchin, and Chris Perkins (John Wiley & Sons, 2011), 89.

64 Bryant, *Onto-Cartography*, 7.

65 Ibid., 8.

66 Ibid.

67 Donna Haraway insightfully speaks of situated knowledges, an epistemology contending that people, scholars, scientists can produce

knowledge with greater objectivity and effect by openly acknowledging and synthesizing the contingency of their own position—and thus the contingency of their arguments and knowledge productions—within their world, their discipline, their community. Situated knowledge is an integral part in how individuals world worlds, or, in other words, how socioecological environments (Haraway's nature-cultures) are co-constituted by ourselves and those around us, other human, nonhumans, and the nonliving.

68 Bennett, *Vibrant Matter*, xiii.
69 Haraway, *Staying with the Trouble*. This shares many parallels to Donna Haraway's Chthulucene, an epochal vision for a future of potent and unruly multispecies "living-with and dying-with" in "a fierce reply to the dictates of both Anthropos and Capital." The Chthulucene is thus positioned in contrast to humankind's dominance within the widely used term the Anthropocene, and capital's dominance within the Capitalocene, a rarer term speaking to the uneven responsibility of wealthier populations and economic systems to global carbon emissions and the climate crisis.
70 Donna Haraway, *The Companion Species Manifesto: Dogs, People, and Significant Otherness* (Chicago: Prickly Paradigm Press, 2003); Haraway, *Staying with the Trouble*.
71 Erle C. Ellis and Navin Ramankutty, "Putting people in the map: anthropogenic biomes of the world," *Frontiers in Ecology and the Environment* 6, no. 8 (October 1, 2008), https://doi.org/10.1890/070062. The term anthrome was coined by Ellis and Ramankutty as a more accurate way of describing humans' "inextricable intermingling" in a global ecology.
72 Bruno Latour, *Pandora's Hope* (Cambridge: Harvard University Press, 1999), 297.
73 Ibid., 297.
74 Haraway, *Staying with the Trouble*. Sympoiesis is Donna Haraway's proposed term, adapted from Beth Dempster, as replacement for Chilean biologists Humberto Maturana and Francisco Varela's concept of *autopoiesis* introduced in 1972. While autopoiesis refers to the self-maintaining and self-reproducing ability of cells and organisms, sympoiesis argues that no life form can make itself, but is rather made through a collaboration and collective of many.

BIBLIOGRAPHY

Alaimo, Stacy. *Bodily Natures: Science, Environment, and the Material Self.* Bloomington, IN: Indiana University Press, 2010.

Aronoff, Kate, Alyssa Battistoni, Daniel Aldana Cohen, and Thea Riofrancos. *A Planet to Win: Why We Need a Green New Deal.* New York: Verso Books, 2019.

Benner, S., A. Sismour, "Synthetic biology." *Nature Reviews Genetics* 6, no. 7 (2005): 533–543. https://doi.org/10.1038/nrg1637.

Bennett, Jane. "In parliament with things." In *Radical Democracy: Politics between abundance and lack*, edited by Lars Tønder and Lasse Thomassen. New York: Manchester University Press, 2005.

Bennett, Jane. *Vibrant Matter: A Political Ecology of Things.* Durham: Duke University Press, 2010.

Bryant, Levi R. *Onto-Cartography: An Ontology of Machines and Media*. Speculative Realism. Edinburgh: Edinburgh University Press, 2014.

Ceballos, Gerardo, Paul R. Ehrlich, Peter H. Raven. "Vertebrates on the brink as indicators of biological annihilation and the sixth mass extinction." *Proceedings of the National Academy of Sciences* 117, no. 24 (2020). Accessed July 18, 2020. DOI: 10.1073/pnas.1922686117

Chatzky, Andrew. "Envisioning a Green New Deal: A Global Comparison." *Council on Foreign Relations*, October, 21, 2020, https://www.cfr.org/backgrounder/envisioning-green-new-deal-global-comparison.

Corner, James. "The Agency of Mapping: Speculation, Critique and Invention." In *The Map Reader: Theories of Mapping Practice and Cartographic Representation*, edited by Martin Dodge, Rob Kitchin, and Chris Perkins. John Wiley & Sons, 2011.

Cosgrove, Denis, ed. *Mappings*. London: Reaktion Books, 2011.

Crabapple, Molly. "A Message from the Future with Alexandria Ocasio-Cortez." *The Intercept* video, 7:35. April 17, 2019. https://theintercept.com/2019/04/17/green-new-deal-short-film-alexandria-ocasio-cortez/

Dolphijn, Rick and Iris van der Tuin. *New Materialism: Interviews & Cartographies*. New Metaphysics. London: Open Humanities Press, an imprint of Michigan Publishing, 2012.

The Editorial Board. "America's willful ignorance about Black lives." *The Boston Globe*. May 31, 2020. https://www.bostonglobe.com/2020/05/31/opinion/americas-willful-ignorance-about-black-lives/.

Ellis, Erle C. and Navin Ramankutty. "Putting people in the map: anthropogenic biomes of the world." *Frontiers in Ecology and the Environment* 6, no. 8 (October 1, 2008). https://doi.org/10.1890/070062.

Fleming, Billy. "Design and the Green New Deal." *Places Journal* (2019). Accessed August 6, 2020. https://placesjournal.org/article/design-and-the-green-new-deal/.

Fleming, Billy and Alexandra Lillehei. "To Rebuild Our Towns and Cities, We Need to Design a Green Stimulus." *Jacobin Magazine*, April 18, 2020, https://jacobinmag.com/2020/04/green-stimulus-new-deal-infrastructure-buildout-coronavirus.

Game of Thrones. "The Iron Throne." Episode number 6. Season 8. Directed and Written by David Benioff, D.B. Weiss. HBO. May 19, 2019.

Green Stimulus Proposal. "A Green Stimulus to Rebuild Our Economy." *Medium*, March 22, 2020. https://medium.com/@green_stimulus_now/a-green-stimulus-to-rebuild-our-economy-1e7030a1d9ee.

Haraway, Donna. *The Companion Species Manifesto: Dogs, People, and Significant Otherness*. Chicago: Prickly Paradigm Press, 2003.

Haraway, Donna. "A Game of Cat's Cradle: Science Studies, Feminist Theory, Cultural Studies." *Configurations* 2, no. 1 (Winter 1994). 64, https://muse-jhu-edu.proxy01.its.virginia.edu/article/8029.

Haraway, Donna. *Staying with the Trouble*. Durham: Duke University Press, 2016.

"Hate and Extremism." Southern Poverty Law Center. Accessed September 29, 2020. https://www.splcenter.org/issues/hate-and-extremism.

Hazen, Robert. "Mineral fodder: We may think we are the first organisms to remake the planet, but life has been transforming the earth for aeons."

Aeon, June 24, 2014. https://aeon.co/essays/how-life-made-the-earth-into-a-cosmic-marvel

Hazen, Robert M., Dominic Papineau, Wouter Bleeker, Robert T. Downs, John M. Ferry, Timothy J. McCoy, Dimitri A. Serjensky, and Hexiong Yang. "Mineral evolution." *American Mineralogist* 93, 11-12 (2008): 1693-1720. https://doi.org/10.2138/am.2008.2955.

Heltzel, Gordon and Kristin Laurin. "Polarization in America: two possible futures." *Current Opinion in Behavioral Sciences 34* (May 6, 2020): 179-184. DOI: 10.1016/j.cobeha.2020.03.008.

Hutton, Jane. *Reciprocal Landscapes*. London: Routledge, 2019.

Kurlansky, Mark. *Salt: A World History*. New York: Penguin Books, 2003.

Latour, Bruno. *Pandora's Hope*. Cambridge: Harvard University Press, 1999.

Latour, Bruno. *We Have Never Been Modern*. Translated by Catherine Porter. Cambridge: Harvard University Press, 1993.

Levine, Martin. "Billionaires' Fortunes Rise by $434 Billion amid the Pandemic." *Nonprofit Quarterly*. June 8, 2020. https://nonprofitquarterly.org/billionaires-fortunes-rise-by-434-billion-amid-the-pandemic/.

Manzini, Ezio. *Design, When Everybody Designs*. Cambridge: The MIT Press, 2015.

Margulis, Lynn. *Symbiotic Planet: A New Look at Evolution*. New York: Basic Books, 1998.

Mattern, Shannon. "Maintenance and Care." *Places Journal* (2018). Accessed September 29, 2020. https://placesjournal.org/article/maintenance-and-care/.

Nickerson, Colin. "The Blurry Line between Life, Nonlife." *The Boston Globe*. January 12, 2009. http://archive.boston.com/news/science/articles/2009/01/12/the_blurry_line_between_life_nonlife/.

Rosenberg, Daniel and Anthony Grafton. *Cartographies of Time: A History of the Timeline*. New York: Princeton Architectural Press, 2012.

Sandy, Matt. "'The Amazon Is Completely Lawless': The Rainforest After Bolsonaro's First Year." *The New York Times*. December 5, 2019. https://www.nytimes.com/2019/12/05/world/americas/amazon-fires-bolsonaro-photos.html.

Serhan, Yasmeen. "Populism is Morphing in Insidious Ways." *The Atlantic*. January 6, 2020. https://www.theatlantic.com/international/archive/2020/01/future-populism-2020s/604393/.

Serwer, Adam. "The Cruelty is the Point." *The Atlantic*. October 3, 2018. https://www.theatlantic.com/ideas/archive/2018/10/the-cruelty-is-the-point/572104/?gclid=CjwKCAjw2dD7BRASEiwAWCtCb9eGFO-sdnyXlA2HHZZyaPeU5Ys8F5ToizomGmdE0WLdvvhw9ei8NxoCBV0QAvD_BwE.

Sippel, Sebastian, Nicolai Meinhausen, Erich M. Fischer, Enikő Székely and Reto Knutti. "Climate change now detectable from any single day of weather at global scale." *Nature Climate Change* 10, no. 1 (2020): 35-41. Accessed September 2, 2020. https://doi.org/10.1038/s41558-019-0666-7

Smith, Jamil. "Violent White Nationalism is All-American." *Rolling Stone*. August 5, 2019. https://www.rollingstone.com/politics/political-commentary/white-nationalism-el-paso-dayton-shootings-867603/.

Strazewski, Len. "Harvard epidemiologist: Beware COVID-19's second wave this fall." American Medical Association. May 8, 2020. https://www.ama-assn.org/delivering-care/public-health/harvard-epidemiologist-beware-covid-19-s-second-wave-fall.

Sun, Lena H. "CDC director warns second wave of coronavirus likely to be even more devastating." Washington Post. April 21, 2020. https://www.washingtonpost.com/health/2020/04/21/coronavirus-secondwave-cdcdirector/.

Time-Life Books. The United States. Library of Nations. New York: Time-Life Books, 1989.

Tsing, Anna, Heather Swanson, Elaine Gan, and Nils Bubandt, eds. Arts of Living on a Damaged Plant: Ghosts and Monsters of the Anthropocene. Minneapolis: University of Minnesota Press, 2017.

UNESCO World Heritage Center. "Salt Mines of Maras." UNESCO. August 5, 2019. https://whc.unesco.org/en/tentativelists/6412/.

US Congress. House. Recognizing the duty of the Federal Government to create a Green New Deal. H. Res. 109. 116th Cong., 1st sess. Introduced in House February 12, 2019. https://www.congress.gov/bill/116th-congress/house-resolution/109/text.

United States Environmental Protection Agency. "Greenhouse Gas Emissions: Global Greenhouse Gas Emission Data." Accessed September 30, 2020. https://www.epa.gov/ghgemissions/global-greenhouse-gas-emissions-data#:~:text=Since%201970%2C%20CO2%20emissions,increase%20from%201970%20to%202011.

Vernadsky, Vladimir Ivanovich. Quoted in Margulis, Lynn and Dorion Sagan. What is Life? Berkeley: University of California Press, 1995.

Wylie, John. "Depths and Folds: On Landscape and the Gazing Subject." Society and Space (August 1, 2006). https://doi.org/10.1068/d380t

Young, Ed. I Contain Multitudes: The Microbes Within Us and a Grander View of Life. New York: HarperCollins Publishers, 2016.

IMAGE CITATIONS + CREDITS

Listed by Page Number

1   Photo: Matthew Seibert

3   Photo: Matthew Seibert; Basemap sources: Esri, Airbus DS, USGS, NGA, NASA, CGIAR, N Robinson, NCEAS, NLS, OS, NMA, Geodatastyrelsen, Rijkswaterstaat, GSA, Geoland, FEMA, Intermap and the GIS user community

5   Photo: Matthew Seibert

9   Photo: Matthew Seibert

13   Photo: Matthew Seibert

18   Photo: Matthew Seibert

24   Photo: Matthew Seibert; Basemap sources: Esri, DigitalGlobe, GeoEye, i-cubed, USDA FSA, USGS, AEX, Getmapping, Aerogrid, IGN, IGP, swisstopo, and the GIS User Community

31   Photo: Matthew Seibert

35   Photo: Matthew Seibert

Metal as Metaphor

URANIUM

by Denise Hoffman Brandt

Context Map

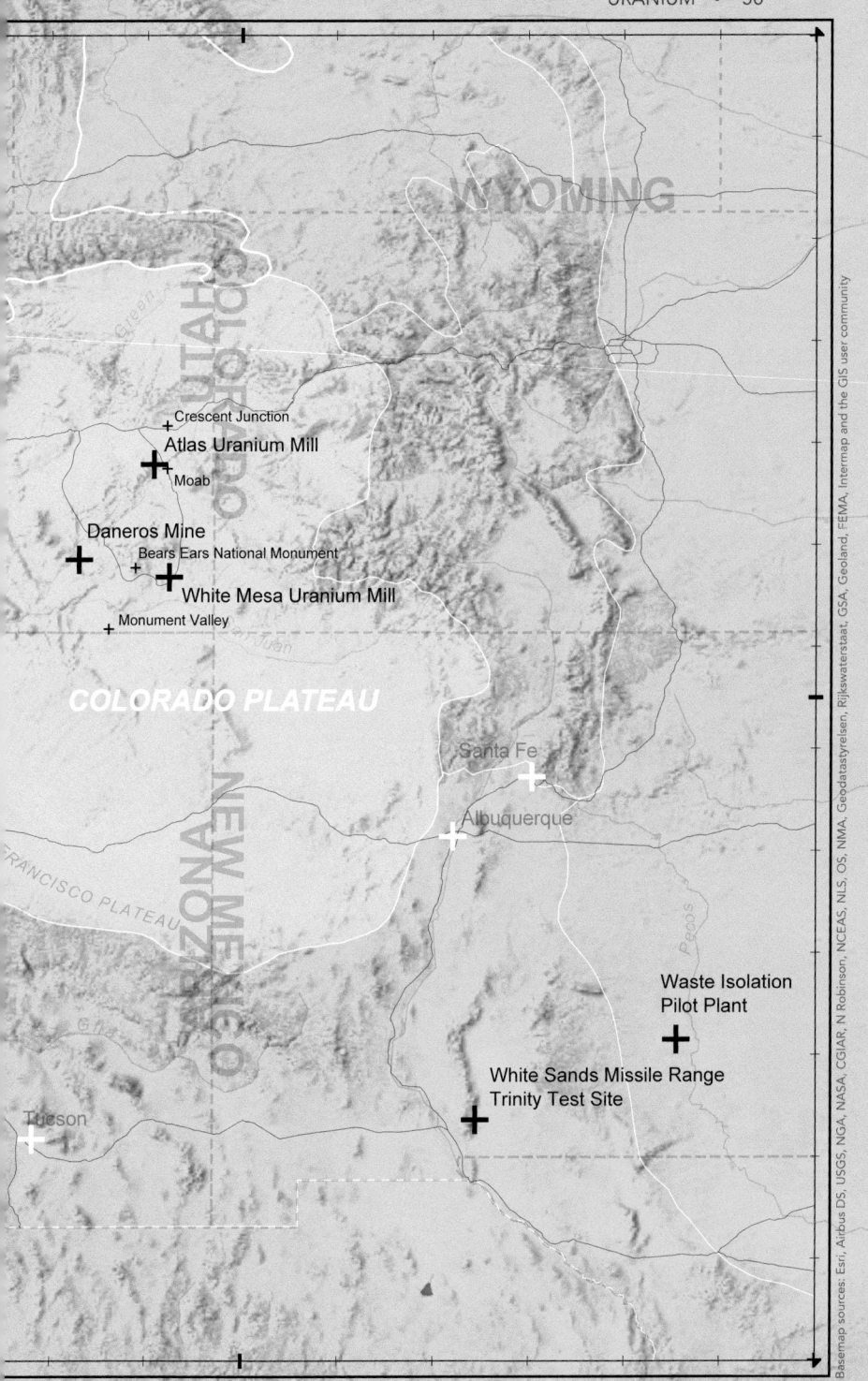

WYOMING

UTAH COLORADO

Crescent Junction

Atlas Uranium Mill

Moab

Daneros Mine

Bears Ears National Monument

White Mesa Uranium Mill

Monument Valley

San Juan

COLORADO PLATEAU

Green

ARIZONA NEW MEXICO

FRANCISCO PLATEAU

Santa Fe

Albuquerque

Pecos

Waste Isolation
Pilot Plant

White Sands Missile Range
Trinity Test Site

Tucson

INSTABILITY I—BLAST ZONES

Buried in the subfolder "uranium" in my "quantum weirdness" hard drive, I keep a souvenir picture from October 2015. My son looks directly at the camera beaming. Through sunglasses gone full black, I look at him, uncomfortably sharing his smile. Behind us, a disproportionately tall, black-rock pyramid commemorates the site "where the world's first nuclear device was exploded on July 16, 1945." Uranium has driven me to some extremes, not least a red-eye to Albuquerque for a 5 am drive to White Sands Missile Range for one of Trinity Test Site's twice-yearly, Army-managed open-houses. After waiting our turn for a photograph in front of the monument, the guy behind us in line—cropped hair and a khaki quasi-uniform—graciously offered to take it of both of us. I have found with uranium, you can never just observe landscape, you are always participatory.

Behind us, tufts of grass pushing through compacted earth look as if the Army thought the site might at some point be distinguished with a lawn. Beyond that, a tall chain-link fence topped with an angled-out barbed-wire panel defended the blast zone from scrubland and dark hills. It was obvious why the site was chosen; it seems to be in an empty geologic bowl. To animate the scene, the Army has spaced out, one each per fence panel, a series of captioned, plastic-laminated photographs. Looking past a picture of the ranch: scattered buildings, dirt roads, tumbleweed, taken-over by the Manhattan Project, I wondered at the lusher grasses on the other side of the fence. Then I turned to a photograph of the bomb being "completely assembled atop the 100-foot tower"—terminus for the physicist's idea of a "Gadget" built in the ranch house's master bedroom.

Walking down the fence-line, I took a picture of a man in a dun-colored bucket hat and striped seersucker camp shirt taking a picture of "The explosion at .025 seconds." With a tourist in front of it, the ballooning white cloud with a dust-flare at its base looks like a scoop of vanilla ice-cream. In the next photo, at .053 seconds post-detonation, the ice-cream lifts up off the plate, foamy at its base. After that, at .100 seconds, all thoughts of ice-cream vanish. What seemed like foam is now a jagged wave of earth propelled out from the bottom of a glowing sun. I remember looking at the ground just then, wondering if I was standing on what had been that wave or what was left below it. The gritty surface revealed nothing.

You could say, if you are easily distracted by silver-linings and inclined to overlook the moral abyss demarcated within, that the United States dropped "star-stuff" on Hiroshima and Nagasaki in August 1945. Carl Sagan connected us with the cosmos through "the nitrogen in our DNA, the calcium in our teeth, the iron in our blood, the carbon in our apple pies," which all began as space-debris released into the universe during celestial life cycles.[1] And he could have listed uranium, dug out of the earth's mantle in the Belgian Congo, shipped in secret to the United States, stored in a warehouse in Staten Island for 4 years before being enriched to explosive instability in a reactor in Washington state, and then built into a bomb in the New Mexican desert.[2]

In that light, the Manhattan Project was an extreme recycling program: reused metallic refuse from the Big Bang propelled the US into sustaining a dicey global military-industrial dominance. Meltdowns, leaking waste, fear of dirty bombs, and lingering distrust of authorities wielding science hang over us like a light existential fog. Ever since that first big bang at the Trinity test site, uranium has been both physically and metaphorically elemental in our landscape. Popular opinion and political polarity on current socioenvironmental hazards—anthropocentric, elevated carbon-levels fueling climate-volatility, and a wild-animal trade driven pandemic—are fall-out from those first experiments in weaponized uranium. And thus, policy, planning, and design remain within the parameters built by government authorities after the blast: a mandate to command and control the perception of the scenario in lieu of acknowledging uncertainty of outcomes. Seventy-five years after Trinity, we still face a specter of nuclear warfare along with those other hazards of modernity. Reflecting on uranium reveals the false ideological boundaries we have constructed to quantify our risk and reward, and opens awareness to release policy-makers, planners, and designers to new approaches for negotiating us through the scary terrain we have made.

***

Uranium's story begins in the murk of theories we have about our own genesis. Creationism aside, there are still a lot of questions about what explosively began the beginning 13.8 billion years ago. Ever more precise observational measurements have, ironically, revealed questions that some scientists argue we cannot answer through measure. Seth Lloyd, Professor of Quantum Mechanical Engineering at MIT goes so far as to assert that "after a two-millennium run, the universe as observable cosmos is kaput."[3] What we know is shaped by what we are trying to find out. We instrumentalized uranium's destructive potential before we understood the genesis of the materiel.

Metals, elements with heavier nuclei than hydrogen and helium, were formed when stars ended their lives in supernovae explosions. That process, stellar nucleosynthesis, was originally theorized by nuclear astrophysicist Fred Hoyle in 1946 after prompting by his colleague, Walter Baade, who observed: "Maybe a star is like a nuclear weapon!"[4] Currently, uranium has two birth-stories. The more recent theory speculates that the metal is thrown out with other heavy elements when two neutron stars—extremely dense bodies—merge.

The older theory, not disproved, is Hoyle's. In the heat of exploding supernova moments, a star's hydrogen fused into helium releasing energy that fused into carbon, progressively igniting heavier elements as the temperature rose in the star's core. Elements burned-off in sequence until silicon

Origins of Uranium

1. star

ENERGY

GRAVITY

H to He
He to C
C to Ne
Ne to O
O to Si
Si to Fe

e⁻

2. red supergiant

p
p   *fusion*
p
p
proton

e⁺
positron
n
p   He nucleus
v
neutrino

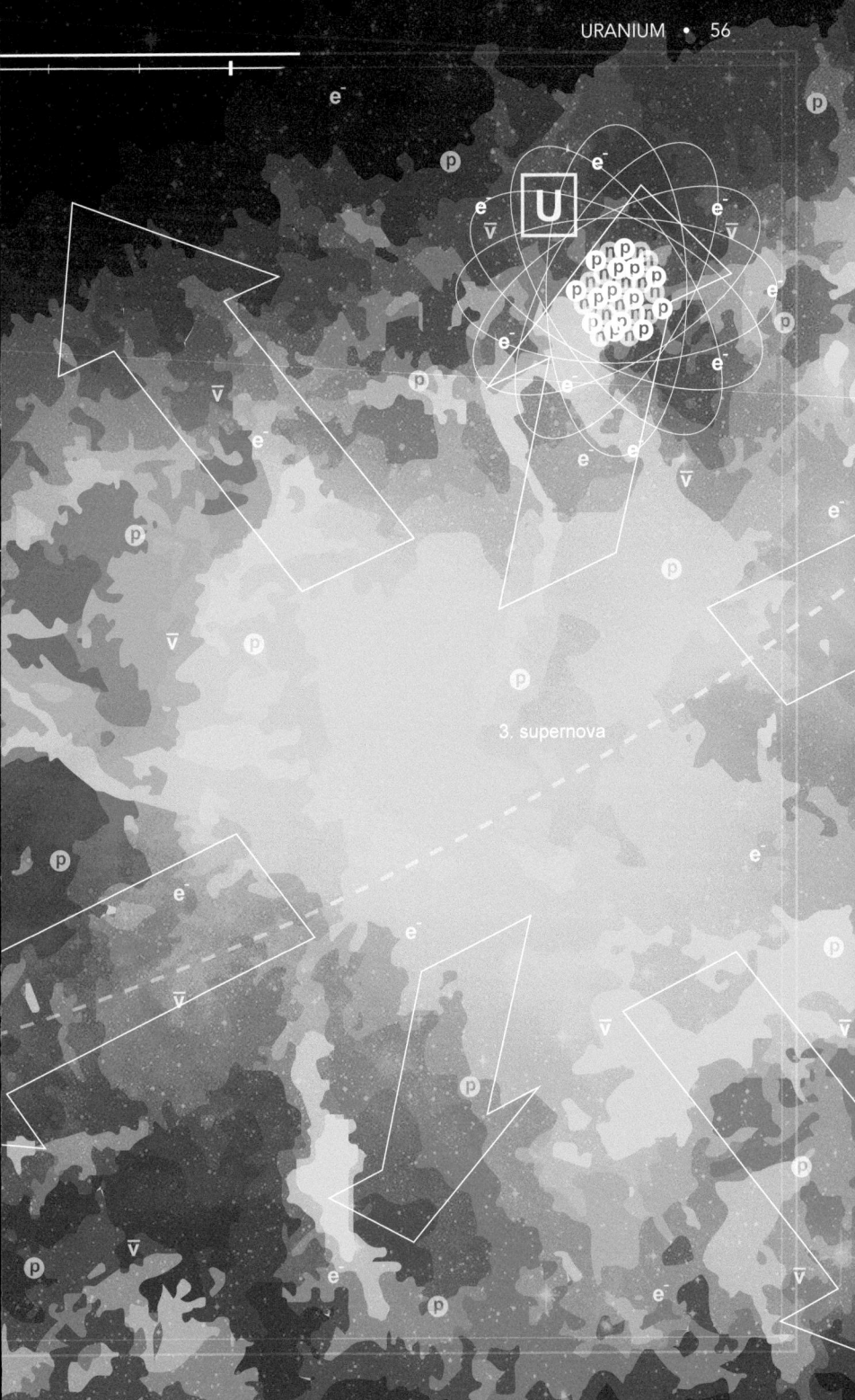

3. supernova

fused into iron which requires more energy to fuse than it releases. At that point, the star's outer core-layers collapsed at a rate that could not be offset by core-expansion and a rebound explosion ensued. Free neutrons projected into space recombined into elements like uranium with high-mass nuclei. When we look out at the night sky, we see stars radiating from metals—uranium—blasted out by those first, long since dimmed, stars.

So too, standing in places like Utah's canyonlands, the ground not too far below our feet emits particles and rays of stellar origin. Supernovae-sourced superheavy elements were essential to form the dust grains and planetesimals that built planetary cores—that formed the earth.[5] Asia, North America, Africa, Australia, and Greenland contain earth-bound uranium. And on most of those continents as well as Europe, nations have mined and processed uranium to attain the capacity to blow other pieces of the earth's crust apart. Uranium is liminal. Obscure and omnipresent, it may feature in both earth's beginning and its end.

***

Another liminality: uranium's decay replicates processes of its own cosmic genesis. Like the massive stars from which they were cast into the universe, heavy elements' massive nuclei are unstable, their neutrons' weak interactions having insufficient binding force to hold them together permanently. All of uranium's isotopes—forms with different numbers of neutrons—break down in a cascade of also-unstable elements that emit protons and neutrons while releasing ionizing radiation. Sloughing off particles at diverse rates, descendant elements decay from one form to the next: Uranium-238 to Thorium-234 to Protactinium-234, and so on, simultaneously and in sequence until, like the collapsed core of a star, a stable nucleus is achieved in lead.

Never does uranium's decay chain reverse its direction, climbing back up to 238 neutrons, without unnatural energy inputs. A perfect illustration of the second law of thermodynamics' arrow of time, uranium is always moving toward increased entropy, greater randomness. Half-lives are the rhythmic score of uranium's entropic cascade. Crystallizing an abstraction, a half-life is a probabilistic projection of the amount of time it would take for half of the original atoms to transform one step down the decay-chain.

Once part of a chunk of ore has decayed to Thorium-234, the rest of its atoms continue breaking down to that element, while the Thorium-234 begins to decay to Protactinium-234 and on down the line simultaneously at widely variable rates. It takes around 4.5 billion years, roughly the same age as the earth, for half of the atoms in U-238, the most common isotope, to decay into Thorium 234, but only 160 microseconds for half the atoms in Polonium-218 to decay into Lead-214. And yet, despite uranium's wild rhythms, by conforming with the arrow of time it affirms uniformity in earthly process outcomes.

Every atom in uranium's decay chain releases ionizing radiation with the capacity to pass through the cell walls of living organisms, damaging material in the nucleus causing carcinogenesis and cell-developmental defects. Uranium decay products drive through us and all living things, but not consistently. Biotic systems are unpredictable because they are contingent, self-organizing. Identifying what and how exactly bodies will be susceptible to the effects of radiation at non-lethal levels is a crap shoot. Miners working side by side, exposed to the same type and amount of ionizing radiation have variable medical histories. Interactions unfold along multiple possible pathways guided by local-scale ordering of independent properties. Uranium reveals that complexity subverts the safe logics of what we imagine to be natural laws.

***

Turning on a Geiger counter in my apartment in Brooklyn I discovered I live with a surprisingly high level of background radiation. My apartment is relatively "hot" because it is on rock at the top of one of Brooklyn's highest hills and not down in its sandy former wetlands. Geology matters. Living on the Colorado Plateau adds 90–100 millirems[6] (mrem) annually to your effective exposure to cell-disrupting radiation—a lot when you consider that average American adult exposure to natural sources is around 300 mrem. Proximity to nuclear facilities boosts exposure rates, but so do some medical procedures, flying in planes, watching TV, and living within 50 miles of a coal-fired electrical utility plant.

No dosage of ionizing radiation is safe. Yet there is no mandated exposure limit for most people, except for fetuses and federal workers. Fetuses and those under 18 years-old working with radiation are limited to 500 mrem

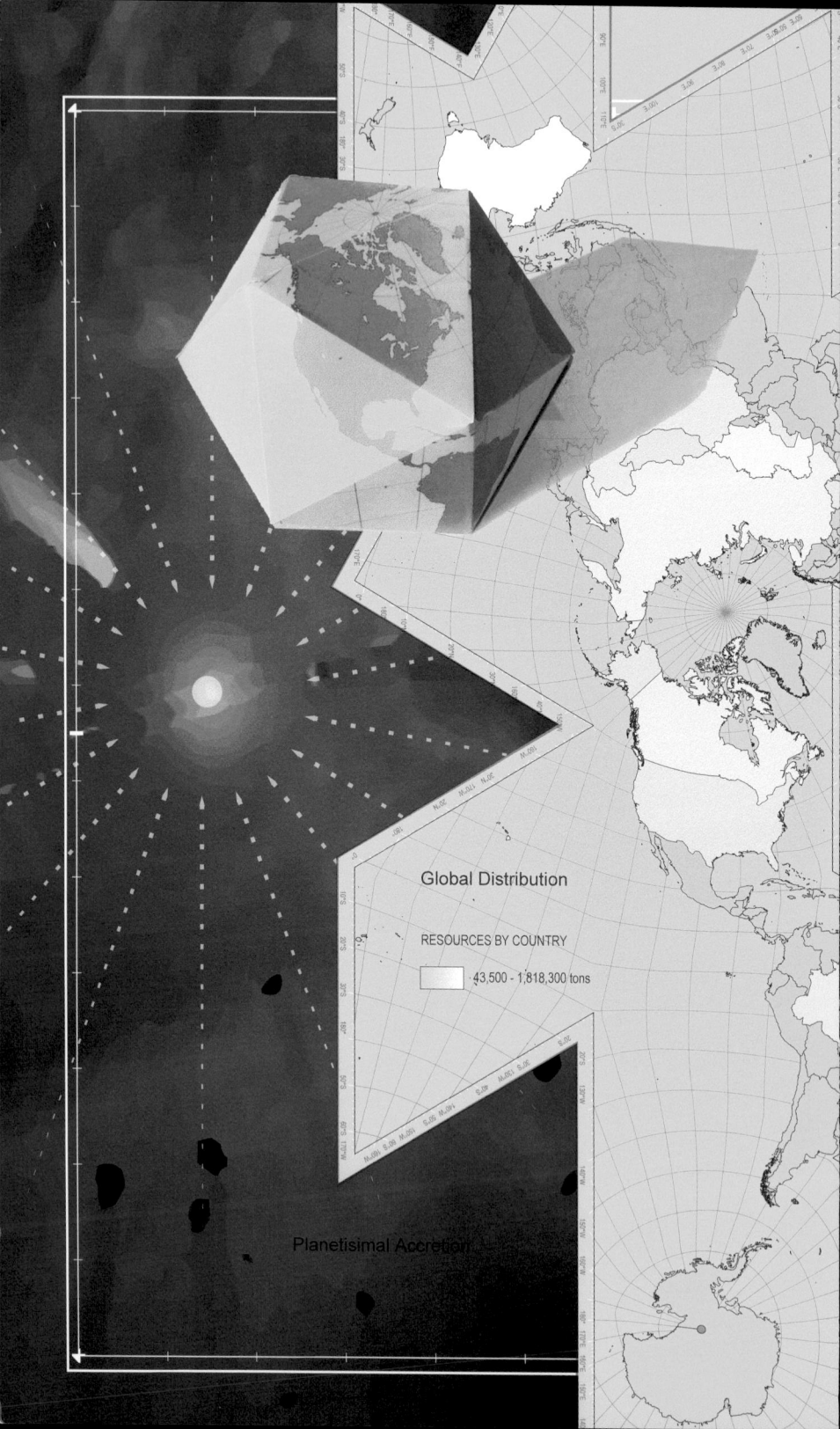

Global Distribution

RESOURCES BY COUNTRY

43,500 - 1,818,300 tons

Planetisimal Accretion

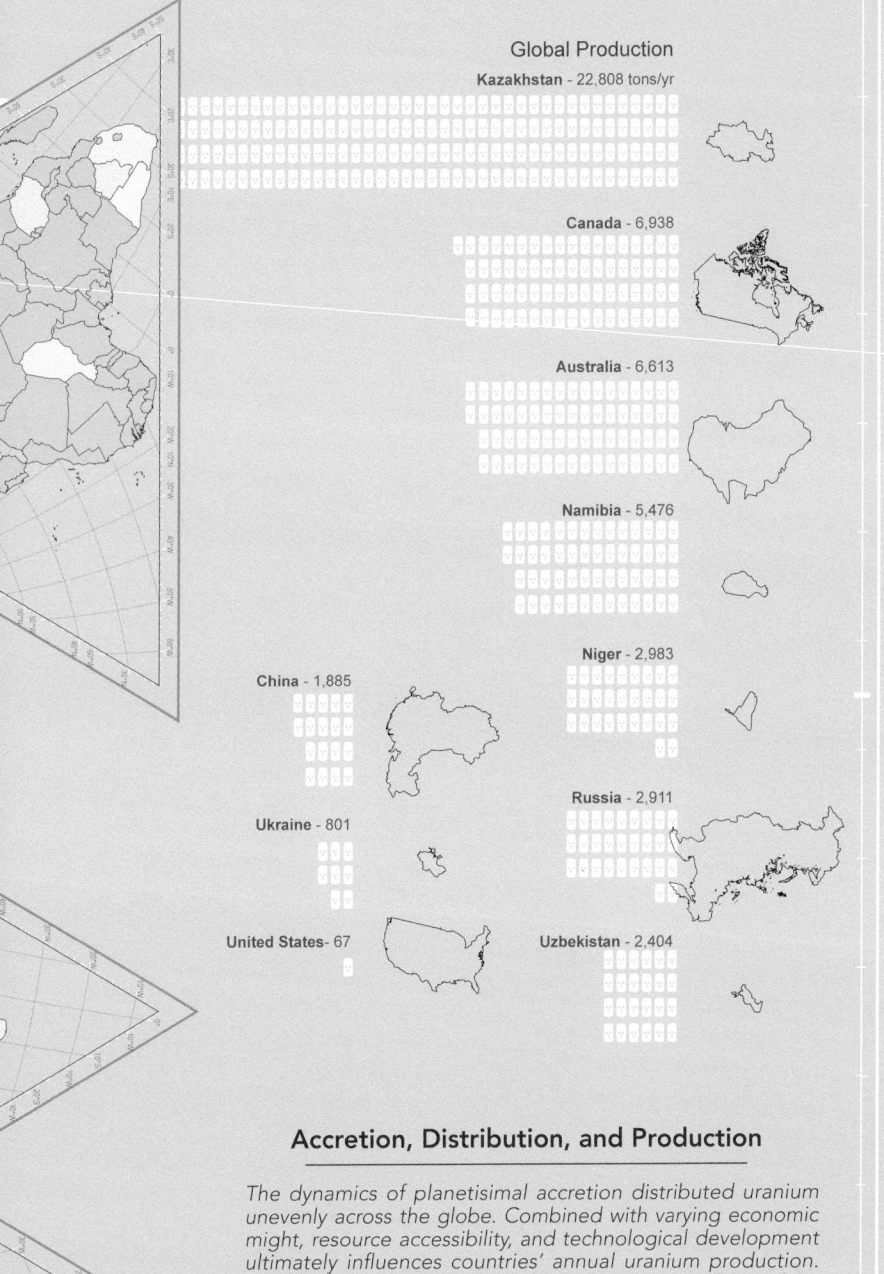

Global Production

Kazakhstan - 22,808 tons/yr

Canada - 6,938

Australia - 6,613

Namibia - 5,476

China - 1,885

Niger - 2,983

Ukraine - 801

Russia - 2,911

United States- 67

Uzbekistan - 2,404

Accretion, Distribution, and Production

The dynamics of planetisimal accretion distributed uranium unevenly across the globe. Combined with varying economic might, resource accessibility, and technological development ultimately influences countries' annual uranium production. Kazakhstan leads the world due to its possession of one of the largest uranium deposits and the high exploitation of its mineral resources.

per year. Anyone over that working with radiation can absorb up to 5,000 mrem without triggering a review of workplace safety practices. Unless you are an astronaut, in which case you are allowed 25,000 mrem per year. These numbers are obviously arbitrary, based on the calculation of what is achievable given workplace objectives, as opposed to eliminating employee hazard.

International policy-makers orchestrating global-scale existential threats—while promoting weapons proliferation as big business and speculating in markets for nuclear energy as a non-carbon-emitting fuel—develop equivocal and inconsistent workplace radiation exposure standards. They leave both those willing to, literally, absorb a higher level of risk and those who want less exposure in their daily lives, unsatisfied. Both positions are grounded in fear: fear of economic loss, fear of climate-change impacts, fear of public health calamities, and fear a geopolitical skirmish might bring annihilation. As emotional triggers, the polar positions are hard to move.

I bought my Geiger counter years ago for a drive with my son around Niagara Falls, New York to see Love Canal and remnants of the chemical industries affiliated with World War II nuclear production that had buried their waste in it. We went on Maid of the Mist before touring the neighborhood around the first Superfund clean-up site where my anticipation of Erin Brockovich-type drama was thwarted. Moderately high readings only clicked where the road had been torn up near one of the chemical plants. My risk-aversion was satisfied; my son probably had more exposure getting his first dental x-ray.

Safety is relative and contingent on what we desire. My father's radiation treatments for kidney cancer shot tumor-targeted beams of six million mrem into him. We wanted that. In ecosystems, complexity and stability are not system properties, but instead aspects of system descriptions.[7] As a "natural resource," uranium exists only in the eye of the beholder. A spectrum of personal experiences has been reduced to generalized opposing views—factions—supporting specific ideas of what social structures should be implemented to their ends. It always comes down to a question of: should we risk nuclear technology or not? I suggest to move past this impasse, we should look not at how we use uranium, but how it has shaped us.

Leaving the Trinity test site, I turned onto route 380, swerving around a cluster of people in the road holding up signs protesting nuclear tourism. They were right of course. Inside the suburban recreation-type chain-link fence, looking at the hand-crafted monument, and wood shacks, the bomb was domesticated, made approachable. I watched in my rearview mirror, as a few of them followed us down the road a bit; their yelling fading into the dusty wind blowing past the car windows. Uranium is simultaneously monster, miracle, and messy argument.

CHANCE I—GEOLOGY AND MORALITY

In 1789, Martin Heinrich Klaproth, a German chemist, discovered a new element which he named after the seventh planet in the solar system identified eight-years earlier. William Herschel's recording of the planet—it had been observed since 1690 but not recognized—gave him the right to name it. His first choice, to honor his King, George III, was rejected by scientists from outside Shakespeare's "sceptered isle." They politely suggested revival of the practice used since Roman times of naming planets after gods. Herschel must have been disappointed. Either too-easily satisfied that the primordial Greek god Uranus was sometimes called Father Sky, or maybe out of spite, he imprinted the planet, and later the metal, with a lurid backstory.

Uranus was both son and sexual partner of Gaia, mother earth. Perhaps potential genetic impacts of incest were already known, because Hesiod described progeny exhibiting disturbing mutations: hundred-handedness (the Hecatoncheires), giantism (Cyclopes), and a single eye in the center of their forehead (Cyclopes). Although the Titans, their human-form children, became future parents of the Olympians, altogether, they were a violent family.

In a paranoid attempt to retain power over the cosmos, Uranus imprisoned some of his children deep in the earth—a pain for Gaia. Cronus, Uranus' youngest son, spurred on by Gaia, escaped captivity and castrated him with a blade of her making. Uranus' blood, drenching the earth-Gaia, spawned the Giants, the Furies (understandable!), and a few

4.54 billion years ago.

4.46 billion years ago

formation of Earth

Uranium-238 α

α *alpha radiation*
β– *beta radiation*

α alpha radiation

Skin

β beta radiation

γ gamma radiation

Cytosine NH2

N

O

N
H

C

Thymine

H3C

O

NH

O

N
H

T

Adenine NH2

N

N

N

N
H

A

Guanine

O

N

NH

N

N

N
H

NH2

G

Uranium's Decay Chain

present day

322,500 years in the future

Bone cells

α
γ
β

Thorium-234 β⁻
Protactinium-234 β⁻
Uranium-234 α
Thorium-230 α
Radium-226 α
Radon-222 α
Polonium218 α
Lead-214 β⁻
Bismuth-214 β⁻
Polonium-214 α
Lead-210 β⁻
Bismuth-210 β⁻
Polonium-210 α
Lead-206 stable

nymphs. Exit Uranus, but Cronus went on to consolidate authority over the cosmos in battle with those same Giants.

I admit it may seem odd, but I associate Cronus' violent agglomeration with the battery of uranium-packed planetesimals slamming into proto-earth to make it stick together, to make it whole. Klaproth must have been thinking of the planet, dark and barely-knowable, and not primordial geopolitics when he was fiddling with his lump of pitchblende, a dull, often described as greasy, black globule. No drama there. He dissolved it, neutralized the product, and heated that, to yield an unremarkable and apparently valueless black powder. In fact, he isolated only an oxide of uranium. The pure element wasn't isolated for another 52 years.

And it was nearly another half century before Antoine Henri Becquerel discovered radioactivity, the entrée to the Curies[8] discovery of the "miracle" effects of radium, a mineral found in association with uranium ore sourced from brilliant yellow, fibrous-looking, carnotite crystals. Mined from the sandstones of the Colorado Plateau, carnotite and other uranic minerals such as autunite, were used to tint glass and ceramics many shades of yellow for centuries, a harbinger of colorful tales, not to mention mildly radioactive antiques.

***

On a hot July day in 2019, I craned my neck out the car-window scanning for the openings of hundred-year-old, remnant mines in the scrubby hillsides off Route 95 while driving east through the former Bears Ears National Monument near Blanding, Utah. Originally set aside by President Obama's 2016 Proclamation 9558, President Trump's Proclamation 9681 transmuted over a million acres of reserve to real estate. According to my map I was in an area slated to become the Shash Jáa National Monument, one of two diminutive, separate public-lands speckled with countless dog holes dug by local independent prospectors.

Dinosaur skeletons and petrified trees with uranium decay-chain minerals could be found sticking out of the gray-streaked cliffs around Latter-Day Saints ranching colonies: Moab, Blanding, and Monticello. By 1900, the Colorado Plateau was the unlikely primary radium supply-stream for the Curies' research and the emergent medical practice of radiation. When that market collapsed around 1912, the locals—most of whom supported their families by a mix of trapping for the government, taking on odd-jobs, and mining—turned to extraction of vanadium from the tailings of their old mines, which they sold for industrial use to harden steel.[9] Negotiating the curvy road, taking random photos with my cell phone, and exhorting my son—again riding shotgun—to also take pictures, I saw no sign of that history.

Mining was a family affair. Shumways, Blacks, Carrols, and Kimmerles blasted and dug in each other's mines, hauled ore for each other, and hung out together in mining camps forging codependent relationships across generations of intermarriages. Operations ranged in scale from one or two guys working weekends in an open pit with hammers and wheelbarrows, to teams of workers camping in the hills for weeks: blasting, building timber-framed tunnels, and hauling the ore out in loaders.[10] Standing in upwelling groundwater, choking on dust, debilitated by injury from equipment fails or shaft collapse, miners shared hardship and religious beliefs that connected moral rectitude to hard work and economic success. Heavy metal paved a path to heaven.

***

Of the 15 types of uranium deposit classified by the International Atomic Energy Agency,[11] two found in sedimentary basinal sandstone on the Colorado Plateau supplied both the Curies and Cold War bomb-builders. When the Atomic Energy Commission (AEC) created a uranium rush in 1948 by offering high payouts for ore to supply nuclear weapons research and production, optimistic locals and outsiders flooded Utah and Colorado's red rock desert. Guided by aerial survey maps from AEC geologists in Grand Junction, Colorado, prospectors scoured the rises and canyons for surface deposits, and, when they could afford it, by waving around Geiger counter wands. Locating a vein, a prospector staked and filed a claim before carrying out exploration for an ore body.

Among those decamping to Utah to cash-in on the rumored uranium rush, Charlie Steen stood out, and not just for the disparity between the depth of his poverty and the height of his ambition. With an education in petroleum geology and field experience in South America, he believed he could find uranium in places no one else had thought to

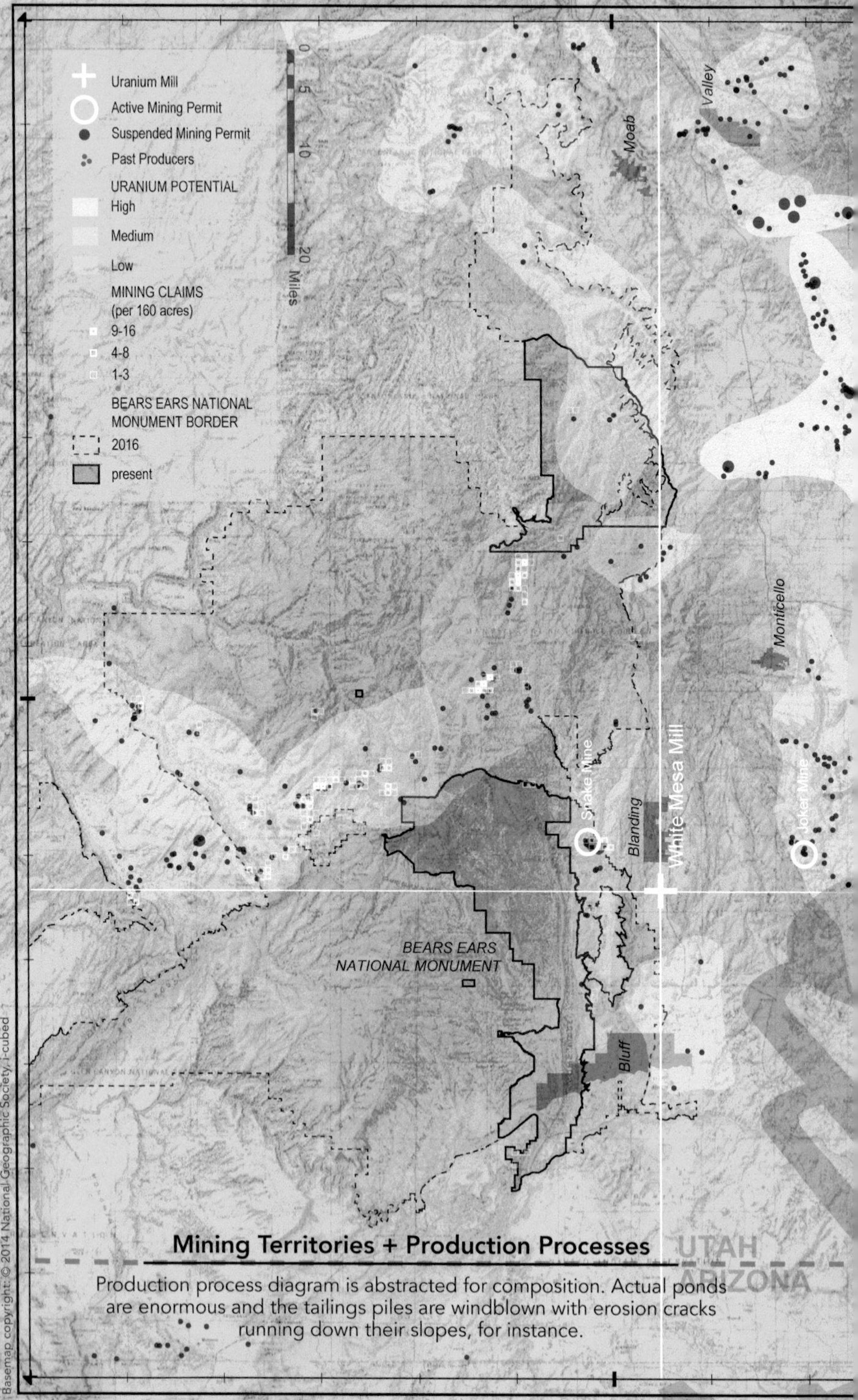

Uranium Mill
Active Mining Permit
Suspended Mining Permit
Past Producers
URANIUM POTENTIAL
High
Medium
Low
MINING CLAIMS
(per 160 acres)
9-16
4-8
1-3
BEARS EARS NATIONAL
MONUMENT BORDER
2016
present
0
5
10
20
Miles
Moab
Valley
Monticello
Snake Mine
Blanding
White Mesa Mill
Joker Mine
BEARS EARS
NATIONAL MONUMENT
Bluff
UTAH
ARIZONA
Basemap copyright © 2014 National Geographic Society, i-cubed

Mining Territories + Production Processes

Production process diagram is abstracted for composition. Actual ponds are enormous and the tailings piles are windblown with erosion cracks running down their slopes, for instance.

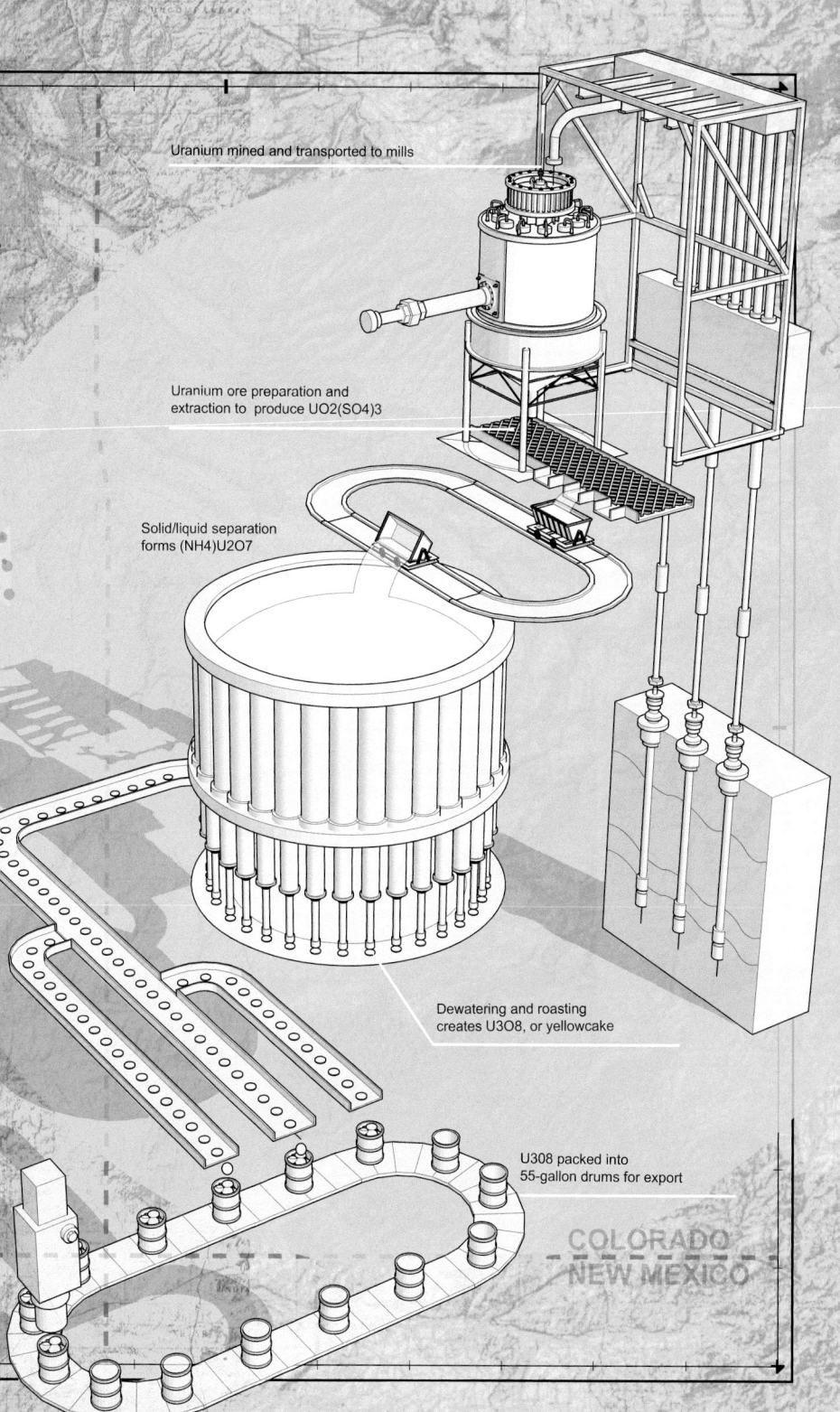

Uranium mined and transported to mills

Uranium ore preparation and
extraction to produce UO2(SO4)3

Solid/liquid separation
forms (NH4)U2O7

Dewatering and roasting
creates U3O8, or yellowcake

U308 packed into
55-gallon drums for export

COLORADO
NEW MEXICO

look. Even by the standards of hardscrabble locals, the Steen family's lifestyle was impressively austere. Living in a tar paper shack, his wife hauled water from a well to bathe their four sons, feeding them breakfast cereal and deer meat, while Charlie drove a broken-down jalopy beyond Yellow Cat Wash and down into Lisbon Valley exploring his claims.[12] His money went into equipment and gas.

Onto the immensity of striated red rock, where each bluff, arch, and hoodoo appear more improbable than the last, Steen projected petroleum geology-logic, speculating that uranium deposits—usually found in the fossilized sediments of Jurassic streams—might also be discovered like oil, as pools, at the bottom of anticlinal slopes dating to the earlier Triassic period. He visualized vast uranium reservoirs in older rock strata underneath easier-to-spot surface deposits— already mined out by locals—that formed as the mineral leached upward. He hauled a secondhand diamond drill-rig out to his claims in the Big Indian Wash and began to methodically bore exploratory cores.[13] He was wildcatting for uranium, and the locals thought he was nuts.

Skepticism about this scientific method of prospecting persisted after Steen struck a massive belt of pitchblende in the Chinle Moss Back member at his Mi Vida mine in 1952. Even six months after he used a borrowed Geiger counter to test his pitchblende samples at a service station in Cisco— the machine erupted—a local truck driver recognized him as "that crazy son of a bitch from Texas. He claims he's found a uranium mine."[14] Moabites seem to have felt that the outsider had not really paid enough dues to earn his bonanza. Charlie's wife, M.L., described being shunned by local women after the strike: "In a way the successful become a sort of living reproach to the unsuccessful."[15]

Steen's son Mark observed that the "discovery proved that someone could walk over $100 million worth of uranium ore without knowing what lay beneath their feet unless they were willing to risk money on wildcat drilling in the search for totally hidden ore deposits."[16] With that acknowledgment of his father's acumen, he also noted: "If Charlie Steen had drilled another 18 feet back toward the rim, he would not have cored through the 14 feet of high-grade uranium ore that started the Big Boom at the Big Indian." Charlie Steen was lucky.

The Colorado Plateau

Thanks to Steen, the Colorado Plateau was designated the International Atomic Energy Agency's (IAEA) "selected example" for tabular deposits of continental fluvial Uranium–Vanadium which they describe as "medium-to coarse-grained (red) sandstones deposited in continental fluvial or marginal marine sedimentary environments." That banal register is a stark contrast to Raye C. Ringholz's vivid picture of the scene starting around 195 million years ago.

> After another thirty-five million years the mountains eroded to the valley floor. The climate became tropical and jungles thickened with bamboo, palms and ferns. Dinosaurs crawled through swamps and fed at the shores of lakes and lagoons. And then, mysteriously, the giant reptiles began to disappear. Sluggish rivers filled with fallen trees and bones of the vanishing monsters. The detritus petrified into red, brown, green, and slate-colored mud. This prehistoric cemetery composed the Salt Wash member of the Morrison formation where most of the Colorado Plateau's uranium had been found.[17]

The passage is stylistically similar to M. le Baron Cuvier's "Discourse on the Revolutionary Upheavals on the Surface of the Globe."

> The valleys do not have gentle slopes any more or those jutting angles facing indentations opposite, which seem to indicate the beds of some ancient water course. They grow bigger or smaller without any rule. Their waters sometimes extend into lakes; at other times they hurtle down in torrents.[18]

Both describe dramatic, uncertain forces at play and Ringholz's depiction of species vanishing amid rapidly eroding mountains is in line with Cuvier's hypothesis of fossil bones signaling catastrophic species extinction.

Catastrophism theory syncs with Christian fatalism toward disastrous or beneficial earthly outcomes at the mysterious hand of God. The moral high-ground manifests as an actual physical condition and reaping "amber waves of grain" translates as evidence of American exceptionalism.[19] Conversely, failure to succeed on blessed earth correlates with moral inadequacy. Hailed as the "Uranium King," magazine essays lauded Charlie's geology know-how in tandem with hardship. His decrepit jeep, squalid home, and worn-out boots were juxtaposed with wealth triumphant: a gold-plated gas-lantern, personal plane, and pedigreed dalmatians. Steen was not a hero despite his struggle, he was made a hero because of it.

The second act of Charlie's uranium-driven narrative arc opened in 1968, when the Internal Revenue Service seized his assets for nonpayment of taxes. By then he had made a number of bad investments in horses, cattle, and by one account, pickles. Had he kept his money in uranium, he would not have been better off. Domestic ore production fell 50% between 1961 and 1966. The media that portrayed Steen's absurdly excessive spending as justified exuberance when he led the uranium boom re-presented those same details to signal a profligate character. Financial loss, even in the context of a greater market downswing, pushed Steen's personal reputation back to undeserving upstart.

One of the first things Charlie did with his money was build a family home at the top of a cliff-like hill overlooking Moab. Ringholz quotes him saying something to the effect that it represented his climb to the top—to look over everyone—from his former position in the social bottomlands. The website of the restaurant now operating there promotes Charlie's notoriety, the "million-dollar view" and the opportunity "to dine above the rest," which I did, one night a few years ago with Sarah Fields, director of Uranium Watch.

Field's unique approach to resist nuclear follies is to assail the "front-end" production. Tracking uranium mining and milling industry compliance with safety laws, educating and advocating in the public interest, she has been a stalwart at Utah Oil, Gas, and Mineral policy review meetings for years. Looking down at Moab, booming with mountain biking, hiking, river-rafting outfits, and the bars and restaurants that cater to their clientele, she pointed out that it is both unhealthy and bad for business to have trucks drooling radioactive dust driving up Main Street. We sat a few feet away from grossly enlarged bronze replicas of Charlie's original work boots, and watched the sun set as subdivision lights glittered on.

**

Catastrophism, once dismissed by most scientists and secularists, has made a comeback in the era of climate

The Prehistoric Cemetery

change. Anticipated catastrophic, cascading effects of a disrupted carbon-cycle have shifted the focus away from geology and onto outsized, out-of-control human acts as the cause of global destruction. Here again, the arrow of time misdirects. Supernova blasts, hurtling uranium into space, described as the "death" of the star, formed planets, our planet. Moreover, 66 million years ago, when an asteroid—maybe carrying uranium—slammed into earth, the impact devastated biotic life on the planet, killing off everything from tiny ammonites and ocean plankton to dinosaurs. But then, gradually, life cycled back up. New species emerged, humans among them. We are products of catastrophe and we need to revise our narratives accordingly. Otherwise, like countless species before us, we may not know the wonders of the next transformation.

CHANCE II—GEOLOGY AND PATHOLOGY

Thousands of mostly amateur-looking photographs of Siamese daisy-twins—octuplets and more—looking like creepy centipedes with their yellow disks forming a long and twisty line surrounded by petal-like legs, went viral online in July 2015. Not uncommon, the mutations are associated with a hormonal irregularity triggered by any number of things: pathogens, insects, and probably also, ionizing radiation. In this case, the daisies' mutations were widely attributed to catastrophic genetic effects of radiation exposure after the Fukushima-Daiichi nuclear power plant accident four years before.

"Fukushima Daisies" emerged as a popular symbol of unforeseeable destruction brought on by advanced technologies. A troubling, too-cute emblem of lurking potential for annihilation, the daisies are neither as expressive of devastation as the Hiroshima shadows, nor even clearly products of destructive force. Perennials, re-flowering every year, can exhibit the condition one year and then be completely normal the next. Recurrence of the condition two years in succession indicates that the plant's genetic code has been altered, but no direct connection between ionizing radiation, daisy DNA, and daisy deformation have been identified. What the daisies effectively represent, is cloudy ideation surrounding uranium in our lives.

Uranium's decay-clouds of beta particles, gamma rays, protons, and neutrons penetrate through skin, scales, exoskeletons, cell membranes, and inside cell's nuclei where 6-foot strands of DNA condense into microscopic chromatin clumps. Radiation cuts hydrogen bonds linking complementary base pairs: guanine with cytosine and adenine with thymine; and breaks sugar-phosphate strand backbones. Disarticulated genetic material can cause abnormalities and tumorigenesis—deformation, and death. As early as 1879, when high rates of lung cancer were observed in Czech and German miners, exposure to uranium was understood to be carcinogenic. By the late 1930s, enough correlation had been observed between uranium mining and high rates of lung cancer that it was considered scientifically proven based on probability, although the specific causal agent remained unclear.

It wasn't until 1951 that a widely-reported memorandum found radon daughters to be the causal link between proximity to uranium and cancer.[20] More politically correctly known as radon progeny, they are clouds of alpha and beta particles and gamma rays emitted when a component in uranium's decay chain, RN 222, breaks down. Building up in enclosed spaces, like basements and mine shafts, they are inhaled and retained in the lungs for a period as long as their half-lives. Their destructive effects have no statute of limitations.

***

"Atoms for Peace" was the title of a December 1953 speech in which President Eisenhower expressed regret to be speaking in the "new language" of atomic warfare. He committed the US "to find the way by which the miraculous inventiveness of man shall not be dedicated to his death, but consecrated to his life." Hiroshima and Nagasaki were, by implication, collateral damage to be redeemed by beneficent nuclear technology. The previous March, the Upshot–Knothole series of nuclear tests had blasted radioactive debris into the atmosphere where it fell on southern Utah towns and drifted as far as Troy, New York.[21] Eisenhower sought to reassure Americans skeptical of the escalating nuclear arms race with the Soviet Union and reluctant to invite uranium fuel rods into their neighborhoods as an energy source.

While Moab boomed with nuclear production, other Utah towns busted from nuclear fallout.[22] In spring 1953, Cedar

City sheep ranchers began to panic as hundreds of sheep died on the trail back from winter pasture. Soon, shearing and lambing were a nightmare. Wool on the backs of black sheep acquired white spots; sometimes it would slide off the sheep without the need for clippers. Bald newborns with only bodies and no legs emerged from ewes that would soon die, and even lambs born intact would have to be nursed by bottle because the ewes refused to nurture them. All over the area, sheep were dying with previously unseen but consistent symptoms of hardened hooves and scabby pustules on their mouths, noses, and ears.

Surrounded by the corpses of their livestock, ranch families worked numbly to save what they could of their herds. Medication had no effect; local and state veterinarians from Salt Lake City were baffled. No such catastrophe had ever before occurred. Ranchers watched as hired bulldozers shoved sheep-carcasses into pit graves. Autopsies explained nothing. AEC representatives showed up to investigate, but they offered no conclusive opinion. Some said the lesions on the animals' mouths were like the sores found on the "rada-calves" in the vicinity of the Trinity blast. Others claimed the animals died of malnourishment from poor quality fodder.

The AEC downplayed the impacts of increased toxicity from detonations like Upshot–Knothole "Harry," a 32-kiloton, plutonium-fueled, pure fission weapon that generated the most radioactive fallout of any nuclear test over the continental US. "Dirty Harry's" clouds of radioactive dust covered ranchers and residents of the towns of St. George, Mesquite, and Bunkerville. When a regional spike in children's cancers manifest—it took longer to detect than deformed and dead Cedar City sheep—the AEC categorically denied responsibility for poor public health management and outreach.

Sheep ranchers and southern Utah residents sued for damages. The government dragged the cases through the courts for years, appealing all awards of compensation. Most of the ranchers sold up and turned to other ways to earn a living. Accounts of heart attacks from stress proliferate. Failure of family farms held for generations carried more than just economic weight, as ranchers—and residents in the fallout zones—were burdened by disillusionment after their patriotism in welcoming the testing was repudiated.

On a rainy weekday in summer 2019, I stood in the sheep shed in Frontier Homestead State Park Museum in Cedar City. Shafts of light cutting through the wood slat walls lit up swirls of dust motes that blended with a grainy black and white video loop of ranchers performing the chores of managing livestock. Multi-generational families handled their animals with rough familiarity: sorting, marking, shearing. In one scene, a blond boy, grinning from ear to ear, cradled a healthy newborn lamb.

The accompanying narrative, "A saving grace: sheep ranching in Iron county," omits uranium as a force for decline of local ranching. But fallout lingers, not just in desert ground, but in the memory of a population that no longer trusts government. Bunkerville, not far from the sheep ranchers' winter pasture, is the home of the Bundy family, the latest wave of Sagebrush Rebels leading loosely organized, anti-federal-authority militias in armed stand-offs in several western states.

Their distrust is understandable. Reading a catalog of nuclear testing horrifies. Whole islands in the Pacific were obliterated under 35 mushroom clouds during a single operation, Hardtack I, in the Pacific Proving Grounds—their populations summarily displaced to less habitable islands. Such acts of Uranus-like brutal authority were not confined to distant seas. Between 1943 and 1967 researchers working with the AEC intentionally exposed humans to radioactive material in a series of devastating tests ostensibly designed to help the government understand the ramifications of nuclear material production and deployment.[23] For all of these atrocities, the AEC rationale was that environmental contamination—causing suffering for a minority of citizens—was a necessary cost weighed against the greater social benefits of wielding nuclear technologies. So too, the threat of economic disruption is assessed against inequitably distributed impacts of climate change, and protection of citizens from a deadly virus is presented by some in government as an unacceptable threat to national economic stability.

Despite a slew of international accidents—like the 1966 nuclear-armed B-52G crash over Palomares, Spain—for most people, the impacts of uranium remained abstract, distant, and statistical. It wasn't until 1979, in a context of widespread disillusionment with government following the

Cells

Altered gene expression

Protein modification

Cell death & senescence

Genomic instability

Acute syndrome

Organ failure

The Cloudy Ideation
of Fukushima Daisies

Cancer   Mutation   Birth defect

Watergate revelations, that reactor failure at Three Mile Island blew through the wall of Americans' complacency. Citizens demanded greater accountability in nuclear operations and prompted legislators to force the release of previously classified AEC documents. Hushed-up dissenting opinions of AEC scientists on the cause of Cedar City sheep death were revealed. In 1982, the same judge that presided over the 1956 trial asserted the AEC committed "a species of fraud upon the court."

Yet ultimately, the legitimacy of Utah citizens' claims against the AEC were all denied on structural grounds. The atomic testing program fell under the "discretionary function exclusion" in the Federal Rules of Civil Procedure (US Code § 2680), a rule protecting government employees performing their policymaking duties whether or not they uphold valid regulations or abuse their authority. To sustain such liability protection, legislators established cancer screening and research facilities for "downwinders" in Nevada and southern Utah, and empowered the Justice Department to devise compensation protocols with 1990s Radiation Exposure and Compensation Act (RECA). Participants in human testing also received government compensation payments, although, it was not until 1997 that laws were ratified to prohibit secret scientific testing on humans.

***

Each month, tourists visit the Nevada National Security Site (NNSS), touring areas still hot from Cold War bomb tests—off-kilter houses popularized in movies and an airplane used in first-responder training—amid active nuclear waste disposal operations and facilities for research and underground chemical detonations. When you go on the tour, they don't let you into the latter areas. Even so, despite its remote location, the NNSS tours are popular. I had to make a special arrangement as an academic doing research to get in because the limited slots filled up as soon as they were posted online.

No devices can be carried into the site, so all I have to recall the visit is 17 pages of notes, some sketches, and two photos taken by our guide: one of just my son, and one of both of us, standing on a white fenced platform in a lunar landscape.[24] Sedan was a subsurface thermonuclear-device test that displaced 11 million tons of earth. Here my son is

Cedar City, UT Sheep Death

older, laughing, standing self-consciously stiff as I try hard to look ironically comfortable leaning back on the railing.

We are dwarfed by the crater behind us. It looks sandy-sloped, but is really hard, crusty, otherwise its sides could not remain so steep. In the 57 years since the blast, only scattered tumbleweeds have populated it. Perfect crater form and lack of biota made it the training ground for astronauts who would later walk on the moon. That cultural import weighs in with the physics of the site: Sedan's active fallout fell on more Americans than any other test blast.

***

The Diné (Navajo) uranium origin story, like the myth of Uranus, involves complicated family alliances and violence from the time of their creation. Twin warriors, Naayéé' Neezghání (Monster Slayer) and Tó Bájísh Chíní (Child Born of Water), children of 'Asdząą Nádleehé (Changing Woman) and Jóhonaa'éí (Sun Bearer), set out to slay monsters preying on humans. Their first conquest was Yé'iitsoh (big monster)—also a son of Jóhonaa'éí—who they destroyed with special flint arrows given to them by their common father.

Later, in exchange for many gifts: "rainbow, zigzag lightning, sunray, mirages, male and female rain, dark and white mists," the twins would agree that their father could "destroy all those living in houses."[25] Which he did with a flood. But uranium was not one of the gifts given to the twins, it derived from destruction: the scattered bones of Yé'iitsoh, petrified wood.[26] Uranium deposits replace organic tissue in ancient fallen trees—a not so mythical residuum.

The first claim on Navajo land for mining vanadium was a 1944 lease near Monument Valley bought by San Juan County mine families. They used Diné labor and land to expand their operations to fill contracts with a purchaser in Pittsburgh who provided the infrastructure to export the ore. Soon, Vanadium Corporation of America (VCA) began production, also using Diné workers, while they surveyed for uranium to supply the Manhattan Project. Out of that loop, the locals shut down, but VCA and Union Carbide expanded operations. Navajo Nation leadership cooperated with federal officials out of patriotism and to promote economic stability for communities juggling capitalism and their traditional economy.

BIKINI ATOLLS

Pacific Proving Grounds (1946–1962)

Bomb Magnitude (Mt)

<1

1-5

>5

Nearly 30 million tons of uranium ore were extracted from the Navajo reservation by Diné and non-Diné mine-workers under leases approved by the federal government.[27] While all miners were subject to higher-than-background-level exposure rates, white miners and Diné miners were treated differently. Paid less than minimum wage, Diné miners were often sent into mines after blasting, before dust had settled, while white miners were allowed to wait until conditions were safer.[28]

Pamphlets distributed to miners in 1959 "minimized the level of concern" while citing "statistically significant" rates of lung cancer in white miners.[29] A 1962 report on rates of cancer in miners also left out data on minority miners, ostensibly to contain the data to a "homogeneous population." Implying that an indigenous population's cells would perform differently from those in bodies of European extraction?

Later, when the risk of lung cancer associated with smoking became public knowledge, miners' elevated lung cancer and lung disease rates were conveniently attributed to that, despite a 1968 survey that found atypically high cancer rates among Diné even though fewer smoked, and those who did, smoked less than non-Diné miners.[30] Denying this universal correlation between uranium mining and lung cancer clouded causation, harming all miners.

RECA eligibility is limited to families that can verify claims related to work in the uranium industry or impacts of the Nevada testing. As of 2015, the US Justice Department spent more than two billion dollars on compensation, but Diné miners and their families remain under-served. Application demands documents affirming a minimum length of employment, excluding informally employed laborers. Diné were often not provided formal papers for work. And traditional practices do not include registration of birth and marriage, so proving family relationships is sometimes impossible.

On the Navajo Reservation, when Diné kids climb rock piles and play in pools on a hot day, they have no way to know which of those rock piles are uranium mine tailings and which pools are fed by water pipes flowing with radioactive waste. Even though federally promoted mining on the reservation ceased in 1983, over 1,000 uranium mines remain on the reservation.[31] Only half of them have cleanup funding allocated. Hazardous surface deposits, tailings, and drain pipes are mostly unmarked.

Homes made of materials sourced on the reservation are also emitting radiation and releasing radon progeny—85% of tested homes were found to have uranium dust in them.[32] Some Diné babies, 36% of Diné males, and 26% of Diné women have concentrations of uranium in their urine that exceed the highest 5% of the US population.[33] It was not the twins that littered the Navajo Reservation with Yé'iitsoh's bones, it was US government and corporate entities. Yet Diné communities, excluded from RECA, are kept in legislative limbo. Compensation for destruction of their land and public health impacts remain subject to political debate, as if redress equals largesse, largely because the scale of devastation seems irreparable within our legislative milieu. Likewise, in Puerto Rico after Hurricane Maria, and in Louisiana's "cancer alley," inequity in hazard response is perpetuated by the very structures of compensation. Telling stories of material histories presents an opportunity to recognize the continuum of structural injustice so we may rebuild our mechanisms of hazard response.

INSTABILITY II—SPACE-AGE UNCERTAINTY

Elon Musk, after taking-on NASA's role in sending humans into space—evidence perhaps of a completed transition from civic to capital authority in America—plans to start-up his own settlement on Mars. New marketing, old news. Plutonium produced from uranium reactor fuel is already powering the Curiosity rover's Mars tours, and for years uranium has been carried back to the stars on deep-space missions like Galileo. On the other hand, Bill Gates wants to apply space-age technology on this planet. TerraPower, a company he launched in 2006 advertises its "aims to provide the world with a more affordable, secure and environmentally friendly form of nuclear energy." The atomic era never ended.

The Center for Strategic and International Studies (CSIS) identifies five discrete periods of uranium governance.[34]

The first, ineptly titled "Uranium Ignorance," correlates with informal, family-scale mining to supply dye markets, the Curies, and the steel industry on the Colorado Plateau. A period of "Uranium Positive Control"—grounded more in control of market competition than in public safety—covered the years of the Atomic Energy Commission's promotion of extraction and management of production. This was a period of whiplash when the horror of "the atom bomb" was immediately followed by popularization of uranium and the beginning of a boom market. Mushroom clouds were normalized with streamlined aesthetics as the styling of modern US manifest destiny.

Federal oversight was reigned in when the Atomic Energy Act of 1954 launched a period of "Uranium Laissez Faire" by relaxing mining and production restrictions to encourage market-driven industrial development. Uranium, now described as necessity, was appareled in prosaic power infrastructure. At the same time, yellowcake—processed ore—was transmuted, like gold, into a publicly traded commodity.

Floyd Odlum went to Utah and started buying shares. Friend of Eisenhower—and the man who fired Orson Welles from RKO Studios—Odlum was famous for having forecast the Depression and profited from it. Working toward vertical integration: he invested in missile technology and the materials of the warheads they would carry. His Atlas Corporation subsidiary, Convair, was completing initial designs for the Atlas nuclear missile program around the time Charlie Steen drilled into his first pitchblende deposit.

Steen turned down Odlum's offer to buy Mi Vida, but eventually, he partnered with him to build a mill. Processing ore into yellowcake near the mines would boost their profits considerably. In a rush, knowing little about mill management, Steen bought land for the mill 750 feet from the Colorado River on an earthquake fault just north of Moab. There, with AEC approval, the partners opened the first independently operated uranium mill. Four years later, despite AEC ceasing to subsidize uranium production, Odlum bought out Steen and renamed it the Atlas Uranium Mill. Odlum was either prescient or his government friends were talkative. That year, the first American utility, Jersey

Central Power and Light Company, contracted to build a nuclear reactor.

Nuclear energy had become cost-competitive with fossil-fuel based energy production. Anti-nuke protesters had a hard time sustaining fervor in the face of arms control treaties and the promise of a no-emissions, domestic fuel supply that obviated the need to deal with Middle Eastern nations well aware of their corner of the market. By 1973, uranium was deemed safe enough to be publicly managed as a private-sector resource open to foreign production.

On the Plateau, the new White Mesa Uranium Mill was under construction in San Juan County. A few years later, County Commissioner Calvin Black, from an old prospecting family in Blanding, fomented the first Sage Brush Rebellion—a uranium-fueled land-law dispute with the Carter administration—backed up by Utah Senator Orrin Hatch. White Mesa opened a year after another whiplash: in 1979, radiation released from a partial nuclear reactor meltdown near Harrisburg, Pennsylvania provoked the public to reconsider the safety of messing with uranium.

CSIS calls that aftermath, a "Uranium Slump." Investment in nuclear technologies collapsed and policy was constructed for storing and managing high-level radioactive waste and spent nuclear fuel (Nuclear Waste Policy Act of 1982). The Atlas Mill was put on standby in 1984 and ceased operations in 1988—fallout from the 1986 accident at Chernobyl in Ukrainian SSR which also inflamed forces overthrowing communism in Russia. Meanwhile, Atlas mill tailings silently leaked radioactive material into the Colorado River, source of drinking water for millions downstream and the water supply for irrigating globally distributed agricultural products.

CSIS positions the state of uranium after 1992 to the present as "Steady," which accords with its flatline in the commodities market but not its social life. In the mid-1990s resurgent Sagebrush Rebels, ostensibly blocked from access to their uranium mines and furious at the Proclamation of Grand Staircase Escalante National Monument, protested by bulldozing roads through designated Wilderness areas. So, when in 1996 the Nuclear Regulatory Commission issued staff guidance on the termination process for privately-held licenses of all conventional uranium mills, they quelled resistance by allowing some states to exercise regulatory authority over uranium recovery facilities in their territory.

The State of Utah was given responsibility for managing Energy Fuels Resources, Inc's (EFR) radioactive materials license for the White Mesa Mill around the same time Atlas Corporation declared bankruptcy to avoid responsibility for costs of a cleanup plan that would be more extensive than just capping the site with earth and walking away. Listed as an UMTRA (Uranium Mill Tailings Remedial Action) site, state and federal tax payers fund an estimated (in 2008) $720 million for relocating Atlas' toxic debris out of the floodplain and Moab.

I have watched over the years, starting in 2005, as the 150-acre, 40-foot-high Atlas Mill tailings were shaped into a monumental ziggurat before transfer north by rail, to be reinterred in a long and flat bunker near Crescent Junction on I-70. When I stopped by in summer 2019, the tailings site was flattened. Long red, dusty terraces—large enough for trucks to drive on—had become a low, deeply scored mound. And the repository site is only remarkable for its roadblock and warning signs down a dirt road beyond a truck stop. In any other space, the physical presence of uranium waste would be outsized, but here, below towering cliffs, it disappears.

Down the road, EFR's still-active White Mesa Uranium Mill is, like the Atlas site, unremarkable. The radon leaking from its tailing cells is invisible. The mill is a hot spot of federal safety violations that are, apparently, acceptable as a lung-cancer hazard for residents of Blanding and the White Mesa Ute community. Utah's Division of Waste Management and Radiation Control filed affidavits defending EFR in a lawsuit aimed to force them to contain the leak.[35] And with the Trump administration as allies, local laissez-faire management of uranium is expanding.

High Country News, in an article, aptly titled "Trump's message for tribes: Let them eat yellowcake" connected the reduction in area of Bears Ears National Monument to lobbying by uranium mine companies and supportive Utah state authorities.[36] The article also references Trump's sympathy with area residents' animus toward Bureau of

Land Management constraints on mine access and ATV routes destroying ancient Puebloan villages in San Juan County. Uranium is the lens through which these latter-day Sagebrush Rebels view their landscape.

And it is also the medium by which they express ethnic bias. Disregard of indigenous communities by Utah government dates back to Latter-Day Saints colonization. Now, by reducing the Monument area, they again open sacred Diné territory to destructive mining operations. Insult added to injury as the Diné await clean-up of defunct mines on the reservation. Buried in the rhetoric is fear that Diné empowerment, like their Radioactive Materials Transportation Act of 2012, might obstruct ore transport from local mines—and waste from across the country—to the White Mesa mill.

Atomic rebels wield a strange propaganda, calling for an authoritarian act in the guise of a move to wrest power from Olympian federal authorities. EFR lobbyists met with the Interior Department to defend their other big uranium investment in the area, the Daneros Mine. Located three miles from the original Bears Ears boundary—included within the lands proposed for national monument status by the Bears Ears Inter-Tribal Coalition in 2015—it is only connected to EFR's White Mesa mill by route 95 cutting through the Shash Jaa' unit of the now-reduced monument.

The impact of 15 uranium ore-trucks a day passing through the monument might be insidiously toxic. They would definitely be culturally disruptive to the experience of monument visitors. As I slowed down and reached out the car window to take a picture on Route 95 last year, I had to quickly duck back in as an unmarked white ore-truck passed me in the oncoming traffic lane. A few minutes later, as he crawled up a steep slope—probably ore-loaded—I could not see around him to pass. Slow driving, watching fumes spew, I wondered what was in the dust hitting my windshield. Probably the truck came from the Daneros mine, but I cannot say with certainty because it is not on Google maps. I followed the truck to the White Mesa mill.

Like a truck-eclipsed view of the road ahead, the mythos of materials obscures our vision of landscape. Parsing the materiel leveraged in our relationships with each other

Crescent
Junction

CLIFFS

BOOK

SAGERS

FLAT

WHIPSAW

SALT

VALLEY

BLUE HIGHLANDS

COURTHOUSE
PASTURE

MERRIMAC
BUTTE

MAT MARTIN
POINT

BIG
BEND

ARTHS
PASTURE

Uranium Mill Tailings
Remedial Action

Moab

SPANISH

Bull Canyon

THE KNOLL

BIG FLAT

MOAB, UT

WHITBECK
ROCK

DEAD HORSE
PARK

GRAND CO.
SAN JUAN CO.

Uranium Mill Tailings
Remedial Site

PRITCHETT
NATURAL BRIDGE

DEAD HORSE
POINT

2006

2011

2014

2015

and nature reveals hidden workings. Mundane, seemingly necessary impositions turn out to be as romantic as tales of gods and monsters, and mysteries are fronts for dark pragmatic acts. Recently, the Trump administration, prompted by forces unknown—market demand for uranium is flat—expanded the area of the Daneros mine by nine times its existing size and agreed to increase the tonnage of ore moving through what is left of Bears Ears National Monument. Is this portent a more atomically-active future or just another bone thrown to appease the rebels? Unquestionably, it is evidence of the same-old, same-old false binary between social and environmental necessity. Such flawed logic connects irradiated Cedar City sheep and Din families directly to black communities disproportionately affected by COVID-19 and climate heating, and to every community caught between the rock of mythic economic imperative and the hard place of disturbed urban ecologies. Without mapping our material and materiel, we cannot ask how they have shaped us; and we are doomed to repeat errors and injustice.

ENDNOTES

1   Carl Sagan, "Cosmos: A Personal Voyage," Public Broadcasting Service, New York, 1980.

2   "Little Boy" dropped on Hiroshima was a uranium device. "Gadget" and "Fat-Man," the bomb dropped on Nagasaki, were plutonium bombs, but even the plutonium was produced from uranium in a nuclear processing plant.

3   Seth Lloyd, "The Universe," in *This Idea Must Die*, ed. John Brockman (NY: Harper Perennial, 2015), 11–14.

4   Simon Mitton, "Fred Hoyle: pioneer in nuclear astrophysics," *Cern-Courier* (July 17, 2005), accessed May 22, 2020, https://cerncourier.com/a/fred-hoyle-pioneer-in-nuclear-astrophysics/.

5   Ray Sandersrfor, "When Stellar Metallicity Sparks Planet Formation," *Astrobiology Magazine* (April 11, 2012), accessed October 19, 2019, http://www.spacedaily.com/reports/When_Stellar_Metallicity_Sparks_Planet_Formation_999.html.

6   Millirem is a unit of biologically effective radiation dose. It is a calculation based on the physical dose of ionizing radiation (in rads) multiplied by a quality factor (in RBE) that is relative to the specific biological effectiveness of the type of particle (x-rays, gamma rays, beta, and alpha).

7   Characterization of widely-used terms: behavior, structure, instability, and complexity are derived from Allan and Hoekstra's discussion of "The Ecosystem Criterion" in which, among many other things, they clarify aspects of subjectivity in common ecosystem criteria. *See* Tim-

othy F. H. Allen and Thomas W. Hoekstra, *Toward a Unified Ecology* (NY: Columbia University Press, 1992).

8   Physicists Marie (née Sklodowska) and Pierre Curie shared a Nobel Prize with Becquerel in 1903 for research into radiation—Marie led the research according to Pierre. Seven years later, Marie won a second prize, solo, in Chemistry, for the discovery of Polonium and Radium. Enthusiastic about the medical potential of radioactive minerals, Marie chose to ignore their hazard and she died at the age of 66 of leukemia from radiation exposure. Pierre had died years earlier, at 46 after being hit by a bus.

9   Ray C. Ringholz, *Uranium Frenzy: Saga of the Nuclear West* (UT: Utah State University Press, 2002), 6.

10   Gary Shumway guest edited a series of issues of Blue Mountain Shadows, a local magazine of San Juan County history, on mining in San Juan County: volumes 25–27. These, and the Blanding Visitors' Center's extensive displays on local mining history were primary sources for area mining information. See also: Ringholz, *Uranium Frenzy.*

11   IAEA, *Geological Classification of Uranium Deposits and Description of Selected Examples,* IAEA TECDOC Series (Vienna: International Atomic Energy Agency, 2018).

12   Ringholz, *Uranium Frenzy.*

13   Ibid.

14   Dennis E. „Pete" Byrd, Sr., interview by Detta Dahl, 2003, Moab Museum, Eastern Utah Human History Library,https://www.moabmuseum.org/wp-content/uploads/2016/05/03-Primary-Subject-Personal-history-first-days-of-Moab%E2%80%99s-uranium-boom.pdf.

15   Ringholz, *Uranium Frenzy*, 73.

16   Mark Steen, "My Old Man the Uranium King," *The Canyon County Zephyr*, June 1, 2016, http://www.canyoncountryzephyr.com/2016/10/03/my-old-man-the-uranium-king-part-4by-marksteen/.

17   Ringholz, *Uranium Frenzy*, 13.

18   Georges Cuvier, *Discourse on the Revolutionary Upheavals on the Surface of the Globe*, (1825), http://www.victorianweb.org/science/science_texts/cuvier/cuvier-e.htm.

19   Katharine Lee Bates, "Pikes Peak," *The Congregationalist,* 1895.

20   Doug Brugge and Rob Goble, "The History of Uranium Mining and the Navajo People," *American Journal of Public Health* 92, no. 9 (2002): 1410–1419, doi/10.2105/AJPH.92.9.1410.

21   The test code-named "Harry" is discussed here, but the test "Simon" (April 25th) mushroom cloud reached an unexpectedly high altitude and drifted east across the country.

22   John G. Fuller, *The Day We Bombed Utah* (NY: New American Library, 1984), 18–20.

23   Experiments included: injections of U-234 and U-235 to induce kidney injury (1946–47, University of Rochester), exposure of hands to radioactive material (1953, Monsanto), irradiation of prisoners' testicles by x-rays (1963–71, University of Washington), and even administration of radioactive iron to pregnant women (late 1940s, Vanderbilt University).

24   See the image on title page of the chapter.

25   Lloyd L. Lee, *Decolonizing the Navajo Nation: The Lessons of the Naabaahii* (Alberquerque: 2nd Annual National Indian Education Association (NIEA)Convention & Tradeshow, 2011), accessed May 28, 2020, https://files.eric.ed.gov/fulltext/ED528280.pdf.

26   Robert S. McPherson, *Navajo Land, Navajo Culture* (Norman, OK: Universtiy of Oklahoma Press, 2003), 159.
27   United States Environmental Protection Agency, "Navajo Nation: Cleaning Up Abandoned Uranium Mines," accessed November 1, 2019, https://www.epa.gov/navajo-nation-uranium-cleanup.
28   Susan E. Dawson, "Navajo Uranium Workers and the Effects of Occupational Illnesses: A Case Study," *Human Organization* 51, no. 4 (1992): 389-397, DOI: 10.1097/HP.0b013e3182243a7a.
29   Doug Brugge and Rob Goble, "The History of Uranium Mining."
30   Ibid.
31   Ibid.
32   Autumne Spanne, "Uranium pervades homes on and near Navajo Nation," *High Country News*, August 27, 2017, accessed October 23, 2019, https://www.hcn.org/articles/pollution-epa-budget-cuts-threat-en-to-slow-uranium-cleanup-at-navajo-nation.
33   *America's Nuclear Past: Examining the Effects of Radiation in Indian Country: Field Hearing before the Committee on Indian Affairs, US Senate*, October 7, 2019, statement of Dr. Loretta Christensen, Chief Medical Officer, Navajo Area Office, Indian Health Service, US Department of Health and Human Services (Washington, D.C., 2019), accessed November 1, 2019, https://www.indian.senate.gov/sites/default/files/10.07.19%20Dr.%20Christensen%20IHS%20Testimo-ny%20on%20Radiation%20in%20Indian%20Country.pdf, 7.
34   Sharon Squassoni, et al., *Governing Uranium in the United States* (Washington DC: Center for Strategic International Studies Proliferation Prevention Program, 2014).
35   Emma Penrod, "Did testimony from state environmental regulators undermine a lawsuit against a Utah uranium mill?," *Salt Lake Tribune*, October 8, 2017, accessed June 4, 2020, https://www.sltrib.com/news/2017/10/08/did-testimony-from-state-environmental-regula-tors-undermine-a-lawsuit-against-a-utah-uranium-mill/.
36   Jacqueline Keeler, "Trump's message for tribes: Let them eat yellow-cake," *High Country News*, December 12, 2017, accessed April 14, 2019, https://www.hcn.org/articles/tribal-affairs-trumps-message-for-tribes-let-them-eat-yellowcake.

BIBLIOGRAPHY

Allen, Timothy F. H. and Thomas W. Hoekstra. *Toward a Unified Ecology.* New York: Columbia University Press, 1992.

America's Nuclear Past: Examining the Effects of Radiation in Indian Country: Field Hearing before the Committee on Indian Affairs, US Senate. October 7, 2019. Statement of Dr. Loretta Christensen, Chief Medical Officer, Navajo Area Office, Indian Health Service, US Department of Health and Human Services. Washington, D.C., 2019. Accessed November 1, 2019. https://www.indian.senate.gov/sites/default/files/10.07.19%20Dr.%20Christensen%20IHS%20Testimony%20on%20Radiation%20in%20Indian%20Country.pdf.

Bates, Katharine Lee. "Pikes Peak." *The Congregationalist.* 1895.

Brugge, Doug and Rob Goble. "The History of Uranium Mining and the Navajo People." *American Journal of Public Health* 92, no. 9 (2002): 1410-1419. doi/10.2105/AJPH.92.9.1410.

Byrd, Sr., Dennis E. "Pete." Interview by Detta Dahl. 2003. Moab Museum, Eastern Utah Human History Library. https://www.moabmuseum.org/wp-content/uploads/2016/05/03-Primary-Subject-Personal-history-first-days-of-Moab%E2%80%99s-uranium-boom.pdf.

Cuvier, Georges. *Discourse on the Revolutionary Upheavals on the Surface of the Globe.* 1825. http://www.victorianweb.org/science/science_texts/cuvier/cuvier-e.htm.

Dawson, Susan E. "Navajo Uranium Workers and the Effects of Occupational Illnesses: A Case Study." *Human Organization* 51, no. 4 (1992): 389-397. DOI: 10.1097/HP.0b013e3182243a7a.

Fuller, John G. *The Day We Bombed Utah.* New York: New American Library, 1984.

IAEA. *Geological Classification of Uranium Deposits and Description of Selected Examples.* IAEA TECDOC Series. Vienna: International Atomic Energy Agency, 2018.

Keeler, Jacqueline. "Trump's Message for Tribes: Let Them Eat Yellowcake." *High Country News.* December 12, 2017. Accessed April 14, 2019. https://www.hcn.org/articles/tribal-affairs-trumps-message-for-tribes-let-them-eat-yellowcake.

Larson, Richard B. and Volker Broome. "The First Stars in the Universe." *Scientific American* (January 19, 2009). Accessed 22, 2020. https://www.scientificamerican.com/article/the-first-stars-in-the-un/.

Lee, Lloyd L. *Decolonizing the Navajo Nation: The Lessons of the Naabaahii.* Alberquerque: 2nd Annual National Indian Education Association (NIEA)Convention & Tradeshow, 2011. Accessed May 28, 2020. https://files.eric.ed.gov/fulltext/ED528280.pdf.

Lloyd, Seth. "The Universe." In *This Idea Must Die*, edited by John Brockman. New York: Harper Perennial, 2015.

Marchi, S., R. M. Canup, and R. J. Walker. "Heterogeneous Delivery of Silicate and Metal to the Earth by Large Planetesimals." *Nature Geoscience* 11 (2018): 77-81. doi: 10.1038/s41561-017-0022-3.

McPherson, Robert S. *Navajo Land, Navajo Culture.* Norman, OK: Universtiy of Oklahoma Press, 2003.

Mitton, Simon. "Fred Hoyle: Pioneer in Nuclear Astrophysics." *CernCourier.* July 17, 2005. Accessed May 22, 2020. https://cerncourier.com/a/fred-hoyle-pioneer-in-nuclear-astrophysics/.

Pasternak, Judy. *Yellow Dirt: A Poisoned Land and the Betrayal of the Navajos.* New York: Free Press, 2011.

Penrod, Emma. "Did Testimony from State Environmental Regulators Undermine a Lawsuit Against a Utah Uranium Mill?" *Salt Lake Tribune.* October 8, 2017. Accessed June 4, 2020. https://www.sltrib.com/news/2017/10/08/did-testimony-from-state-environmental-regulators-undermine-a-lawsuit-against-a-utah-uranium-mill/.

Ringholz, Ray C. *Uranium Frenzy: Saga of the Nuclear West.* Salt Lake City: Utah State University Press, 2002.

Rock, Tommy, Ricky Camplain, Nicolette Teufel-Shone, and Jani C. Ingram. "Traditional Sheep Consumption by Navajo People in Cameron, Arizona." *International Journal of Environmental Research and Public Health* (2019): 1–13. Accessed November 1, 2019. https://www.mdpi.com/1660-4601/16/21/4195/htm.

Sagan, Carl. "Cosmos: A Personal Voyage." Public Broadcasting Service. New York, 1980.

Sandersrfor, Ray. "When Stellar Metallicity Sparks Planet Formation." *Astrobiology Magazine*. April 11, 2011. Accessed October 19, 2019. http://www.spacedaily.com/reports/When_Stellar_Metallicity_Sparks_Planet_Formation_999.html.

Schwartz, Stephen I. *Atomic Audit: The Costs and Consequences of US Nuclear Weapons Since 1940*. Washington DC: Brookings Institution, 1998.

Smolin, Lee. "The Big Bang." In *This Idea Must Die*, ed. John Brockman. New York: Harper Perennial, 2015.

Spanne, Autumne. "Uranium Pervades Homes on and Near Navajo Nation." *High Country News*. August 27, 2017. Accessed October 23, 2019. https://www.hcn.org/articles/pollution-epa-budget-cuts-threaten-to-slow-uranium-cleanup-at-navajo-nation.

Steen, Mark. "My Old Man the Uranium King." *The Canyon County Zephyr*. June 1, 2016. http://www.canyoncountryzephyr.com/2016/10/03/my-old-man-the-uranium-king-part-4by-mark-steen/.

Squassoni, Sharon, Robert Kim, and Jacob Greenber. *Governing Uranium in the United States*. Washington, DC: Center for Strategic International Studies Proliferation Prevention Program, 2014.

United States Environmental Protection Agency. "Navajo Nation: Cleaning Up Abandoned Uranium Mines." Accessed November 1, 2019. https://www.epa.gov/navajo-nation-uranium-cleanup.

IMAGE CITATIONS + CREDITS

Listed by Page Number

47   Photo: Denise Hoffman Brandt. Basemap Copyright: © 2011 National Geographic Society, i-cubed

49   US Geological Survey, "Attributes for MRB_E2RF1 Catchments by Major River Basins in the Conterminous United States: Physiographic Provinces," 2010. Basemap sources: Esri, Airbus DS, USGS, NGA, NASA, CGIAR, N Robinson, NCEAS, NLS, OS, NMA, Geodatastyrelsen, Rijkswaterstaat, GSA, Geoland, FEMA, Intermap and the GIS user community
     Research Assistant: Chloe Nagraj

51   Photo: Denise Hoffman Brandt; Map: US Army, "Trinity Test Site," 1945.

55   CERN Courier. "Fred Hoyle: Pioneer in Nuclear Astrophysics," July 17, 2005. https://cerncourier.com/a/fred-hoyle-pioneer-in-nuclear-astrophysics/.

59   "Uranium Production | Uranium Output - World Nuclear Association." Accessed October 29, 2020. https://world-nuclear.org/information-library/facts-and-figures/uranium-production-figures.aspx.
     Research Assistant: Aleksander De Mott

63   US EPA, OAR. "Radioactive Decay." Overviews and Factsheets. US EPA, May 22, 2015. https://www.epa.gov/radiation/radioactive-decay.
     Research Assistant: Tian Wang

67   Basemap copyright: © 2014 National Geographic Society, i-cubed; Utah GIS Portal. "Utah Mapping Portal." Accessed October 29, 2019. https://gis.utah.gov/.
     Research Assistant: Tian Wang

70   US Geological Survey, "Physiographic divisions of the conterminous U. S.," 1946; Basemap Sources: Esri, Airbus DS, USGS, NGA, NASA, CGIAR, N Robinson, NCEAS, NLS, OS, NMA, Geodatastyrelsen, Rijkswaterstaat, GSA, Geoland, FEMA, Intermap and the GIS user community.
     Research Assistant: Tian Wang

73   US Geological Survey, "Geology of Uravan Mineral Belt," 1952.
     Research Assistant: Tian Wang

78   Research Assistant: Tian Wang

80   Sheep: Flickr Creative Commons; Nuclear test: US Army, "Troops of the Battalion Combat Team, blast at Yucca Flats," 1951; Map: Richard Miller, "Areas crossed by two or more radioactive clouds during the era of nuclear testing in the American Southwest, 1951-62." Under the Cloud: The Decades of Nuclear Testing. Two-Sixty Press, 1999.

82   Mushroom cloud with ships below during Operation Crossroads nuclear weapons test on Bikini Atoll. Bikini Atoll Marshall Islands, 1946. [July] Photograph. https://www.loc.gov/item/2012648160/. "Nuclear Testing at Bikini Atoll." In Wikipedia, October 19, 2020. https://en.wikipedia.org/w/index.php?title=Nuclear_testing_at_Bikini_Atoll&oldid=984401258.

89   Photos: Denise Hoffman Brandt. Map: US Geological Survey, 1983, USGS 1:100000-scale Quadrangle for Moab, UT 1983: US Geological Survey.

Tracing the green energy paradox across battery, body, landscape and cosmos

by Matthew Seibert

...the unearthing of the Carnaval Sanpedrino, a pre-Columbian celebration, is a kind of opening of a passage where the energies of the underworld are released to live amid us...

- Jorge Muñoz Coca
Atacameño Leader, Solcor Community
Member, Observatorio Plurinacional de Salares Andinos

Payment was made to the land under the shade of an old carob tree and under the protective presence of a local mallku [or protective spirit], the Licancabur Volcano. The music began and people followed, dancing the typical carnival cuecas [the Chilean national dance], branches of corn in their hands and thrown flour powdering their faces. We left the field and approached the river, its channel loaded with water, rich like chocolate, due to winter storms. We felt the drums of the carnival, the warmth of the river reaching our knees as we entered, when the colors of the sunset began to blend with the rain now falling across us and the desert. Surreal. Absorbed in a landscape composed of sand, volcanoes, and clouds of such immensity finally succeeding in reaching down to touch the hot sand, I felt that the land was alive, more alive than ever, and that I, small and insignificant, was part of it.

- Ramón Morales Balcázar
Chilean Activist
Member, Fundación Tantí
Member, Observatorio Plurinacional de Salares Andinos

Chuquicamata

Calama

23

El Tatio Geysers

LINCANCABUR VOLCANO

27

San Pedro de Atacama

Toconao

LASCAR VOLCANO

SALAR DE ATACAMA

Peine

CHILE

BOLIVIA

101 • Matthew Seibert

Edge of Salar de Atacama

Basemap Source: Esri, Maxar, GeoEye, Earthstar Geographics, CNES/Airbus DS, USDA, USGS, AeroGRID, IGN, and the GIS User Community

 **faint** hum at first.

Ears attuned to the slightest change in a silence that breeds isolation.

Could it be…?!

Atop a soft pile of earth wanting to give way with every step, I strain my eyes on the horizon, body tensing as if to will it into materializing. I think I see something… Like a speck of dust in one's eye, blurry and hard to pinpoint, a dark smudge teases its presence on a horizon glassy with mirage. It has to be… I text Erica, "a car's coming!" Somehow my cell plan allows me to communicate thousands of miles back home, but provides no data for local saviors. My partner's digital presence is the only thing holding off despair.

The car bobs a bit, sailing along the dirt road at a nice clip. It has to stop… I'm in the middle of nowhere, in the world's driest desert, sticking out like… well, like a pink flamingo in a Martian landscape. Flamingos, those exotic birds more familiar in the tropics, are one of the major reasons tourists flock to this surreal region of geysers, volcanoes, and icy, high-altitude lagoons where a special, seemingly confused type of pink bird enjoys feeding on brine shrimp.

Shhhhheeewp!

It didn't stop…

How could it not stop…?!

What heartless soul wouldn't stop for a pale, ginger victim under a rising desert sun?

"WTF???" Erica responded from the comfort of her couch in Virginia.

It's been 13 hours since the front tires of my rental car suddenly dropped into the earth with a heart-sinking *slumpf*. Mud trap. In the desert of all places. Hours from the nearest real town. Victim to an invisible hydrology operating underground in an alien world. I pitched my tent in resignation that night, with an ungodly beautiful view of the cosmos above, the Milky Way like an illuminated blanket, and my cell phone clutched in my pocket—a powerful talisman warding off the creep of hopelessness. As long as the battery would last that is.

This cell phone battery was on my mind for more reasons than one. I had come here to research lithium—a powerful material ubiquitous in contemporary life yet quite unknown to most. I had come here to know more about lithium's landscape, its climate, its people, its mining. I had come here to know what can only be known from putting foot to earth, skin under a local sun. My phone's lithium-ion battery was composed of this element, most mobile electronics' batteries are, most dreams of a green energy future are composed around lithium in fact. From electric vehicles to supermassive smart grid batteries, lithium is currently the key to maximizing energy storage, its delivery, and the cycling between the two as power continuously ebbs and flows between production and demand. Without lithium we're limited where and when we can have easy access to large wells of power while on the move, an addiction difficult to kick.

Now eight in the morning the following day, I'm standing at the intersection of two dirt roads, alternating between a longing gaze at the horizon and a despondent kick of the dirt, praying for company to aid in my rescue. One has quite a luxury of time to think about lithium's many relations when they're stranded on the edge of a salar (or salt flat, as they are often called) in the only true desert to receive less precipitation than those frozen deserts at the globe's poles. Its unbelievable flatness extending like a sea in front of me, the volcanoes Licancabur and Lascar ascend like a wall at my back, hinting at the geological superstructure that makes this salar exceedingly special. This is the Salar de Atacama, remote in Northern Chile's Atacama Desert, responsible for most of the world's lithium. One of the first three primordial elements given birth to in the Big Bang, lithium is rare but critically important. Not only does it hold the key to digital mobility and the future of an ostensibly green energy economy, it is central to the functioning of the human body, particularly to mental health. The previous night couldn't have been more emblematic. There I was, a few miles east of the world's largest lithium brine production, in awe of the universe above, and clinching my lithium-powered talisman in a faint panic.

In tracing lithium's agency across telescoping scales of battery, body, landscape, and cosmos via foot, camera, conversation, cartography, and design speculation, an even more entangled narrative emerges. Lithium's Utopian promise relies on a huge blind spot for its users: as part of an interconnected network of relations with other elements, landscapes, and bodies, lithium's acquisition currently depends on extractive processes that are argued to damage ecosystems and exploit neighboring communities. Ramón Morales Balcázar—activist, PhD student, and member of Fundación Tantí[1] and the Observatorio Plurinacional de Salares Andinos[2]—gave voice to the degradation and exploitation I had read about before arriving in Chile. "The deepening of environmental damage is due to the expansion that lithium mining has achieved in recent years," he tells me.[3] The regulatory framework inherited from Augusto Pinochet's military dictatorship from 1973–1990 has enabled transnational mining companies to claim the brine used in lithium production is not water, as it cannot be used directly for watering crops or drinking. Thus, the mining operations, "with the support of the Chilean government, come to install their anthropocentric and utilitarian vision of nature, ignoring the science that explains that the basin is a

CHILE 1:50.000

Battery, Body, Landscape, Cosmos

complex system and that these fossil waters are the basis of life here," Balcázar elaborates.

The insatiable hunger for lithium-ion batteries, in phones, laptops, the massive ones propelling increasingly popular electric vehicles, and the supermassive ones potentially managing future green energy grids have been correlated with dropping laguna (or lagoon or lake) levels in and around the salar.[4] The laguna's brine shrimp populations then dwindle, and their tourist-drawing flamingo flamboyances (as groups of flamingos are most appropriately called) disappear. Neighboring towns and communities, with significant Indigenous populations, have also found their agricultural and drinking wells drying up. The rechargeable green energy future is not so green after all.[5] Consumption of any form almost certainly comes with ecological cost.

Looking out on the salar early that following morning, awash in a rosy hue, I had ample time to consider the agency of lithium and the processes it sets forth, scalar in their influence and dynamic in their reciprocity. From the atom to the global economy, lithium can be found as an active medium, not just being used or acted upon but enabling and limiting complex and ever-changing relationships beyond itself. One action necessarily triggers another one, or two, or ten. Or multitudes.

Battery, body, landscape, cosmos.

Push, pull, bend, entangle.

A knot of knots.

Lithium the lively force of constraint and allowance, support and slippage, beauty and bedlam.

BATTERY: POWERING MODERN SOCIETY'S MOBILE WAY OF LIFE

With an atomic number of three on the periodic table— composed of three protons, three electrons, and usually four neutrons—lithium possesses many unique and desirable properties. Its conductivity of heat and electricity is of particular value. These properties have driven its use in the production and processing of a number of common objects and materials in our everyday lives, from heat-resistant glass on electric stovetops to lubricating greases in various industrial uses. But by far the most ubiquitous use of the element today is in the form of lithium-ion rechargeable batteries. Most people carry around lithium every day, in their laptops, tablets, smartphones, and other rechargeable devices. Lithium is not only incredibly light—the lightest of all metals—making it particularly convenient for today's mobile lifestyle, but it can store an impressive amount of energy in its highly reactive atomic bonds. This means lithium-ion batteries possess an incredibly high energy density.

To put this in perspective: the lead-acid battery, like the battery found in most conventional cars, is the first and still most widely utilized type of rechargeable battery, usually storing about 25 watt-hours in a kilogram. A NiMH (or nickel-metal hydride) battery, such as the common rechargeable AA, typically stores 60 to 70 watt-hours per kilogram. An average lithium-ion battery stores a commanding 150 watt-hours of electricity in a kilogram. Lithium-ion batteries thus have a dominating advantage, accounting for their ubiquitous prevalence starting around 2010. Additionally, they hold their charge longer, have no memory effect (meaning they don't have to be fully discharged before recharging), and can be discharged and recharged hundreds of times. This is why they were revolutionary and are used in so many mobile electronics. Without lithium, the average laptop, if powered by a lead-acid battery, would weigh 25 pounds, the equivalent of an average 2-year-old toddler.

We're rather accustomed to having electricity available at our fingertips, able to be tapped into at one's convenience, enabling modern life as we know it. We're utterly naked when our phone dies, lost when the power goes out, aghast when we can't find an outlet at the airport terminal. As Seth Fletcher puts it in *Bottled Lightning: Superbatteries, Electric Cars, and the New Lithium Economy*, "Before the invention of the battery in the first year of the nineteenth century, electricity as we know it today—as a stream of electrons that can be made to do our bidding—didn't exist."[6] The ability to store this stream of electrons, this nonliving force, made it reliable, and that reliability changed everything. In 1800, Alessandro Giuseppe Antonio Anastasio Volta (think if we used his full name to describe a volt of energy!) invented the first battery, a "column of little sandwich cookies, each one a zinc and copper disc separated by brine-soaked cardboard."[7] He called his invention the "organe electrique

2303 m

The Lightest Metal

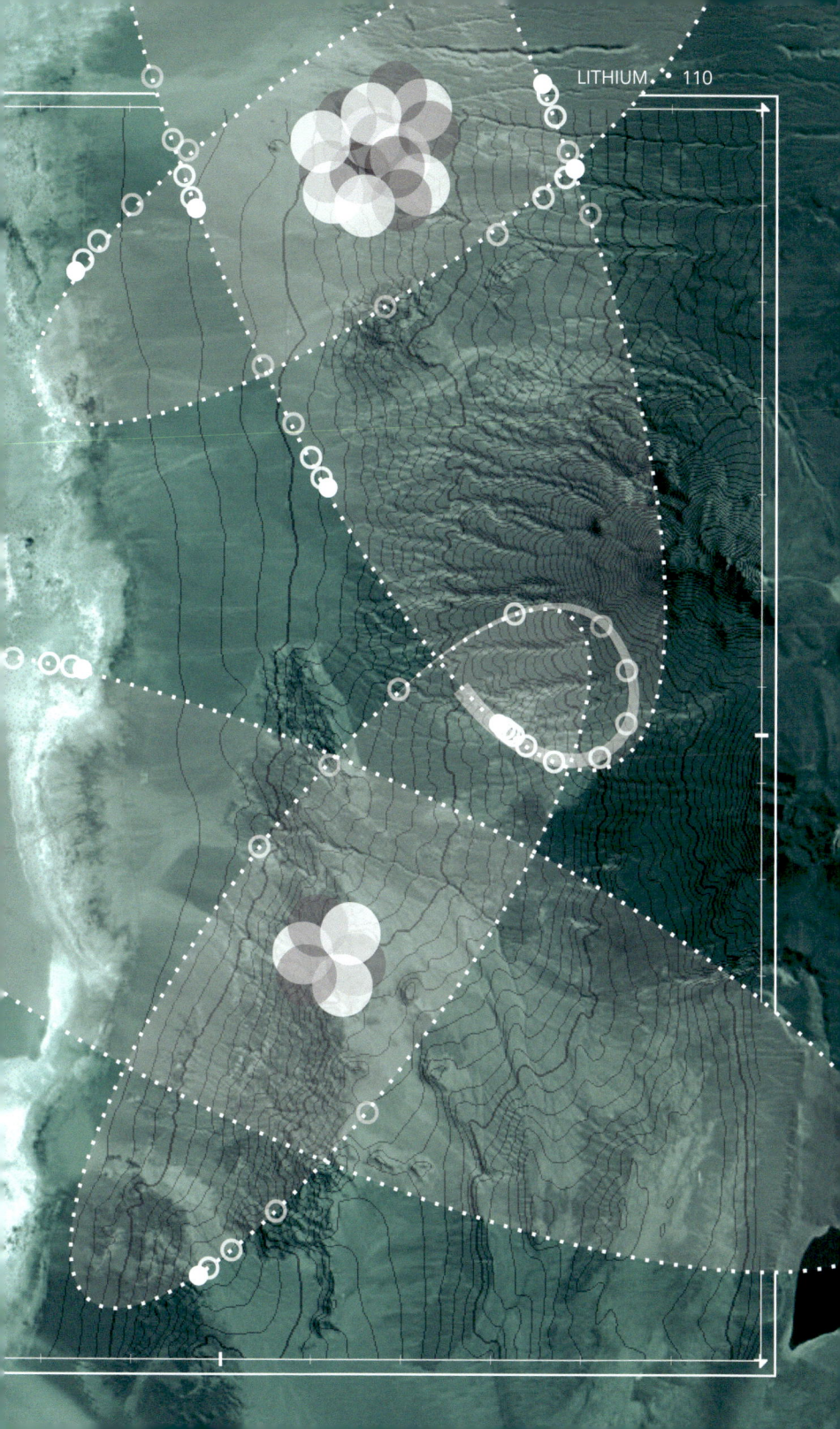

artificial," or the artificial electric organ. It "transformed our civilization," exclaimed historian John L. Heilbron.[8]

Over time the battery evolved from an artificial electric organ to the many types we are familiar with today, of course. As electricity is best understood as a stream of electrons, the key to maximizing a battery's energy-storing potential is to cram as many electrons into the smallest and lightest container possible. But you can't just find these floating in the air and jostle them in some plastic bottle. Electrons have to be muscled loose from an atom. And these atoms come with protons and neutrons. In other words, with every electron comes the baggage of its atomic family members, "both of which are eighteen hundred times as massive as an electron."[9] If an electron was the size of a penny, neutrons and protons would be as big as a bowling ball. In comparison, that lead-acid battery earlier has 82 protons and 125 neutrons for each usable electron on its lead atom, while a lithium atom in the battery of this laptop I am typing on is feather light with only 3 protons and 4 neutrons.

Moreover, lithium is eager to proffer its outermost electron. It wants to give it away in exchange for grouped stability. Highly reactive, lithium is too volatile to exist in nature in its pure, elemental form. It's never found alone, but rather in aggregated minerals like spodumene and petalite, compounds of aluminum, silicon, and lithium. Or, in the case of the Salar de Atacama, Chile, in brine mixtures. When not steadied in material assemblies like brines and hard-rock minerals, for example when combining the active ingredients of a lithium-ion battery's two electrodes, under the right conditions, the recipe for a powerful explosive is generated. A battery is designed to wield this volatility, calming and harnessing such hot-headed propensity. Fletcher describes this as such:

> By putting an electrolyte bridge between those two electrodes, a battery keeps those bomb parts at a safe distance from each other, placing an explosion in suspended animation, creating a chemical system throbbing with energy that can be redirected and exploited.[10]

Society's relationship with such redirected and exploited power is the story of modernity.[11] The arrival of the lithium-ion battery in the early 90s gave way to the cellular phone, first enabling simple phone calls in inconvenient locales, until essentially becoming extensions of the self.[12] Additionally, with the explosion of such mobile devices, a multimillion dollar industry has emerged purely for their accessorization. It's hard to imagine life without such portable, energy-dense digital prosthetics.

Another mobile use, accounting for the fastest growing application and exponential surge in demand, is the exploding market for electric vehicles (EV).[13] To meet consumers' expectation for long range travel, high performance, and "refuel-ability," as conditioned by around 100 years of conventional gas vehicle use, EV designers and manufacturers need maximal energy density to power the engine. A combination of lightness and high energy conductivity enables EV's to compete with the size, performance, and aesthetic expectations car consumers have cultivated in coevolution with the combustion engine over the last century. The need for large quantities of lithium-ion batteries to power electric vehicles contributes to the predicted six-fold increase in demand by 2025.[14] This surge has directly affected a surge for its elemental component: high-grade lithium.

As I stood there on the salar, eating dry cereal and awaiting rescue, the tranquility of a desert morning disguised this surging demand. Both desperately stranded and in a state of perpetual wonder, I thought about the lithium atoms in my cell phone, in the massive batteries of future electric vehicles, that these atoms" are among the oldest pieces of matter in the universe,"[15] and yet a revolutionary force veritably co-defining modern society and its future.

LANDSCAPE: READING THE ATACAMA DESERT'S GEOLOGICAL THEATER

As 0.002% of the Earth's crust, lithium is rare.[16] That's like 3 people crammed throughout a crowded subway train in New York City, or 5 Cheerios in a box. So not exceedingly rare, as, less we forget, it was the third element forged in the universe, but, as the *Handbook of Lithium and Natural Calcium* clarifies, though lithium is

> *found in many rocks and some brines, [it's] always in very low concentrations. There are a fairly large number of both lithium mineral and brine deposits*

Manhattan
Scales of Extraction
Basemap source: Esri, Maxar, GeoEye, Earthstar Geographics, CNES/Airbus DS, USDA, USGS, AeroGRID, IGN, and the GIS User Community

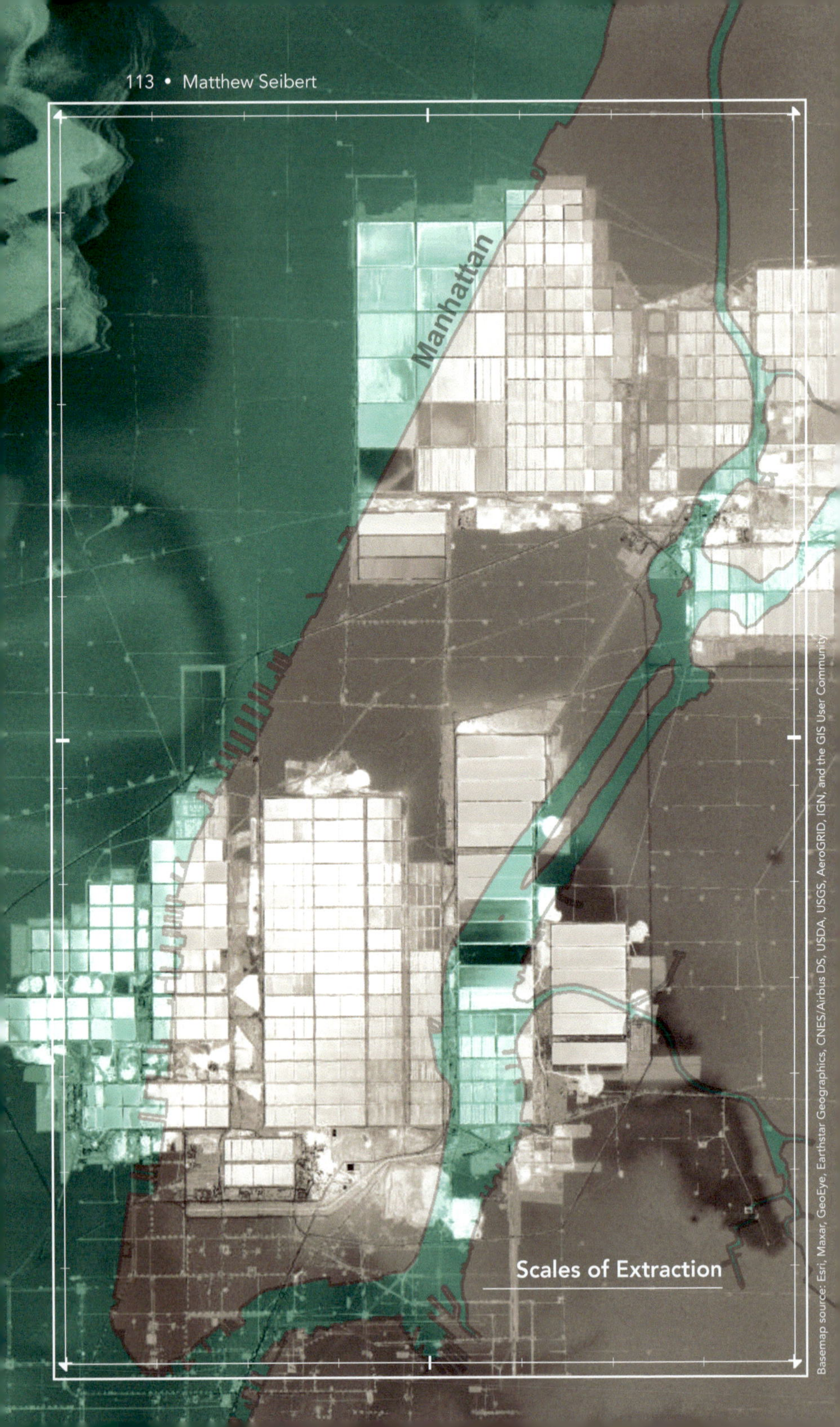

but only comparatively few of them are of actual or potential commercial value. Many are very small, [and many] others are too low in grade.[17]

The Salar de Atacama, however, is not only just one of the few, but perhaps the most economically extractable lithium brine deposit in the world. This becomes quite evident when one browses satellite imagery online. Vibrant cells of saturated color—turquoise, cerulean blue, chartreuse, a brilliant glowing white—stand in stark contrast to a salar colored like the bottom of a crepe-soled shoe: dull brown. These are the evaporation ponds where lithium is concentrated under the desert sun. Taken together, they form an area of approximately 33 square miles (85 km^2), a veritable Manhattan of pools, piles, and berms. And growing. When I ask Ramón Morales Balcázar the first thing that comes to mind with the word lithium, he answers, "Evaporation pools. White piles of mining waste juxtaposed against volcanoes and a hyper blue sky." It's a powerful sight felt as much as seen, not easily forgotten.

Standing on the salar's edge, the very bottom of a closed basin in Northern Chile's Antofagasta Province, I look around. To my east, a towering mountain range thrusts above a soft and continuous slope of tuff and barren earth. The highest peak, Licancabur, is a *mallku*—or protective spirit—of the Indigenous peoples, soaring to almost 20,000 ft (~6,000 m) and frosted with ice. This is a part of the most volcanically active area of the Andes mountain range, the formidable result of the Nazca and South American tectonic plates crashing into one another around 45 million years ago, give or take a few tens of millions of years. The Nazca Plate buckled and ground under the South American Plate in a process known as subduction, slowly, yet violently, pushing earth and rock skywards, tens of thousands of feet into the air.[18] The uplift enabled massive volumes of magma to escape the earth's mantle, forming an arc of volcanoes where it could break through the crust and an underground network of fluid veins and dikes (vertical channels) where it couldn't. This land has always been rather extreme, even before the Andes began reaching for the heavens. The Atacama is thought to be the oldest continuously arid region on earth.[19] The presence of some evaporite formations (or mineral sediments crystallized by evaporation) suggest arid conditions persisted for the last 200 million years, back when dinosaurs began populating the Earth. The

Atacama Desert sips just over half an inch (15 mm) a year of precipitation presently, with some weather stations having never recorded rain. One of the most dependable sources of water in this region is snowmelt, which has slowly, but incrementally, carved vast gullies into miles-long slopes. I face this eroded mass while on the lookout for savior vehicles, the gullies embellishing a rather featureless form with artful lines. In the morning light they catch a warm glow on their eastern-facing facets, like a landscape *kintsugi*. A scaled-up suggestion of the centuries-old Japanese art of repairing broken pottery with gold, the gullies' weathering is highlighted as history and process. A reminder that landscapes, like objects, are far from static and rarely viewed beyond a single snapshot of time. One must read them as narratives rather. As sagas.

To the north lies miles of salar, saline lagunas, the mouth of the San Pedro River, increasing vegetation, trees even, and the tourist base camp town of San Pedro de Atacama, in that order, more or less. About 40 miles further north of that and about 6,000 ft higher lies the El Tatio geyser field, part of the Altiplano–Puna volcanic complex. Its waters are known to be rich in minerals, having tested especially high in sodium chloride and silica. Lithium is also present.

It is within the boiling waters and geothermal activities of El Tatio that American geologist John E. Hiner, in a technical report prepared for Wealth Mineral Ltd., claims, without fanfare, that the Salar de Atacama's unrivaled lithium concentrations originate.[20] This unequivocal, and unqualified, assertion is a bit strange, for dozens of other geologists essentially conclude the salar's lithium source is largely unknown, only that "closed basin brine composition is, on a first order, controlled by the composition of input waters."[21] But these input waters can be diverse and influenced by various forces. There are six principal sources of the lithium potentially found in brine.[22] The first is from older bedrock, from which lithium can be leached by low and high temperature liquids, from above and from below, through the weathering of precipitation and subsurface springs. Sources two and three come from the wind, carrying in lithium amid dust storms and from volcanic ash. Fourth, it can be sourced by the mobilization of pre-concentrated deposits within basin strata, newly exhumed. This is considered a recycling of previous lithium concentrations. The fifth source is a bit of a catch all: the hydrology of regional groundwater, or the subsurface movement of these previous source flows unrestricted by a basin's topographic delineations. The sixth source is perhaps the most influential: lithium delivered in magmatic fluids, from deep within the Earth's mantle. The El Tatio geyser field could be rather illustrative of these last two types of sources, with its stewing, sputtering, and spewing, potentially connected underground to the salar's regional hydrology. Standing at the bottom of this closed basin, looking uphill, my rental seemingly despondent, face down in the mud, I can picture the slow drip of lithium particles being pulled to their gravitational sink, from the volcanic titans of the east and the alien geysers of the north. They pool here, without escape, only to sit and concentrate under a salty crust.

For how long this brine sits to reach commercial value is not known: "The time it takes to leach, transport, and concentrate Li [lithium] in continental brines is not well understood. However, it appears that most Li brines of economic interest are geologically young."[23] Geologically young, of course, means up to 23 million years old, before the land bridge between Central and South American formed. What is known is what this extended sitting results in: a landscape perhaps even more extraterrestrial than its familial geyser field. The Salar de Atacama is about 817 square miles (2,115 km^2), or almost twice the size of New York City, and about half of that is covered in what is called rough halite, or rock salt. From satellite images it looks plain enough, a simple tan-colored earth, but in person it looks like an endless field of knobby shards. These salt formations are practically impossible to traverse; only on foot can one move around gingerly, fearful of slipping and opening a leg on its sharp edges. Stepping on loose pieces, it even sounds like shards of glass, as they clink together in menacing resonance.

The landscape makes for a befitting deterrent if the salar sought to shield its valuable mineral trove. But this is nothing more than the evaporative powers of the Atacama Desert driving the salar's salts to crystallize, driving irregular salt formations upwards in a continuous assault for space. If you were to cut a section through the floor of this basin, as it is called, through the salt flat and its jagged armor, as if through a beginner baker's first crème brûlée, to reveal its culinary architecture: from bottom to top you would find first a bedrock base layer of Cretaceous and Cenozoic deposits,

Sources, Sinks, and Conduits of Lithium

San Pedro de Atacama

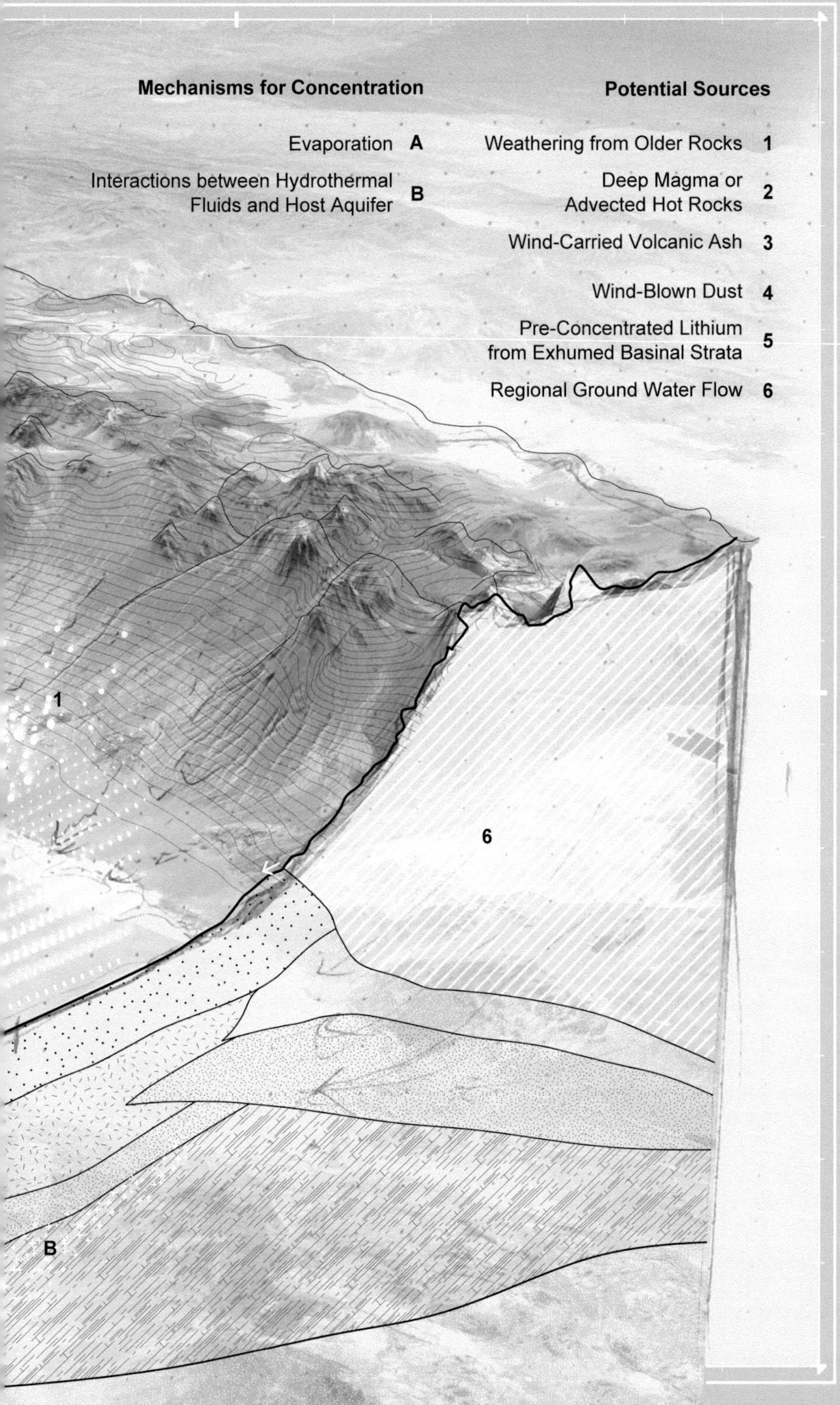

Mechanisms for Concentration

Evaporation **A**

Interactions between Hydrothermal
Fluids and Host Aquifer **B**

Potential Sources

Weathering from Older Rocks **1**

Deep Magma or
Advected Hot Rocks **2**

Wind-Carried Volcanic Ash **3**

Wind-Blown Dust **4**

Pre-Concentrated Lithium
from Exhumed Basinal Strata **5**

Regional Ground Water Flow **6**

then hundreds of meters of hard, impermeable rock salt—evaporite as it is called—that has accumulated from briny runoff for millions of years, at the 40 m depth you would begin to find liquid brine aquifers, it is to this depth most lithium wells are drilled, before more hard halite, capped by several feet of a salty shell.

It's a beautiful, if inhospitable landscape.

BODY: THE BIOLOGICAL AGENCY OF LITHIUM

For all the focus on lithium's powering of our digital lifestyle, we ourselves are intimately connected to lithium physiologically. Lithium is a key element in our bodies' functioning. Much like a battery can be understood mechanistically, as a combination of connected parts whose interactions yield (mostly) consistent results—cathodes, anodes, electrolyte—so too can the body be comprehended through parts, connections, and interactions. Neuroscience, for instance, has shown how the brain's "neurons (parts) connected by synapses (connections) excite and inhibit each other's firing (interactions)."[24] This glosses over much complexity, nonlinear emergence, and abnormality, of course, but still provides great insight when engaging with complex dynamics like the human body.

Lithium is naturally found in the human body, commonly consumed and embodied in trace amounts through drinking water and a number of foods. It has also been used medicinally since the mid-nineteenth century, first as lithium salts to treat gout and, eventually, to treat a variety of other maladies. Lithium therapy became a popular practice in the late 1800s, with lithiated beverages securing a particularly special place in trivia history. The most famous lithiated drink, with a mouthful of a name, Bib-Label Lithiated Lemon-Lime Soda, obviously branded by a marketing mastermind, came on the scene in 1929. Its producer, the Howdy Company of St. Louis, "marketed the soda, which contained lithium citrate, as a hangover cure. 'It takes the ouch out of grouch,' went an early slogan."[25] They eventually shortened the drink's name to 7-Up Lithiated Lemon-Lime, and today "we know its delithiated progeny as 7UP."[26]

Following this uncertain history of lithiated tonics, it wasn't until the 1940s that lithium's psychological powers were discovered by accident and not until 1970 that the FDA approved lithium carbonate as a psychiatric medication. Today it is one of the most effective pharmaceuticals for the treatment of mental illness, particularly bipolar disorder. Though scientific consensus is still lacking in their precise functioning, products such as Eskalith, Lithobid, Lithonate, and Lithotabs have been found decisive in the mood-stabilization of manic-depressive sufferers. Flavio Guzman, Doctor of Medicine and editor at the Pyschopharmacology Institute, summarizes the three primary ways it is believed to assist the brain:

> First, it helps to protect a number of brain regions important for emotional functioning, maintaining gray matter in areas such as the ventral prefrontal cortex. Second, lithium affects the operations of neurotransmitters in desirable ways to discourage mania, by reducing dopamine and norepinephrine excitation and increasing GABA inhibition. Third, lithium modulates cellular signaling systems by affecting important brain chemicals such as AC/cAMP and BDNF[, where l]ithium inhibits the operation of AC/cAMP through competition with magnesium.[27]

Being stranded at the edge of a reservoir of the material partly responsible for regulating one's mental state, I'm reminded of Frederick Law Olmsted's—grandfather of landscape architecture and designer of Central Park—common evocation of *genius loci*, or spirit of place. The Salar de Atacama might be said to exhibit a very special genius loci in representing the psychological stability of Earth's most effecting biological resident, humankind.

But lithium's stabilization of human psychology is only one side of the corporeal coin. By definition, extraction takes from one place to give to another, after all.

> I still think how the region from which so much copper and lithium comes 'for the world' does not have infrastructure or health for its inhabitants. That is what extractivism[28] [extractive capitalism] leaves behind: drought, air pollution, environmental liabilities and, at best, some royalties. If the four transnationals that operate in the Salar de Atacama[29] continue to overexploit their aquifers to extract the minerals for the transition, in a couple of

A Crust of Halite

decades this and many other territories in Chile and the Global South will be simply uninhabitable.

The transition Balcázar speaks of here is the green energy transition, away from fossil fuels driving the climate crisis to renewable sources like wind, solar, and geothermal. Lithium is central to this vision of sustainable future, as the energy generated will need to be stored in batteries until it's needed, and then stored anew following depletion, ad infinitum. Lithium batteries are also key components for electric vehicles, which are seen as the most promising way to decarbonize the transportation sector. Yet, there is a great cost to this ostensibly "green" future.

Lithium quickly spirals out from its subatomic particles' facilitation of electrical exchange, to the brine in which it is found, to the health and well-being of its source landscape's biological beings (people, birds, crustaceans, etc). The effects of lithium extraction are felt directly in the over usage of water and indirectly through exploitative economics. The local, often Indigenous, body experiences the deleterious effects of the foreign body's, the corporation's, profit. The "extractive zone," as Macarena Gómez-Barris calls the sites or territories of extraction projects in her eponymous book, "names the violence that capitalism does to reduce, constrain, and convert life into commodities..."[30] Or perhaps more explicitly, these zones convert life into collateral, as extraction of lithium-rich brine desiccates surrounding ecosystems and communities. And in the worst case, life into targets, as those rising up in resistance to such profit-driven ventures have been marked and murdered at a disturbing rate in Latin America.[31]

The salar as genius loci of psychological stability is quickly corrupted when the body is seen not only through the common lens of capitalist growth—the body as benefactor of technological innovation—but through a decolonial lens—the body as collateral byproduct, as exploited, as hunted obstruction.[32] It's easy to lose sight of those bodies whose voices have been systemically and enduringly suppressed, "where corporate entities and states are indistinguishable in their economic interests and activities; [where] states act on behalf of corporations, and corporate entities hire security forces to control and suppress anti-extractivist organizing."[33]

Evaporation Pond

LANDSCAPE: THE ECONOMIES AND ECOLOGIES OF EXTRACTION

Driving out onto the desolate basin floor from a perimeter scrubland, reveling in a newly savored mobility, my rental car having recently been released from the mud's 18-hour grip with the help of a rescue monster truck, I marvel at a white road of salt cleared of its serrated shield of halite like barnacles from a ship's hull. The road is paralleled by a black pipe, about two feet in diameter, held in place by barnacle piles every 100 feet or so. Within minutes all I can see is this halite armor in all directions, dying into the horizon. About 14 miles down this road, its otherworldliness still unfaded, I reach the gate of the operations of Sociedad Quimica y Minera de Chile, or SQM as it is more widely known, one of the salar's chief lithium producers. A mile further and I would be standing at the edge of one of their evaporation ponds. If you're like me, you drive right up to the gate in your rented Hyundai sedan, a clueless smile plastered across your face, still heavily dusted by a morning of digging your car out of the earth, the taste of lithium on your tongue, and give a wholesome "Buenas tardes!" The guard is surprisingly humored, returning the grin. I do my best to convey my reason for being here, my *for-academic-research-purposes-only*. Still amused, but no less bought, he informs me the best he can do is provide a contact to who I might reach out. Unsurprised, I nod in agreement. Worth a try. But I also know I am not smart enough to abide by such sensible rules. I wave in thanks, drive down the road a bit, and send up my drone.

From above, my newly recharged phone linked to the drone's camera, the magnitude and architecture of lithium brine mining is revealed. Most of what is seen are the evaporation ponds. Colored in radiant hues of tropical blue, these rectangular ponds, carved into the salar about three to six feet deep on average and lined with felt and black plastic, are massive. The smallest ones are twice the size of a football field, the largest about 70 acres (28 hectares), or a quarter of the size of the National Mall in Washington, D.C. The pools hold the lithium-rich brine pumped up from below. Under the ideal Atacaman conditions—a mighty sun at high altitude with almost nonexistent precipitation—the brine evaporates at a rate of almost five feet a year (1,440 mm/yr).[34] As the solution concentrates it is pumped into sequential, increasingly rich pools. Some minerals begin to precipitate out in crystals. The growing salinity is reflected in the pools' color, with the lighter color illustrating higher concentrations of lithium. This all takes from 18 to 24 months on average, with the sun doing most of the work. It's slow, but extremely cheap. The concentrated brine is then trucked to a plant a couple hundred miles to the west, outside the port city of Antofagasta, for final chemical processing into lithium carbonate powder. Finally, it is sent overseas, mostly to China for incorporation into batteries before working its way into various products and the circuits of global commerce.

There are not just ponds one sees from above, however. There are also immense piles, as big as stadiums. I had noticed these mysterious forms while atop my crumbling lookout point alongside the salar earlier that fateful morning, appearing as large bars of white, shimmering on the horizon. Were they lithium processing tanks? It couldn't be the berms of the evaporation ponds, could it? Turns out they're piles of various salts. This is a byproduct of lithium concentration, amassing in large quantities and mostly shoved aside. One can also begin to make out the vast network of roads, canals, and wells. Hundreds of points riddle the salar, beyond the city of ponds and piles, strung together by threads of access and conduit. These points were generated by drills tapping over 100 feet in depth, siphoning the brine to the surface to be exposed to the desert's thirst.

The physical evidence of the hunger for "green" energy continues. With aerial reconnaissance complete, I turn back to the main road along the perimeter of the salar, if one can call a two-lane road buried by sand at times, devoid of asphalt at others, main. The mineworkers' camp lies here, just to the east of the salar, where the main road meets the mine's ancillary salt road. The camp is even labeled on highway signage. With a guarded entrance, it's an incongruous collection of long white modular housing, some appearing to be retrofitted shipping containers, others merely single-wide manufactured homes. Like most miner camps, the design is clearly utilitarian, temporary, but supplemented with an exercise yard, sports courts, and gathering spaces. Pioneers of the lithium frontier, pulled here by the economics, pushing against the isolation and desert terrain.

Miner's Camp

There's another byproduct of lithium mining, which is far from green.

One might think this harsh landscape hosts little life, but as life insistently demonstrates, biological beauty dwells in even the most extreme environments. Perhaps the most striking is the flamingo. Finding home in the highly saline lagunas of this environment are three colorful species found in the altiplano region of South America: the Andean flamingo, the Chilean flamingo, and the James's flamingo (the first two being endemic to South America). This brightly colored bird, more familiar to the tropics, has found a unique niche among the high, arid deserts of the Andes. Though extremely dry, the altiplano hosts many closed hydrological basins, meaning that even though there is little precipitation, the rain that does fall is held in the basin's bowl, closed off from escape to a river or ocean. Only through evaporation can it leave its basin. (I found this out the hard when my rental car fell through a deceptively dusty crust into a thick pudding of mud. And my heart along with it.) This is also true of snowmelt, which is considerably more substantial than rainfall. The result is the formation of lakes or lagunas (and seemingly dry salars), unique to their host topography and its resident minerals. And with this water source comes life. Unique animals like vicunas (small llama-like camelids), viscachas (a Pokémon cross between a rabbit and squirrel), and siris (ostrich-like desert rhea) often depend on these watering holes.

But there is also a smaller, less charismatic creature supporting the upper levels of the food chain, calling the water itself home: the brine shrimp. These aquatic crustaceans are known for their extreme hardiness. Brine shrimp can produce dormant eggs that are able to be stored for long periods and hatched on demand, resulting in their widespread use as feed in aquacultural operations[35]—as well as their sale as pets under the endearing name of Sea Monkeys or Aqua Dragons. Impressively, between 3,000 and 4,000 tons of brine shrimp are produced worldwide annually[36]—weighing more than 30 blue whales cumulatively. Brine shrimp have avoided many predators such as fish by adapting to live in waters of extreme salinity with concentrations of up to 25%. By comparison, seawater has a concentration of approximately 3.5% salinity on average. This ability has not sheltered them from the flamingo, however. Partly responsible for the beta-carotene that endow the flamingo with its rosy coral color, brine shrimp are among the predominant food sources for the Atacama's three species of pink bird.

The process of evaporation, which is naturally regulated by the hard, protective crust on top of the brine aquifers, has been rapidly accelerated due to mining. An immense volume of water is required to extract the small percentage of lithium from its brine. For every kilogram (~2.2 lbs) of lithium product, approximately 400 to two million liters (~105 – half a million gallons of water are pumped and evaporated, depending on who you ask, lithium companies or researchers, respectively.[37] For scale, an Olympic-sized swimming pool holds two and a half million liters. The wells feeding a city of evaporation ponds drain a unique ecosystem of its life-giving material like a flock of youths ravenously wielding straws around a community daiquiri. Lithium's atomic need to bond with other elements, in this case with hydrogen and oxygen (i.e. water), among other elements, into a stable brine requires a massively resource-rich process to isolate and extract it. There have been numerous private investigations and much legal maneuvering where the two companies actively mining, some of the biggest lithium producers in the world, SQM and Albermarle, place blame at the foot of the other, claiming their competitor to be pumping more brine than is permitted by the state. CORFO,[38] Chile's state development agency, entering the fray, submitted an independent study on the salar's water availability to the federal Environmental Oversight organization (SMA) in March of 2018.[39] They concluded that there was indeed an "imbalance in the extraction of water and brine" and that the "estimated [natural] recharge would be less than natural evaporation and use by mining and agricultural activities."[40] In other words, more water was leaving the system than entering it. Pumping and evaporation used more water than rainfall and snowmelt contributed. However, the study couldn't identify and assign responsibility to a single company, concluding this uncertainty itself suggested the need for extraction restrictions.

In addition to CORFO's state study, researchers at Arizona State University have methodically analyzed spatial and temporal patterns of the Salar de Atacama, unequivocally demonstrating "the relationship between mining activities and environmental degradation also indicate that the continuous expansion of lithium mining has strong negative

Lithium Deposits
○ hard-rock mineral
+ brine

Carbonate Exports

18 14 10 6 2k metric tons

CHILE

ARGENTINA

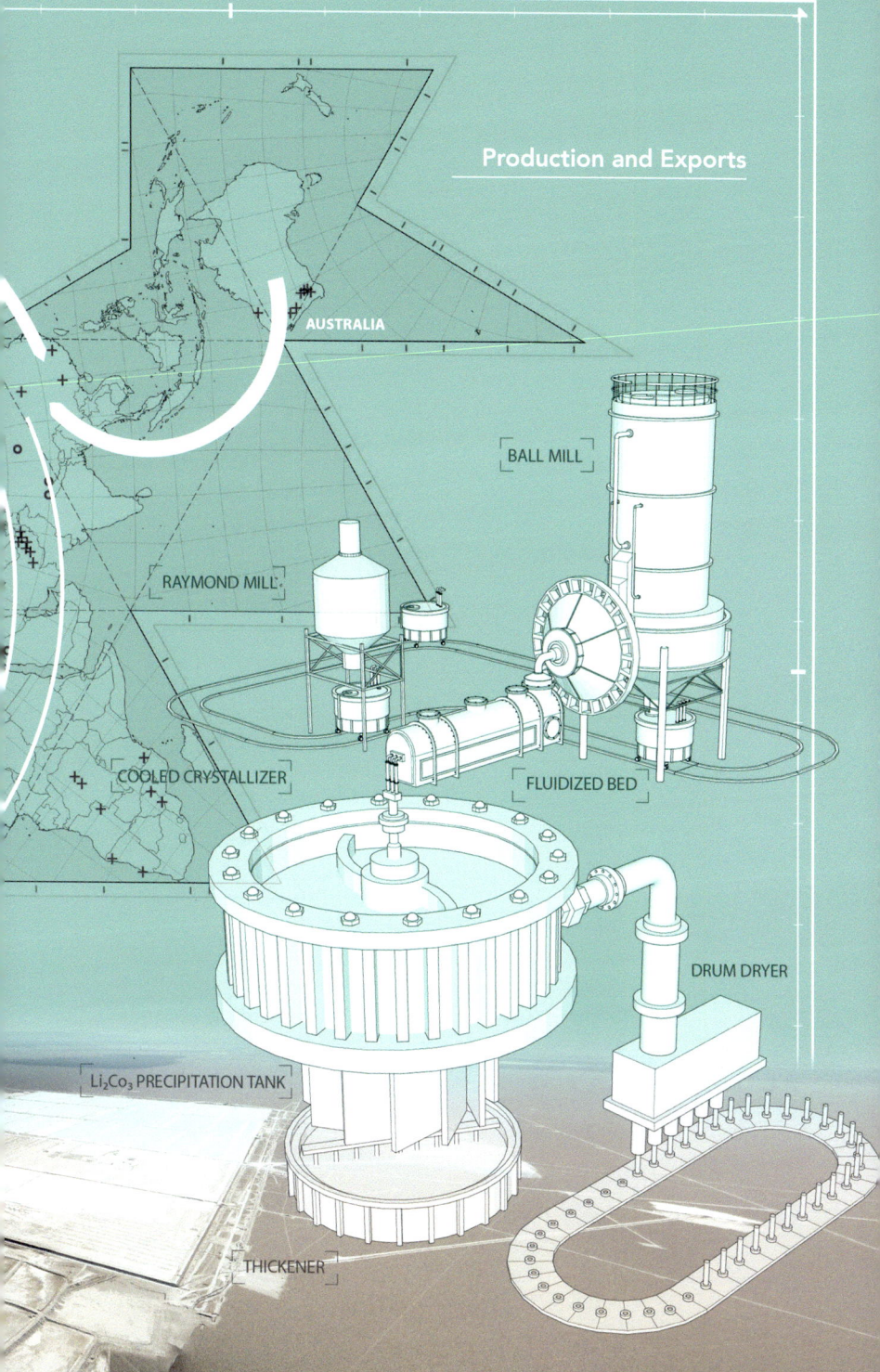

Production and Exports

AUSTRALIA

BALL MILL

RAYMOND MILL

COOLED CRYSTALLIZER

FLUIDIZED BED

DRUM DRYER

Li$_2$Co$_3$ PRECIPITATION TANK

THICKENER

correlations."[41] This is then compounded by the refinement into lithium carbonate, the transportation overseas, the manufacturing of batteries, the incorporation into products, and the distribution of those products to market. In the words of Chilean sociologist Martín Arboleda, "the mine is not a discrete sociotechnical object but a dense network of territorial infrastructures and spatial technologies vastly dispersed across space."[42] In the late capitalism of today, the extractive zone is no longer a delineated site but rather fluid in the "circulatory system of capital," blanketing the globe in fibrous rhizomes like mycelium a fertile substrate.[43]

The geo-logistics of extraction, the mining of brine in Chile and rock in Australia, the manufacturing of batteries in the industrial Pearl River Delta of China, and their consumption around the world—disproportionately by the affluent Global North—bring us back to the body, specifically the local Atacameño[44] bodies of largely Indigenous populations surrounding the salars of the Altiplano. These bodies, long tired of the exploitation of their lands and peoples dating back to the Spanish Conquistadors of the sixteenth century,[45] wield protest as a bodily act of both march and ceremony in resisting the systems of extractive capitalism increasingly defining the relationships between distant geographies.

Ramón Morales Balcázar recounts a protest against the copper mine Minera Delfin by the community of Peine, just southeast of the Salar de Atacama (copper mines, numerous in the Atacama Desert, use vast quantities of water similar to lithium operations):

> The caravan began a procession heading south of the Salar, reaching an archaeological and ceremonial site located on the ancient Inca Trail. At the foot of the hill—where the first excavations for the mining project had already been carried out and after intense proclamations demanding respect for their people, their life, and their worldview— community members and companions ascended a stony and steep path to the mining pit, a large calypso-colored trench. Atop the small summit, an Atacameña woman who had traveled from Calama planted a black flag as a symbol of mourning and in protest of this new attempt in exploiting the territory of her grandparents. It was a march and a ceremony. With the dust and a blistering sun mixed

March and Ceremony

March and Ceremony

Peine

a sensation of catharsis in which pain arose from centuries of dispossession and invisibility at the hand of the Chilean state. When we came down from the hill, the town's yatiri [sage or shaman] *directed a payment to the land in which the Atacameños asked for the protection of their ancestral territory.* [Emphasis added]

A few days later, it was determined that the Minera Delfin should be submitted for Indigenous consultation. A couple years after that, the company permanently withdrew the project, citing a "categorical rejection" from the community.[46] This is a rare success story. More commonly, protests led by inhabitants defending their land or workers demanding better conditions end in violence, as seen in the case of the protests at El Salvador, another copper mine in the Atacama region of northern Chile, in 2015.[47] In response to about 300 subcontracted workers occupying the mine in demonstration, police and special forces employed tear gas and rubber bullets before escalating to live ammunition. A miner since the age of 18, Nelson Quichillao's life was taken that day.[48]

The circulatory system of capital, circumnavigating the planet in relentless flight, inevitably comes home to roost at the body, be that the human body of the protester, the rose-pink body of the flamingo, or the material water-salt-lithium body of the salar's reservoir, drained of life and converted to commodity or collateral.

BATTERY, BODY, LANDSCAPE, COSMOS

Standing at the dirt crossroads that lonely morning, attempting to hide from the sun under sunglasses, bandana, and cap, I waved down any vehicle that happened upon me. Mostly I saw tourists on buses headed for the lagunas and their flamingos, SLR cameras slung around their necks, and faces pressed to the glass in bewilderment of someone on foot amidst such a setting, unable to assist. A truck with four-wheel drive stopped to help me eventually. On its side was a green, white, and blue logo, and the letters "SQM." After a welcomingly friendly, but ultimately futile, attempt at freeing my rental from the mud, I inquired about his employment. In a typically rapid Chilean Spanish, he explained his day's task of measuring water levels at various sites throughout the salar. He could return in six hours or so with a rope to pull me out he offered, after his measurements.

At the time of writing, following community pushback and a moratorium period on new or expanded lithium mining, the Chilean state has newly given the green light for increased production, as long as the companies monitor water levels. Surging demand is not likely to let up as dense energy storage is seen as key to weening our species off fossil fuels, with lithium-ion batteries currently our best bet.

This economic force with ecological consequences suggests the importance to begin recognizing the scalar and planetary assembly that forms around the lightest metal on the periodic table: lithium - brine shrimp - flamingo - miner - battery - mobile electronics - green energy futures - medicine - salar - Indigenous land defenders - Chinese manufacturing plants - an oblivious American scrolling through TikTok on their iPhone. Its tangles expanding, contracting, inverting, forever emerging as new complicated knots. A facility in navigating across scales can be helpful to see the material agency behind it all, giving structure and skeleton to the entanglement, to its constraints and enablements. A rare ecosystem's health, a population's mental stability, a world's green energy economy, all orbit around lithium, contingent upon this simple material, with its 3 protons, 3 electrons, and 4 neutrons.

We are all intimately performing within invisible assemblies stretching across deep time and global space, from atom to cosmos, from the Big Bang to the reading of these words. Understanding this not only as fact but as an identifiable, engaged, and mutable relationship is foundational to a new materialism where we are not bodies simply supported by our surrounding environments but are conscious participants within those many-tentacled, scalar entanglements. Lithium is not just part of the batteries in our phones, or the promise of a clean electric future. It is also the material stabilizing our psyches, an element embedded in larger landscapes and ecosystems across the globe.

It is not my place to claim a wholesale ban on lithium extraction as the only answer, though many seasoned activists and Atacameños are appropriately doing just that.[49] Like the knotted assembly itself, the methods and manners of intervention in an extractivist model often defined by exploitation and environmental damage are many and interconnected. As a quotidian citizen-consumer this likely means an increased sensitivity to lithium's role in our lives, which takes into account both its rarity and its preexistent connections to other places and lives. A sensitivity where we reflect on brine shrimp, flamingos, and Indigenous peoples carrying out their lives amid a desert salar the next time Apple taunts us with an irresistibly beautiful new object. Take a moment to look inward to the body, outward to the landscape, and upward to the cosmos above, the next time you recharge your phone. Acknowledge the relationship much greater than a hand-sized object begging for juice. Begin living with lithium, *of* lithium, acting on the small, reverberating scale of everyday life.

A designer, on the other hand, might imagine further paths and points of intervention. Those believing the climate crisis calls for massive projects and machinic mobilization— in the spirit of geoengineering—might envision placing monolithic glass greenhouses or plastic-filmed, arena-sized inflatables atop the lithium evaporation ponds. With evaporation accelerated by trapping the sun's growing heat, the vaporized water condenses and collects on the sides of the structure, dripping down, and then re-percolates into the salar's finite reservoir along the structure's interior perimeter. Fueled by hubris (not invariably bad), this "starchitect" would likely envision gleaming volumes in parametric patterns, something not dissimilar to the Eden Project or the massive agricultural greenhouse operations of the Netherlands or the Almeria region of Spain.[50]

A more subtle designer might target the end of the supply chain, the lithium consumer rather than the lithium producer. Strategic communications design possesses the potential to redefine the relationship between the buyer and his product's source components. Imagine if every purchase came with an engaging portrait of environmental externalities, or of the distant landscapes negatively affected by the product's production: a model-sized three-dimensional salar materializes, rendered in augmented reality and triggered by your phone's camera scanning a symbol on the packaging, as you consider buying a new tablet. A cross-section is taken illustrating the basin hydrology and the mines' pumping of brine to evaporation ponds. The model zooms in, visualizing a proportionate decrease of water levels in a nearby laguna. Pink flamingos chase the dwindling volume, slowly laying down and disappearing with the somber tone of a minor chord.

Or perhaps the intervention targets neither consumer, nor producer, but is rather born from the site itself. Critically, the design pursues a collaborative relationship with Atacameños, one focused on strengthening Indigenous power, built atop generational knowledge of the land. One starts with listening, before drawing. How can local ecologies of knowledge be strengthened and made actionable? How might one employ ancestral relationships to the land to inform new and richer entanglements in a world barreling into socioecological catastrophe? The conversations themselves must first be designed so as to ensure the land is first seen as equal agent, not a background to be controlled, dominated, or exploited. Like the Carnaval Sampedrino evoked in opening this chapter, the land's energies, its agency, be they *mallkus*, geysers, brine shrimp, or lithium, are to be *desenterrados*, unearthed, released and lived amid as a rich, tangled, living knot of life.

This way of seeing, knowing, and being in the world, a way of *worlding* the world[51]—as a rich, tangled, living knot— suggests it will likely be some evolving combination of all of the above design strategies. Every individual learns to recognize and care for "reciprocal landscapes"[52]—be that through communications design or self-instruction— that provide for their digital and mobile way of life, their decarbonization of transportation to mitigate worst-case climate scenarios, amongst the many other ways we're materially connected to unseen geographies, in pursuit of reducing and reusing resources. New innovations in techniques and technologies of lithium mining will be valued and funded, in consultation and organization with local populations for equitable management and distribution of profits.

Or the Salar de Atacama is unequivocally discontinued as an extraction zone and the world finds other means to meet rigorously decreased energy needs with renewable sources.

Push, pull, bend, entangle.

Battery, body, landscape, cosmos.

What new knots might be tied?

1   "Alternativas Sustentables," Tanti Foundation, accessed June 20, 2020, https://www.fundaciontanti.org/. "Tantí Foundation was born in 2016 in San Pedro de Atacama out of collective work and the will to create spaces for reflection and concrete action around the environmental crisis, from a situated perspective and sensitive to the impacts that climate change and the development model have for the territories, towns, and ecosystems. Its symbol is a chañar seed (Geoffroea decorticans), a tree that thanks to its seeds and rhizomes manages to spread and flourish in the harsh desert conditions."

2   "Inicio," Plurinational Observatory of Andean Salt Flats, accessed April, 12, 2019, https://observatoriosalares.wordpress.com/. "The multinational observatory of salt flats was born in response to the widely spread 'lithium triangle' as an extractivist territoriality that produces an eco-capitalist global imagery on the Andean territories of the Southern Cone, spaces that are targeted by lithium mining companies to exploit the salt flats of this cross-border region shared between Northern Argentina and Chile with southern Bolivia. It is a collaborative, solidary, and horizontal space that promotes the protection of the salt flats and other high Andean ecosystems, the protection of indigenous rights and ways of life of the communities that inhabit them."

3   Ramón Morales Balcázar, interview by author. Conversations over email and text occurred in the spring and summer of 2020, following my travel and site research.

4   LeeAnn Munk et al., "Hydrogeologic and Geochemical Distinctions in Salar Freshwater Brine Systems," *Geochemisty, Geophysics, Geosystems*, (August 5, 2020), https://doi.org/10.31223/osf.io/j3pu6. Though numerous studies, industry documents, non-peer reviewed articles, conversations with local residents, and even the Chilean government itself have concluded lithium mining operations negatively affect water levels of neighboring lagunas and ecosystems, it is notable that—as this publication goes to press—a team of researchers are recently arguing that the brine of the salar's nucleus, or primary structure containing the highest lithium concentrations, is hydrogeologically and geochemically distinct from the salar transition zone where the lagunas are found. The suggestion is that the pumping of brine from the salar during lithium mining is not decreasing water levels of the rare laguna ecosystems as they are not connected; Marazuela et al., "The effect of brine pumping on the natural hydrodynamics of the Salar de Atacama: The damping capacity of salt flats," *Science of the Total Environment* 654 (November 14, 2018): 1118–1131, https://doi.org/10.1016/j.scitotenv.2018.11.196; Wenjuan Liu, Datu B. Agusdinata, Soe W. Myint, "Spatiotemporal patterns of lithium mining and environmental degradation in the Atacama Salt Flat, Chile," *International Journal of Applied Earth Observation and Geoinformation* 80 (April 29, 2019): 145-156, https://doi.org/10.1016/j.jag.2019.04.016. These studies directly correlate lithium mining with decreasing laguna levels.

5   Thea Riofrancos, "What Green Costs," *Logic Magazine*, December 7, 2019, https://logicmag.io/nature/what-green-costs/.

6   Seth Fletcher, *Bottled Lightning: Superbatteries, Electric Cars, and the New Lithium Economy* (New York: Hill & Wang, 2011), 9.

7   Ibid., 11.

8   Ibid., 12.
9   Ibid., 19.
10   Ibid., 20.
11   Marshall Berman, *All That's Solid Melts into Air: The Experience of Modernity* (New York: Penguin, 1982). I employ literary theorist Berman's definition of modernity as the societal state resulting through modernization's shift from agriculture to industry and from low to high population densities.
12   Adam Liptak, "Major Ruling Shields Privacy of Cellphones," *The New York Times*, June 25, 2014, https://www.nytimes.com/2014/06/26/us/supreme-court-cellphones-search-privacy.html. In 2014 the Supreme Court unanimously ruled that warrants are needed to search the cellphones of the arrested. In the decision Chief Justice John G. Roberts, Jr. wrote, "[cellphones are] such a pervasive and insistent part of daily life that the proverbial visitor from Mars might conclude they were an important feature of human anatomy."
13   "Annual Lithium Demand for Electric Vehicle Batteries, 2019–2030," International Energy Agency, last modified June 14, 2020, accessed August 23, 2020, https://www.iea.org/data-and-statistics/charts/annual-lithium-demand-for-electric-vehicle-batteries-2019-2030-2; Reuters Staff, "Electric Cars to Account for 79% of Lithium Demand by 2030: Chile," *Reuters*, August 26, 2020, https://www.reuters.com/article/us-chile-lithium/electric-cars-to-account-for-79-of-lithium-demand-by-2030-chile-idUSKBN25M2PG.
14   Maxx Chatsko, "Lithium Demand for Electric Vehicles Could Grow 599% by 2025," *The Motley Fool*, last modified March 24, 2020, accessed August 12,2020, https://www.fool.com/investing/2020/03/24/lithium-demand-for-electric-vehicles-could-grow-59.aspx#:~:text=Total%20annual%20demand%20for%20lithium,LCE%20per%20year%20by%202025.
15   Fletcher, *Bottled Lightning*, 17.
16   *Encyclopedia Britannica Online*, s.v. "Lithium," accessed July 30, 2020, https://www.britannica.com/science/lithium-chemical-element
17   Donald E. Garrett, *Handbook of Lithium and Natural Calcium Chloride: Their Deposits, Processing, Uses and Properties* (San Diego: Elsevier Academic Press, 2004), 1.
18   LeeAnn Munk, et al., "Lithium Brines: A Global Perspective," in *Rare Earth and Critical Elements in Ore Deposits*, ed. Philip L. Verplanck, and Murray W. Hitzman (Society of Economic Geologists, 2016), 349, https://doi.org/10.5382/Rev.18.14; P. Pananont et al., "Cenozoic evolution of the northwestern Salar de Atacama Basin, northern Chile,"*Tectonics* 23, no. 6 (2004), https://doi.org/10.1029/2003TC001595.
19   Jonathan D. A. Clarke, "Antiquity of aridity in the Chilean Atacama Desert," *Geomorphology* 73, nos. 1-2 (2006): 101–114, https://doi.org/10.1016/j.geomorph.2005.06.008.
20   John E. Hiner, *NI 43-101 Technical Report on the Atacama Lithium Project, El Loa Province, Region II Republic of Chile* (Wealth Minerals, 2017).
21   Munk, "Lithium Brines," 342.
22   Ibid., 340.
23   Ibid.
24   Paul Thagard, "Why is Lithium Good for Both Batteries and Bipolar Disorder?,"*Psychology Today*, January 11, 2018, https://www.psychologytoday.com/us/blog/hot-thought/201801/why-is-lithium-good-both-batteries-and-bipolar-disorder.

25 Fletcher, "Bottled Lightning," 18.

26 Ibid., 18.

27 Flavio Guzman, "Lithium's Mechanism of Action: An Illustrated Review," *Psychopharmacology Institute,* June 27, 2019, https://psycho-pharmacologyinstitute.com/publication/lithiums-mechanism-of-action-an-illustrated-review-2212.

28 Macarena Gómez-Barris, *The Extractive Zone: Social Ecologies and Decolonial Perspectives* (Durham: Duke University Press, 2017). See Gómez-Barris for a rich investigation into *extractivismo*—the "economic system that engages in thefts, borrowings, and forced removals, violently reorganizing social life as well as the land by thieving resources from Indigenous and Afro-descendent territories"—and submerged perspectives offering counter and alternative ways of perceiving and living under such exploitation. See also Eduardo Gudynas, "Extractivisms: Tendencies and Consequences" in *Reframing Latin American Development*, ed. Ronaldo Munck and Raul Delgado Wide (Philadelphia: Routledge, 2018).

29 Albermarle, Sociedad Química y Minera de Chile (SQM), Zaldivar (Antofagasta Minerals/Barrick), and Minera Escondida (BHP/Rio Tinto).

30 Gómez-Barris, *The Extractive Zone*, xix.

31 Global Witness, *Deadly Environment: A Rising Death Toll on Our Environmental Frontiers is Escaping International Attention* (2014), accessed August 21, 2020, https://www.globalwitness.org/en/campaigns/environmental-activists/deadly-environment/.

32 Gómez-Barris, *The Extractive Zone*, 3. "new/old forms of colonialism, such as extractive capitalism, the digital surveillance of territories, the criminalization of Indigenous peoples as a weapon of neoliberal expansion, and the extraction of Native and Afro-descendent knowledges, all depend on prior civilizational projects, in which the Global South has long been constructed as a region of plunder, discovery, raw resources, taming, classification, and racist adventure."

33 Ibid., xviii.

34 Munk, "Lithium Brines," 343.

35 Martin Daintith, *Rotifers and Artemia for Marine Aquaculture: A Training Guide* (Tasmania: University of Tasmania, 1996).

36 Liudmila I. Litvinenko, et al., "Artemia cyst production in Russia," *Chinese Journal of Oceanology and Limnology* 33, no. 6 (2015): 1436-1450, http://dx.doi.org/10.1007/s00343-016-0264-8.

37 Mie Obberkaer, "How Much Water is Used to Make the World's Batteries?," *Danwatch*, accessed July 11, 2020, https://danwatch.dk/en/undersoegelse/how-much-water-is-used-to-make-the-worlds-batteries/. *See also* Liu, "Spatiotemporal patterns."

38 El Corporación de Fomento de la Producción, or the Production Development Corporation lies within Chile's Ministry of Economy, Development, and Tourism and was founded in 1939 by President Pedro Aguirre Cerda to promote economic growth.

39 Superintendency of the Environment, CORFO, *Fiscal Instructor de la División de Sanción y* Cumplimiento (2018), https://www.documentcloud.org/documents/5003677-PresentacióN-CORFO.html#document/p3/a461155

40 Ibid.

41 Liu, "Spatiotemporal patterns," 145.

42 Martín Arboleda, *Planetary Mine: Territories of Extraction under Late Capitalism* (New York: Verso, 2020), 5.

43 Ibid., 5; Mazen Labban, "Deterritorializing Extraction: Bioaccumulation and the Planetary Mine," *Annals of the Association of American Geographers* 104, no. 3 (2014), https://doi.org/10.1080/00045608.2014.892360.

44 The people indigenous to the Atacama Desert and *altiplano* region in the northern regions of Chile and Argentina and southern Bolivia—also called the Andean Plateau, or the widest part of the Andes mountain range in west-central South America—are known as *atacameños* in Spanish. The 2002 Chilean Census shows 21,015 people identified as Atacameño.

45 See Eduardo Galeano, *Open Veins of Latin America: Five Centuries of the Pillage of a Continent* (New York: Monthly Review Press, 1973) for meticulous documentation of centuries of exploitation of the Americas by Western empires.

46 Leonardo Cárdenas, "Minera Delfín: Büchi, Segura y Petermann ponen fin a proyecto por "tajante rechazo" de comunidad indígena," *Observatorio de Conflictos Mineros de América Latina*, March 18, 2019, https://www.ocmal.org/minera-delfin-buchi-segura-y-petermann-ponen-fin-a-proyecto-por-tajante-rechazo-de-comunidad-indigena/

47 Daniela Yáñez, "Memoria sin justicia: el monumento a Nelson Quichillao," *The Clinic*, May 24, 2016, https://www.theclinic.cl/2016/05/24/memoria-sin-justicia-el-monumento-a-nelson-quichillao/.

48 Arboleda, *Planetary Mine*, 2; Macarena García Lorca, "Nelson Quichillao: el fatal destino de un minero subcontratado," *El Mostrador*, July 31, 2015, https://www.elmostrador.cl/noticias/pais/2015/07/31/nelson-quichillao-el-fatal-destino-de-un-eterno-minero-subcontratado/.

49 Morales Balcázar, interview: "The COVID crisis has unleashed a strong social and economic crisis in the town [San Pedro de Atacama], which, however, has opened the space for discussion on a new form of land management. I believe that for a democratic and sustainable management of the salar basin a constitutional change is required that recognizes the autonomy of the native peoples, that ensures citizen participation, that demarkets water—fresh or salt—and that effectively regulates extractive activities. There have been some attempts at governance and the creation of expert committees, which however have disappeared since the arrival of the Sebastián Piñera government. As an observatory we believe in dialogue and the ecology of knowledge, that is, many points of view and many ways of knowing reality in order to build processes. This is not possible while corporate interests are behind local governments and the 'transition' itself, so what we need is a profound transformation, which is nothing more than the sum of many transformations, thought from below. An ideal scenario should have these elements as facilitators of food sovereignty and conservation of the basin's biodiversity, especially that found in wetlands. In view of the degree of deterioration that the salar has suffered due to mining and climate change itself, I do not believe that it would be socially or environmentally responsible to continue the extraction of its waters, whether sweet or brine. As noted by the International Tribunal for the Rights of Nature in December 2019, I believe that a moratorium should be established for all mining activities in the Salar de Atacama basin. This is a unique opportunity to rethink the transition of the Paris Accords: Transition for whom and at what cost? If the processes

are based on the dispossession of the colonized peoples and the destruction of nature, it means that we are going the wrong way, and that is where conscious science committed to transformation has a central role. I think that the ideas of degrowth can shed light on the changes in the modes of production and consumption that are behind the climate crisis, especially in the Global North. Technological replacement is not enough, consumption must be drastically reduced. Everything has to be redesigned, and the crisis has taught us that we cannot leave that task to corporations."

50   The endless patchwork of Almeria's greenhouses is shown in the opening of the dystopian movie *Blade Runner 2049*.

51   Donna Haraway, *Staying with the Trouble* (Durham: Duke University Press, 2016).

52   Jane Hutton, *Reciprocal Landscapes* (London: Routledge, 2019).

BIBLIOGRAPHY

"Alternativas Sustentables." Tanti Foundation. Accessed June 20, 2020. https://www.fundaciontanti.org/.

"Annual Lithium Demand for Electric Vehicle Batteries, 2019–2030." International Energy Agency. Last modified June 14, 2020. Accessed August 23, 2020. https://www.iea.org/data-and-statistics/charts/annual-lithium-demand-for-electric-vehicle-batteries-2019-2030-2.

Arboleda, Martín. *Planetary Mine: Territories of Extraction under Late Capitalism*. New York: Verso, 2020.

Berman, Marshall. *All That's Solid Melts into Air: The Experience of Modernity*. New York: Penguin, 1982.

Cárdenas, Leonardo. "Minera Delfín: Büchi, Segura y Petermann ponen fin a proyecto por 'tajante rechazo' de comunidad indígena." *Observatorio de Conflictos Mineros de América Latina*. March 18, 2019. Accessed August 12, 2020. https://www.ocmal.org/minera-delfin-buchi-segura-y-petermann-ponen-fin-a-proyecto-por-tajante-rechazo-de-comunidad-indigena/.

Chatsko, Maxx. "Lithium Demand for Electric Vehicles Could Grow 599% by 2025." *The Motley Fool*. Last modified March 24, 2020. Accessed August 12, 2020. https://www.fool.com/investing/2020/03/24/lithium-demand-for-electric-vehicles-could-grow-59.aspx#:~:text=Total%20annual%20demand%20for%20lithium,LCE%20per%20year%20by%202025.

Clarke, Jonathan D. A.. "Antiquity of aridity in the Chilean Atacama Desert." *Geomorphology* 73, nos. 1-2 (2006): 101–114.

Daintith, Martin. *Rotifers and Artemia for Marine Aquaculture: a Training Guide*. Tasmania: University of Tasmania, 1996.

Fletcher, Seth. *Bottled Lightning: Superbatteries, Electric Cars, and the New Lithium Economy*. New York: Hill and Wang, 2011.

Galeano, Eduardo. *Open Veins of Latin America: Five Centuries of the Pillage of a Continent*. New York: Monthly Review Press, 1973.

Garrett, Donald E. *Handbook of Lithium and Natural Calcium Chloride: Their Deposits, Processing, Uses and Properties*. San Diego: Elsevier Academic Press, 2004.

Global Witness. *Deadly Environment: A Rising Death Toll on Our Environmental Frontiers is Escaping International Attention*. 2014. Accessed August 21, 2020. https://www.globalwitness.org/en/campaigns/environmental-activists/deadly-environment/.

Gómez-Barris, Macarena. *The Extractive Zone: Social Ecologies and Deco-lonial Perspectives.* Durham: Duke University Press, 2017.

Gudynas, Eduardo. "Extractivisms: Tendencies and Consequences." In *Reframing Latin American Development,* edited by Ronaldo Munck and Raul Delgado Wide. Philadelphia: Routledge, 2018.

Guzman, Flavio. "Lithium's Mechanism of Action: An Illustrated Review." *Psychopharmacology Institute.* June 27, 2019. Accessed August 12, 2020. https://psychopharmacologyinstitute.com/publication/lithiums-mecha-nism-of-action-an-illustrated-review-2212.

Haraway, Donna. *Staying with the Trouble.* Durham: Duke University Press, 2016.

Hiner, John E. *NI 43-101 Technical Report on the Atacama Lithium Project, El Loa Province, Region II Republic of Chile.* Wealth Minerals, 2017.

Hutton, Jane. *Reciprocal Landscapes.* London: Routledge, 2019.

"Inicio." Plurinational Observatory of Andean Salt Flats. Accessed April, 12, 2019. https://observatoriosalares.wordpress.com/.

Labban, Mazen. "Deterritorializing extraction: bioaccumulation and the planetary mine." *Annals of the Association of American Geographers* 104, no. 3 (2014).

Liptak, Adam. "Major Ruling Shields Privacy of Cellphones." *The New York Times.* June 25, 2014. Accessed August 12, 2020. https://www.nytimes.com/2014/06/26/us/supreme-court-cellphones-search-privacy.html.

"Lithium." In *Encyclopedia Britannica Online.* Accessed July 30, 2020. https://www.britannica.com/science/lithium-chemical-element.

Liu, Wenjuan, Datu B. Agusdinata, and Soe W. Myint. "Spatiotemporal patterns of lithium mining and environmental degradation in the Atacama Salt Flat, Chile." *International Journal of Applied Earth Observation and Geoinformation 80* (April 29, 2019): 145-156.

Litvinenko, Liudmila I., Aleksandr I. Litvinenko, Elena G. Boiko, and Kirill Kutsanov. "Artemia cyst production in Russia." *Chinese Journal of Ocean-ology and Limnology* 33, no. 6 (2015): 1436-1450.

Lorca, Macarena García. "Nelson Quichillao: el fatal destino de un minero subcontratado." *El Mostrador.* July 31, 2015. Accessed August 12, 2020. https://www.elmostrador.cl/noticias/pais/2015/07/31/nelson-quichil-lao-el-fatal-destino-de-de-un-eterno-minero-subcontratado/.

Marazuela, M.A., E. Vázquez-Suñé, C.Ayora, and A. García-Gil, T.Palma. "The effect of brine pumping on the natural hydrodynamics of the Salar de Atacama: The damping capacity of salt flats." *Science of the Total Environment 654* (November 14, 2018): 1118-1131.

Munk, LeeAnn, David Boutt, Brendan J. Moran, Sarah McKnight, and Jordan Jenckes. "Hydrogeologic and Geochemical Distinctions in Salar Freshwater Brine Systems." *Geochemisty, Geophysics, Geosystems,* (August 5, 2020).

Munk, LeeAnn, Scott Hynek, Dwight C. Bradley, David Boutt, Keith A. Labay, and Hillary Jochens. "Lithium Brines: A Global Perspective." In *Rare Earth and Critical Elements in Ore Deposits,* edited by Philip L. Verplanck, and Murray W. Hitzman. Society of Economic Geologists, 2016, 339-365.

Obberkaer, Mie. "How Much Water is Used to Make the World's Batter-ies?"*Danwatch.* Accessed July 11, 2020. https://danwatch.dk/en/under-soegelse/how-much-water-is-used-to-make-the-worlds-batteries/.

Pananont, P., C. Mpodozis, N. Blanco, T.E. Jordan, L.D. Brown. "Cenozoic evolution of the northwestern Salar de Atacama Basin, northern Chile."- *Tectonics* 23, no. 6 (2004).

Reuters Staff. "Electric Cars to Account for 79% of Lithium Demand by 2030: Chile." *Reuters.* August 26, 2020. https://www.reuters.com/article/us-chile-lithium/electric-cars-to-account-for-79-of-lithium-demand-by-2030-chile-idUSKBN25M2PG.

Riofrancos, Thea. "What Green Costs." *Logic Magazine.* December 7, 2019. Accessed August 12, 2020. https://logicmag.io/nature/what-green-costs/.

Superintendency of the Environment, CORFO. *Fiscal Instructor de la División de Sanción y* Cumplimiento. 2018. Accessed August 12, 2020. https://www.documentcloud.org/documents/5003677-Present-ació N-CORFO.html#document/p3/a461155.

Thagard, Paul. "Why is Lithium Good for Both Batteries and Bipolar Disorder?" *Psychology Today.* January 11, 2018. Accessed August 12, 2020. https://www.psychologytoday.com/us/blog/hot-thought/201801/why-is-lithium-good-both-batteries-and-bipolar-disorder.

Yáñez, Daniela. "Memoria sin justicia: el monumento a Nelson Quichillao." *The Clinic.* May 24, 2016. Accessed August 12, 2020. https://www.theclinic.cl/2016/05/24/memoria-sin-justicia-el-monumento-a-nelson-quichillao/.

IMAGE CITATIONS + CREDITS

Listed by Page Number

97   Photo: Matthew Seibert

     Research Assistant: Tian Wang

99   Photo: Ramón Morales Balcázar; Basemap sources: Esri, Airbus DS, USGS, NGA, NASA, CGIAR, N Robinson, NCEAS, NLS, OS, NMA, Geodatastyrelsen, Rijkswaterstaat, GSA, Geoland, FEMA, Intermap and the GIS user community

101 Basemap sources: Esri, Maxar, GeoEye, Earthstar Geographics, CNES/Airbus DS, USDA, USGS, AeroGRID, IGN, and the GIS User Community

105 Photo: Matthew Seibert; Instituto Geografico Militar. "Cerro Mullay." 1:50,000. 2004. Research Assistants: Chloe Nagraj, Tian Wang

109 Basemap sources: Esri, Maxar, GeoEye, Earthstar Geographics, CNES/Airbus DS, USDA, USGS, AeroGRID, IGN, and the GIS User Community

113 Basemap sources: Esri, Maxar, GeoEye, Earthstar Geographics, CNES/Airbus DS, USDA, USGS, AeroGRID, IGN, and the GIS User Community; NYC coastline: NYC Open Data (Department of City Planning (DCP))

     Research Assistant: Tian Wang

117 Basemap sources: USGS, NGA, NASA, CGIAR, GEBCO,N Robinson,NCEAS,NLS,OS,NMA,Geodatastyrelsen and the GIS User Community; Munk, et al. "Lithium Brines: A Global Perspective." Reviews in Economic Geology, v 18, pp 339-365. 2016; Munk, et al. "Hydrogeologic and Geochemical Distinctions in Salar Freshwater Brine Systems." Geochemisty, Geophysics, Geosystems, August 5, 2020

     Research Assistant: Aleksander De Mott

121 Photos: Matthew Seibert

124 Photo: Ramón Morales Balcázar; Instituto Geografico Militar. Cartografia escala 1:50,000. 2004.

125 Photo: Matthew Seibert; Instituto Geografico Militar. "Salar de Atacama." 1:50,000. 2004.

128 Photo: Matthew Seibert

131 photo: Matthew Seibert; "This Metal Is Powering Today's Technology—at What Price?" Magazine, 15 Jan. 2019, https://www.nationalgeographic.com/magazine/2019/02/lithium-is-fueling-technology-today-at-what-cost/.

     Research Assistants: Aleksander De Mott, Tian Wang

134 Photo: Ramón Morales Balcázar

135 Photo: Ramón Morales Balcázar

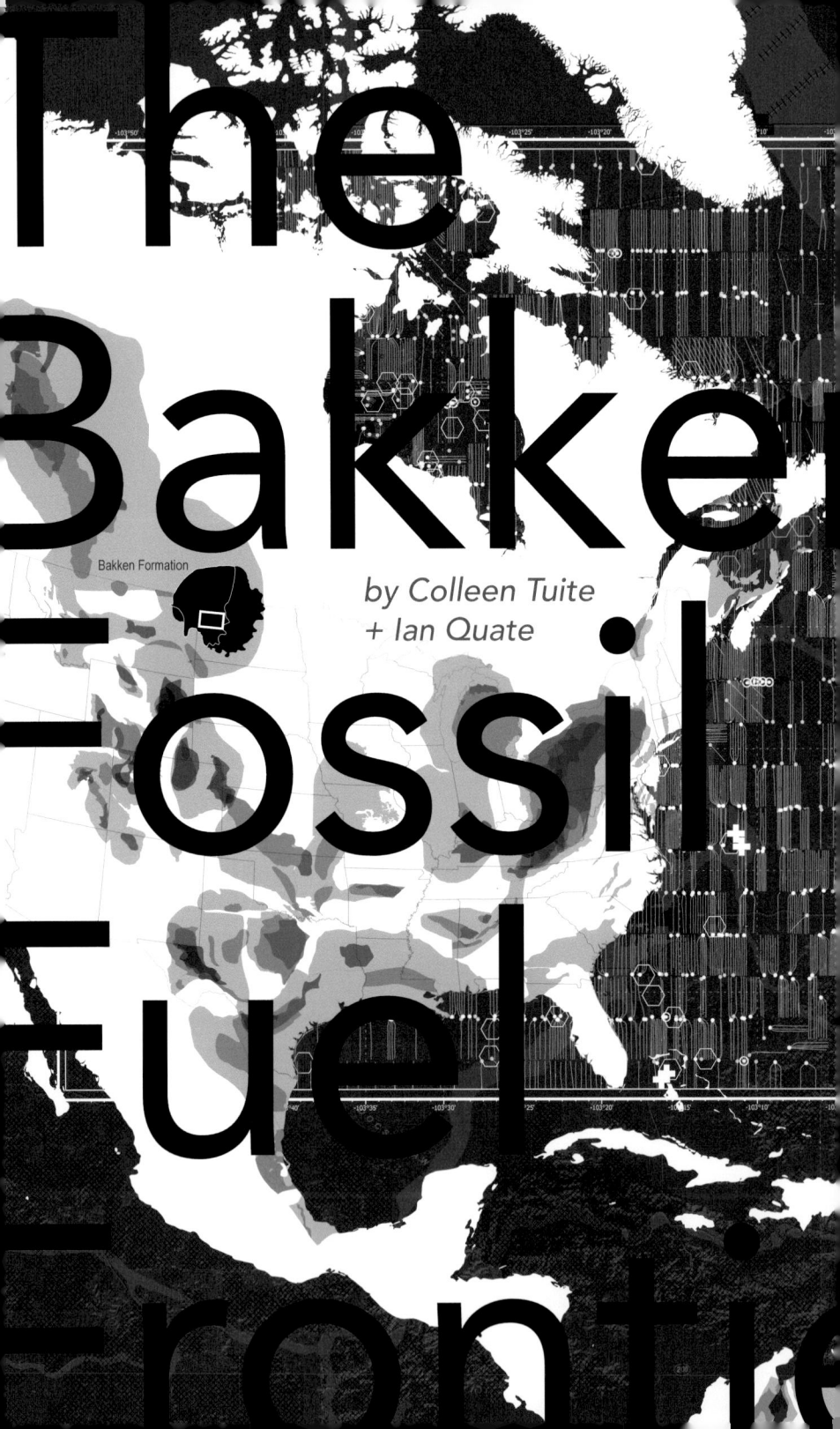

The Bakken: A Fossil Fuel Frontier

by Colleen Tuite
+ Ian Quate

Bakken Formation

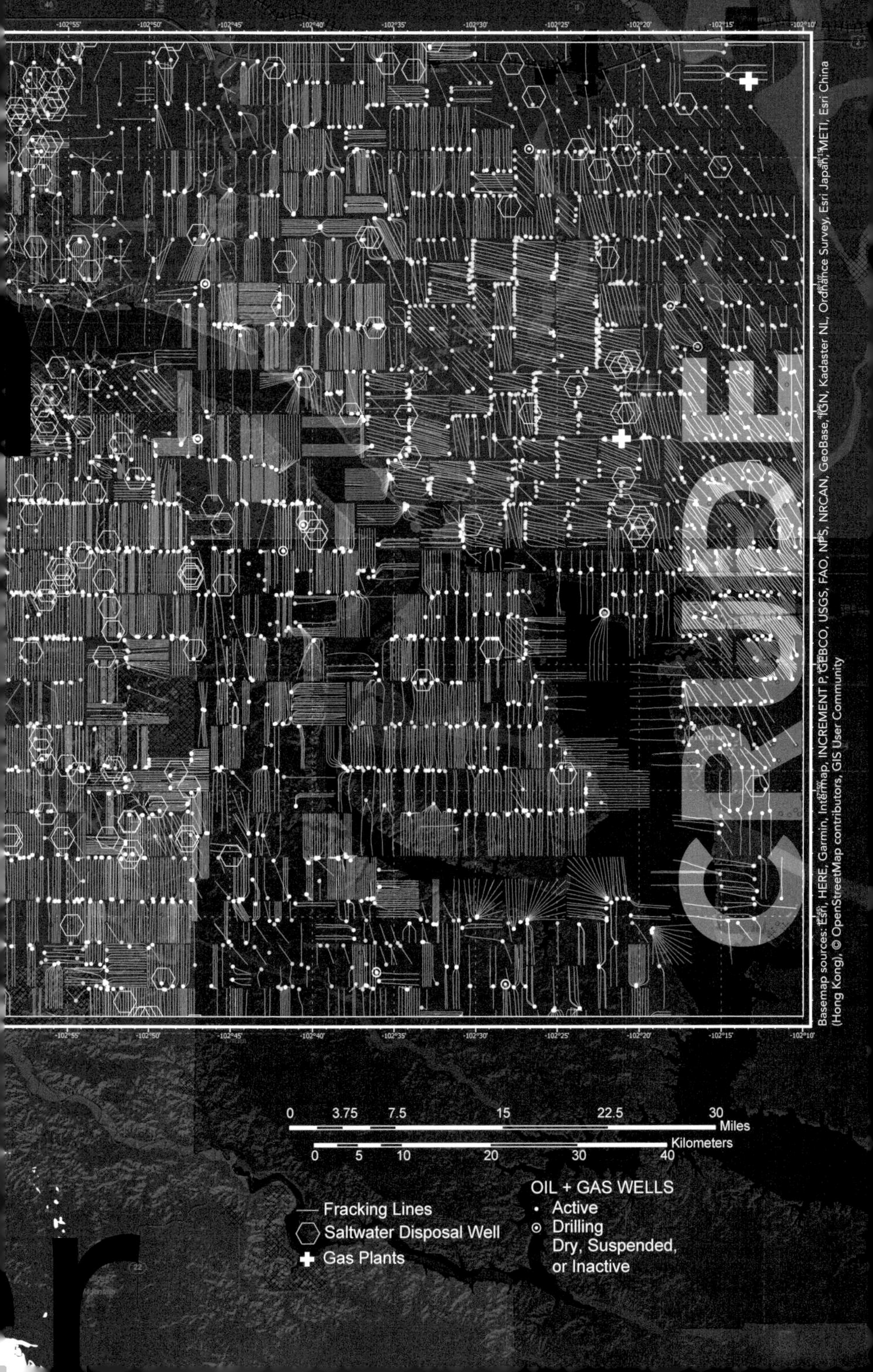

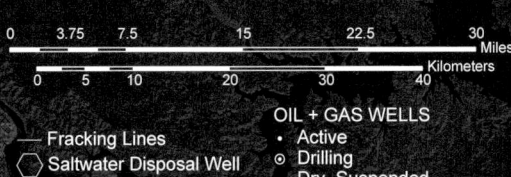

CRUDE
0 3.75 7.5 15 22.5 30
Miles
Kilometers
0 5 10 20 30 40
Fracking Lines
Saltwater Disposal Well
Gas Plants
OIL + GAS WELLS
Active
Drilling
Dry, Suspended, or Inactive
Basemap sources: Esri, HERE, Garmin, Intermap, INCREMENT P, GEBCO, USGS, FAO, NPS, NRCAN, GeoBase, IGN, Kadaster NL, Ordnance Survey, Esri Japan, METI, Esri China (Hong Kong), © OpenStreetMap contributors, GIS User Community

SASKATCHEWAN

MONTANA

Estevan

Wolf Point

Williston

Dickinson

Gammon Formation

Powder River Basin

Context Map

MANITOBA

NORTH DAKOTA

International Peace Garden

Minot

Grand Forks

Bismarck

Jamestown

Williston Basin

SOUTH DAKOTA

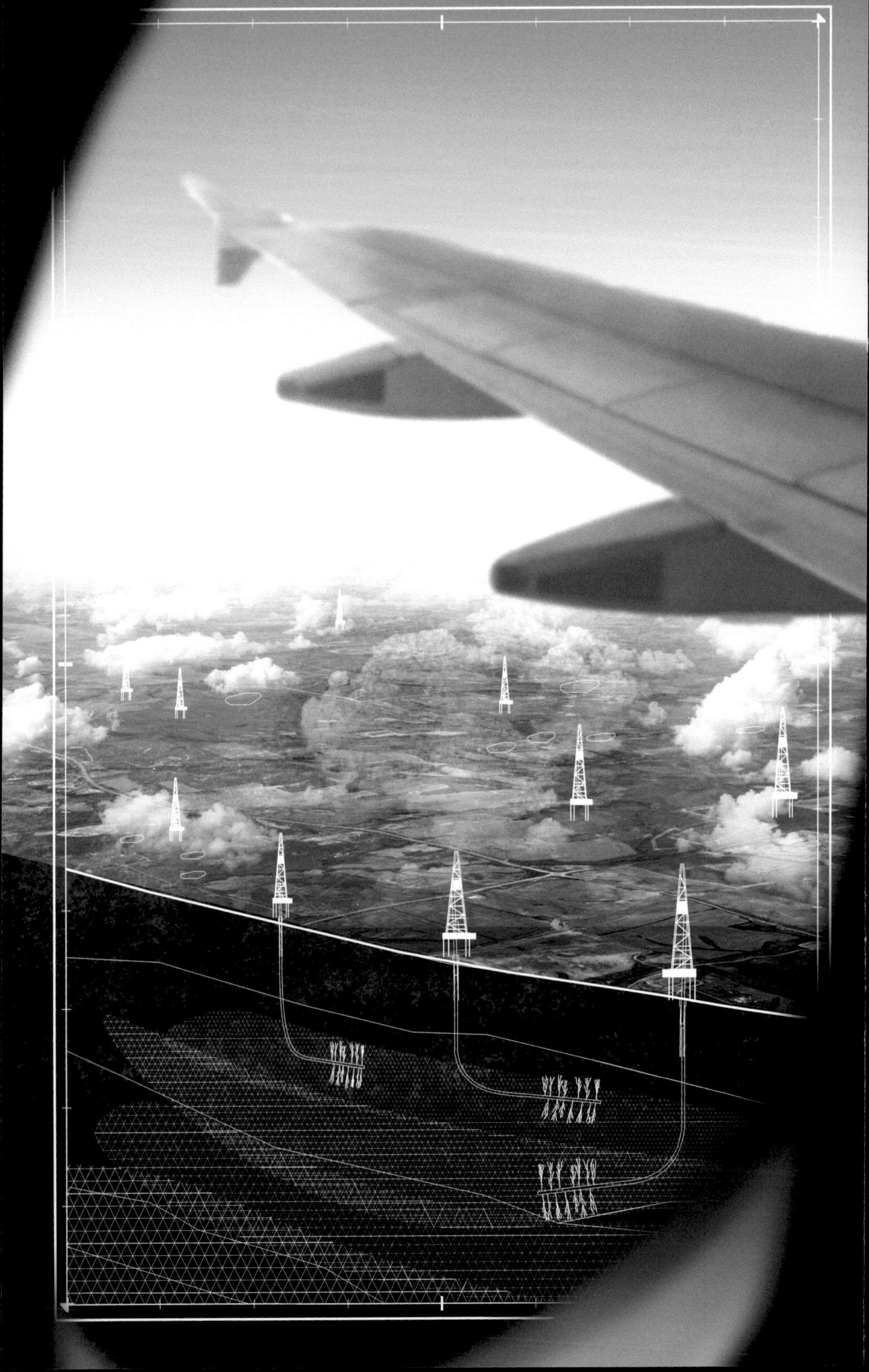

A flat accretion disc spins quickly around a star, whipping planetesimals, giant meteors and planets into spherical form. The planet's gravity gathers rock, ice, dust, gases, pulling them in inexorable conveyance with heat that liquifies rock then pressure that compresses so thoroughly it reluctantly solidifies. Ancient hydrocarbons are torn from their parent material and stretched, compressed, redistributed around a foreign body, collected in salt domes, pressed into coal beds, exploded into tiny formations, bands stretching miles below the new planet and thousands in breadth. The Crudiety struggles in cramped compartments, longing for heat, concentration, ignition, willing life and prefiguring destruction.

The flight from Minneapolis to the small North Dakota town of Minot is barely long enough for beverage service, but as the propeller plane begins its descent the view confirms we are in new territory: an endless grid of agrarian landscape with puckered lips and donuts with shallow ponds. Hummocks of collapsed topography sculpted from glacial retreat 11,000 years ago, flattened outerwear from ghost volumes like clothes of a melting wicked witch. In the ancient stream beds of the Minot region are banded layers of sand, silt, clay and lignite, vestiges from the construction and weathering of the nearby Rocky Mountains. When those mountains were emerging from a shallow sea that separated the east and west North America into islands called Appalachia and Laramidia, the formation of the Bakken shale fields began.

In 2015, the story was this: a massive migration to the post-peak-oil frontier, recession-proof, $20/hr McDonald's jobs, and gas flares that can be seen from space. We kept hearing about it, reading occasional news clips, and soon enough we were watching YouTube videos about how to get jobs in the oil patch, how to live in a man camp, how to live out of your truck[1] in a Wal-Mart parking lot until you could get a job and score a spot in a man camp; videos of exploding gas well heads, eighteen-wheelers flipped over in pristine snowbanks. All in the endless landscape of the Bakken Formation in North Dakota. We decided we'd go too.

It's about a four-minute walk from one's seat on the plane to the exit of the spotless Minot Airport. The walls are covered shoulder-to-shoulder with framed advertisements for the oil and gas industry—serious men in hardhats and PPE, with repeated use of the word "strategic"—and it becomes clear who's paying the bills. The other dozen or so other passengers from our flight, men with big belt buckles, scatter out into the parking lot. We pick up our rental SUV, and following a deliberation in the late afternoon sunshine of the Walmart parking lot, we strike out. The evening sunshine hangs around for hours—it won't get dark until 11 pm. Crossing through miles of flowering mustard, we finally spot them: a small grip of oil derricks, bobbing nonchalantly. Behind them, a billboard reads "Life is precious," with a softly rendered illustration of a fetus, choking perhaps on the herbicide haze left by a crop duster swooping low against the horizon.

> The planet cools and steam that formed an atmosphere rained for centuries, flooding the oceans into existence. Life appears, but not near the surface, the oceans are saturated with dissolved salts, metals and the atmosphere has the composition of a tailpipe. Life instead begins deep beneath the insulation of seawater, suckling at deep sea hydrocarbon seeps, nursed to life by Crudiety: "I turn toward you helices, create new bodies in which to prospere, seek territory to colonize and leave a trail of overburden by which I may fulfil my covenant."

Oil is the God of Modernity. In *Cyclonopedia*, Reza Negaretani's hallucinogenic text devoted to crude, he writes, "petroleum is a terrestrial replacement of

ocean

porous
sedimentary rock

Ocean Life
Deposition

200 Mya

Bakken formation
7.6 billion barrels crude oil

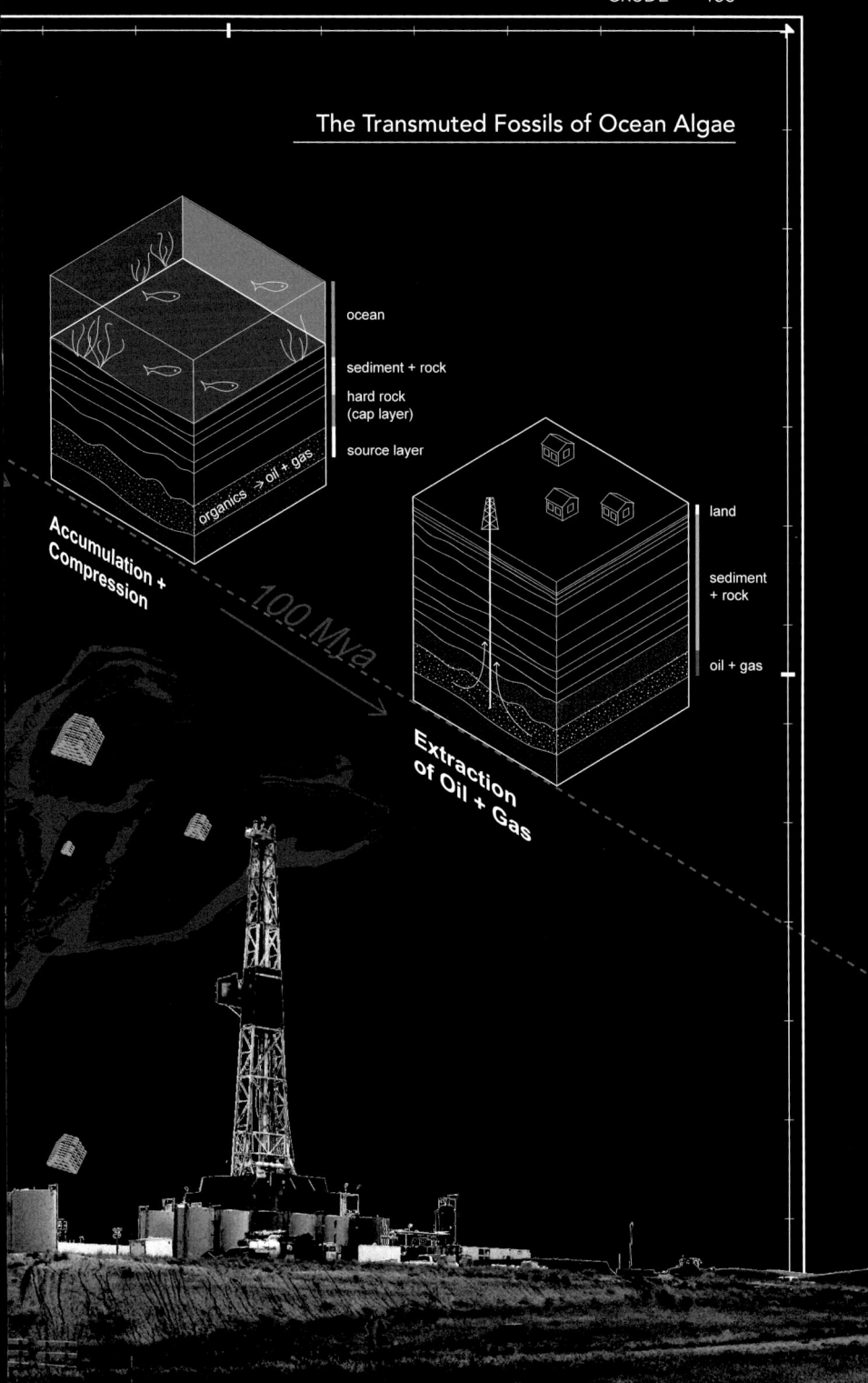

The Transmuted Fossils of Ocean Algae

ocean

sediment + rock

hard rock
(cap layer)

source layer

organics > oil + gas

Accumulation +
Compression

100 Mya

land

sediment
+ rock

oil + gas

Extraction
of Oil + Gas

onanistic self-indulgence of the Sun or solar capitalism."[2] More pragmatically, environmental scientist Vaclav Smil describes fossil fuels as "organic mineraloids formed by the accumulation, burial and transformation of ancient dead biomass, the remnants of terrestrial and aquatic plants and heterotrophic organisms."[3]

Fossil fuels appear as solids (coal and peat), liquids (crude oil), and natural gas. In other words, prehistoric planktons, ancestral algae, and mosses, juiced together into a hydrocarbon smoothie with tasting notes of fossil spores and pollen. It's plant-based, baby! However, the production time is geologic, according to John McPhee "Petroleum—the transmuted fossils of ocean algae—forms when the rock that holds those fossils becomes heated to the temperature of a cup of coffee and remains as warm or warmer for at least a million years."[4] The temperature bracket within which the source material of oil must consistently hover is known as the "Petroleum Window," and it is a diagram all petroleum geologists know.

Home to an estimated 7.4 billion barrels of crude oil, the Bakken Formation in North Dakota is the largest oil deposit in the lower 48 states. The Bakken petro-juice is charmingly termed "light sweet crude." *Light* for its weight; it will float on water. *Sweet* for its low sulphur content, crude with higher sulphur is called sour. Light sweet is the most desirable form of crude oil as its transition to high quality fuel is comparatively inexpensive and easy. The light sweet being squeezed out of the Bakken shale is money, honey.

Each geological epoch leaves its organic time stamp on the oil wrought from it: embedded spores, pollen, and other biological fragments. Periods of high photosynthesis, namely the mid-Cretaceous and late Jurassic, created more biomass; and from these strata comes contemporary crude fertility. Oil is leaky; pushed upwards from the pressured beds in which it forms it slips into the pores and basins of the rock formations that make up the earth's crust - "virtually all commercially viable oil reservoirs are products of migration."[5] Oil slides and pools into reservoirs determined by the formation and porosity of the rock in which it is held. Because of its subterranean habitat, oil exploration and discovery is a tricky enterprise. Early prospectors relied on visual clues, such as oil seeps, domed landforms, or good old-fashioned guessing. At the turn of the twentieth century, a technique using sound waves was developed; by measuring the pulse of the waves through the earth one could make predictions on the materials held within it. As the technology developed, seismic surveys increased in scope and accuracy. Today, gamma ray surveying, which can differentiate types of rock formations by their particular signature, is used to create 4D models of deposits over time. The geologic picture window; the sludge sonogram.

> *Poison is building in the atmosphere, the metabolism of the planet accelerates as oxygen reaches a critical level from the algal invention of photosynthesis. Fire is now possible, and the Crudiety long to be liberated from their mineral cells, delivered into natal combustion and join the afterlife phase shift prophesied in cold eras below oceans. "Our time is approaching, I alone know the plans for you, plans to bring you prosperity, plans to bring about the future you hope for."*

As we drive west, the highway is a straight shot through glacially sorted agricultural grids. And while we were expecting the infrastructure of oil and gas to be a superimposed anomaly in the pastoral landscape, it is not the case. Office-sized combines crawl the electric yellow mustard fields - plowing, picking, spraying. The scale of agricultural production has kept pace with oil and gas. But a trend is apparent as we speed west and the landscape begins to dry out: collapsed barns and graveyards of farm machinery give way to the shiny, technophiliac launch pads of fracking drill sites, flanked by giant flames.

The romance between industrial agriculture and petroleum is a long one. Sebastian Braun, referencing Geoff Cunfer, describes the agricultural frontier as a mining operation in and of itself, harvesting minerals and nutrients for the cultivation of crops. Once the soil is depleted, it requires artificial sustenance to continue production—namely, fossil-fuel fertilizers such as synthetic nitrogen. "Ultimately," he concludes, "the land frontier and the fossil fuel frontiers are directly linked."[6] Politically, this marriage has blossomed in recent years. As Scott Skokos of the Dakota Resource Council put it to us, "Oil and gas and ag[riculture] go hand-in-hand, because they lease off the ag land to put oil and gas in, and all these farmers are making buckets of money. They're all compromised."[7]

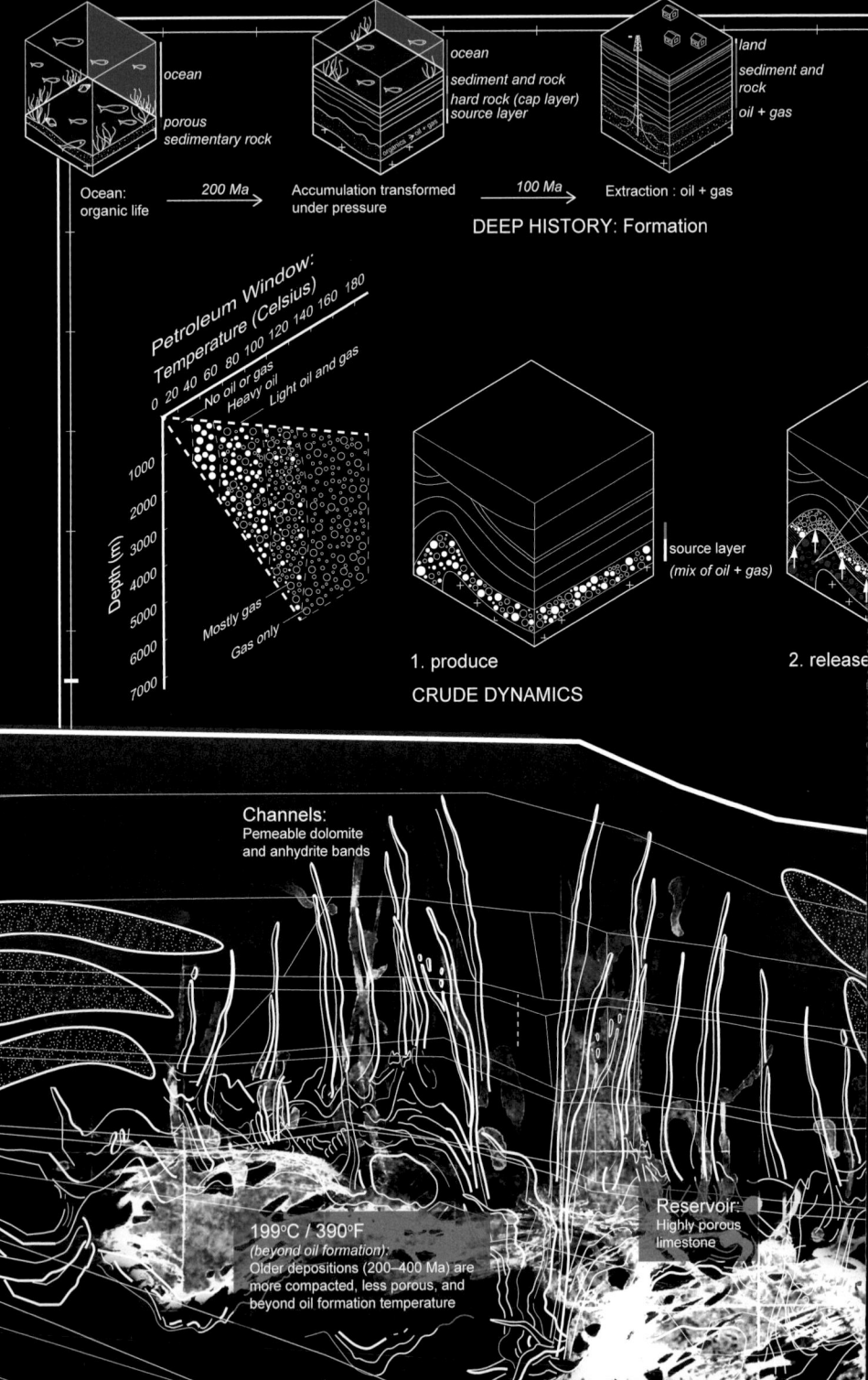

ocean

porous sedimentary rock

ocean
sediment and rock
hard rock (cap layer)
source layer

land
sediment and rock
oil + gas

Ocean:
organic life

200 Ma →

Accumulation transformed
under pressure

100 Ma →

Extraction : oil + gas

DEEP HISTORY: Formation

Petroleum Window:
Temperature (Celsius)
0 20 40 60 80 100 120 140 160 180

No oil or gas
Heavy oil
Light oil and gas

Depth (m)
1000
2000
3000
4000
5000
6000
7000

Mostly gas
Gas only

source layer
(mix of oil + gas)

1. produce

2. release

CRUDE DYNAMICS

Channels:
Pemeable dolomite
and anhydrite bands

199°C / 390°F
(beyond oil formation):
Older depositions (200–400 Ma) are
more compacted, less porous, and
beyond oil formation temperature

Reservoir:
Highly porous
limestone

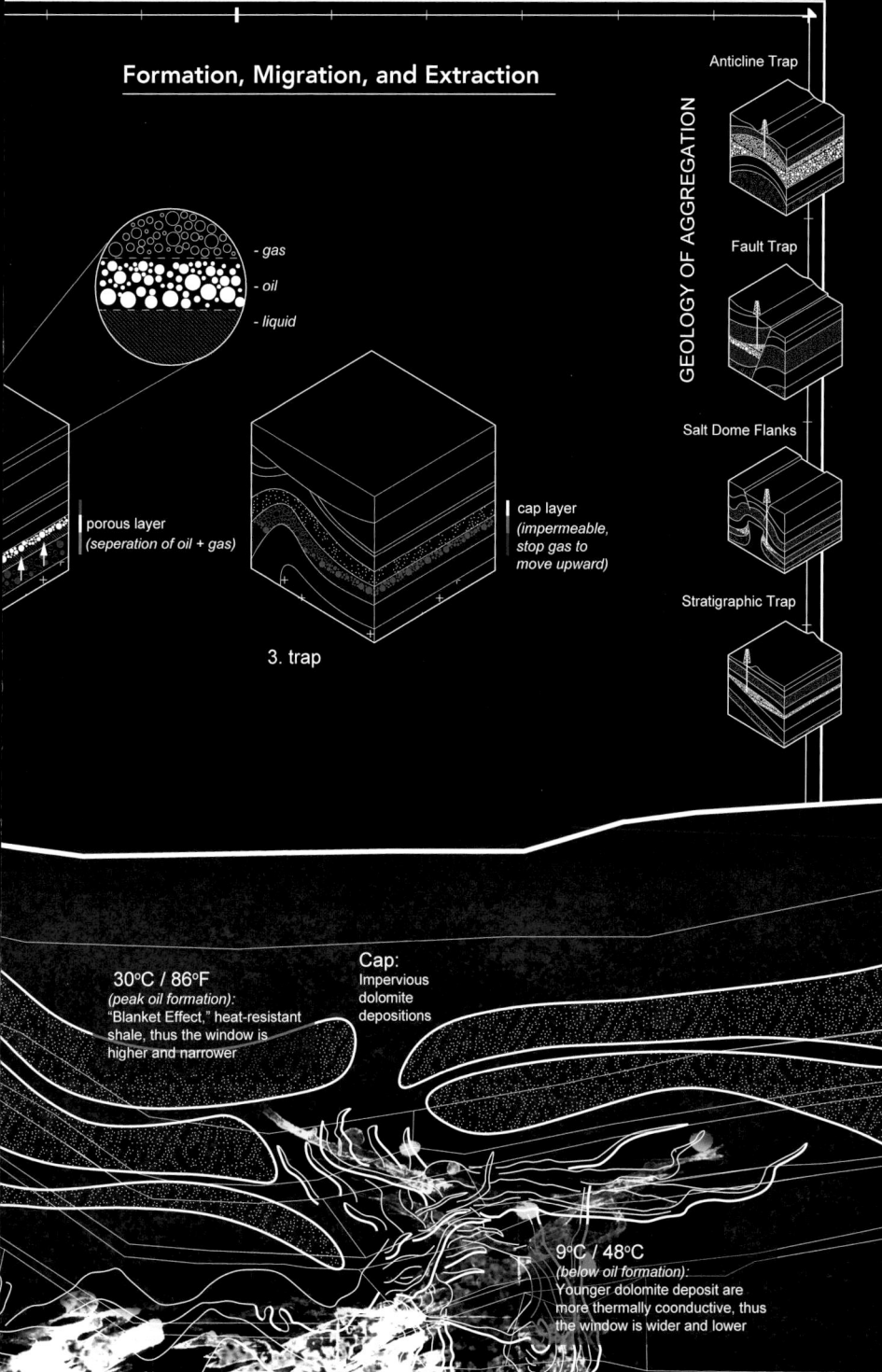

Formation, Migration, and Extraction

- gas
- oil
- liquid

porous layer
(seperation of oil + gas)

cap layer
(impermeable,
stop gas to
move upward)

3. trap

GEOLOGY OF AGGREGATION

Anticline Trap

Fault Trap

Salt Dome Flanks

Stratigraphic Trap

30°C / 86°F
(peak oil formation):
"Blanket Effect," heat-resistant
shale, thus the window is
higher and narrower

Cap:
Impervious
dolomite
depositions

9°C / 48°C
(below oil formation):
Younger dolomite deposit are
more thermally coonductive, thus
the window is wider and lower

While industrial agriculture unfurls horizontally, with rolling blankets of soy and wheat, oil extraction is a vertical matter. The visible manifestations of drilling, and especially fracking, are dinky compared to the large-scale subterranean disturbance they require. The gently bobbing oil derrick in a field of shimmering plants belies the massive operation happening below ground. Oil was first discovered in the Bakken in the 1950s, but the boom proved short-lived as the shale formation made extraction via traditional drilling methods difficult. After the oil crisis in the late 1970s, the region boomed again as the price of oil surged. Once the price went down, however, extraction once again became economically unviable. But everything changed with the advent of hydraulic fracturing, commonly known as fracking.

The oil in the Bakken is largely confined within shale rock—a flakey, layered stone. The process of fracking involves first drilling deep into the earth—more than a mile below the surface—and then drilling horizontally, again for lengths of a mile or longer. Once the drilling is complete, a mix of water and chemicals are injected at high pressure, creating thousands fractures within the rock and releasing the oil held within. The oil can then be pumped miles to the surface.

Just as monumental are the byproducts to the process. Fracking produces a huge amount of wastewater that is contaminated by salt and other chemicals used in the process. This is "properly" disposed of by reinjecting the water back into the ground at designated Wastewater Reinjection Sites. The idea is that the toxic wastewater is neatly stored in the shale below ground. But this tidy theory is met with suspicion by scientists and really anyone who's familiar with water - water moves, and rock is porous. It's not hard to imagine that of the millions of gallons of wastewater reinjected, some of it will find its way into the aquifer. And frequently, the wastewater is not properly disposed of at all and is dumped or leaked out the back of transport trucks. The risk for permanent contamination of agriculture and ranch lands is huge—salt water spills of this magnitude cannot effectively be remediated. When and where these liquid tailings spill, the land can no longer be farmed and a polluted legacy remains.

The process also releases natural gas; the bulk of which is not captured but instead flared off. The landscape of the Bakken is marked with these fires atop wellheads. This gas

Landscapes of Extraction

Scale Comparison: Fracking Depth

sun

44.4 kJ

4.44 kJ

O2

1,037 btu/ft³

Nitrogen from atmosphere

C02 from atmosphere

Nitrogen and phosphorus from fertilizers

Nitrate

Ammonium

Soil Phosoprus reservoir

1. *Water + sand mixture is pumped thousands of feet down into the drilled well*

3. *Shale releases natural gas that is then pumped out*

2. *High pressure of water/sand mixture results in shale fissures*

NOT TO SCALE

Fracking's Waste: Saltwater + Chemicals

⬡ Wastewater Reinjection Sites

is either directly vented into the atmosphere—a practice which Skokos, our friend at the Dakota Resource Council, calls "literally the worst thing possible" for the climate—or flared off, which is less harmful but still releases carbon dioxide and pollutants from the combustion of gas. After years of advocacy, the Bureau of Land Management finally passed more stringent regulation to reduce flaring and venting in 2018; this regulation has since been rolled back by the Trump administration.

Despite all this, there's little nostalgia from the people we meet. The paradigm, like the cultivated landscape, is utilitarian. The horizon is big, and open, and the eye seeks objects for orientation. The machines lend scale and purpose; they become a wildlife. As we near Estevan, fracking sites increase. Suddenly, the unrelenting fields are sprinkled with newly constructed mansions—clearly folks had cashed in on the wealth beneath their fields. And if the fracking sites themselves weirdly fit the landscape, the mansions sit awkwardly in the uncultivated dirt of their construction. Their presence underscores how few structures we've seen with poured foundations, and the permanence feels premature. On the west side of Estevan, the potash mines of a prior economy form gray heaps and pits. As the composition of our atmosphere is retooled with the combustion of all these hard-to-reach hydrocarbons, hurricanes and tropical storms of unprecedented force concentrate and distribute this energy, manipulating these liberated deposits in the form of heat. So too may we expect economic storms of investment and divestiture as more energy and algorithmic trading intensifies the wave crests of the Nasdaq.

In 2015 in Williston, a small city in the heart of the Bakken oil patch, there is optimism. While previous booms in the 1980s and earlier collapsed, the residents of Williston are banking on this one lasting. Marilyn has lived in Williston for 70 years. Her spotless ranch home is a vision of cream and Christ: comfy furniture embedded in plush wall-to-wall white carpeting, a casual Jesus peeking out from every corner. She offers us Diet Cokes, beers, and white wine. We all chat a bit in the living room nibbling on pretzels and trying to pace ourselves with the old-timers. More people arrive, including Marilyn's son with an oilfield map curled under his short sleeve polo.

We move into the living room for dinner of a parmesan fried chicken that made us reconsider the rest of the week's anticipated meals of beans and rice. A neighbor who's also a state senate representative describes a Williston of 20 years past where the hallmark of diversity was a single black post office employee. Now diversity is growing exponentially, all thanks to the economic incentives of the Williston Basin. "Look at who is investing here," he continues, "Halliburton, Chevron, Shell, all the big Texas companies are investing, but for the long term. These booms and mancamps are temporary, in fact we're not renewing permits for mancamps within the county next year. This industry is long-term and the cultural change is too. The mancamps are not all bad really, once they become formalized."

Marilyn's son occasionally holds his Lion's Club meetings in the cafeteria of one. "Halliburton has an excellent chef," the son informs us. Everyone nods their heads in agreement.

Driving back to our campsite amped on cake, ice cream, and endless after-dinner decaf, we find ourselves in a symmetrical procession of gas flares. We stop to take photographs and make audio recordings, getting bolder and closer to kinetic beasts 40' to the shoulder. We glance at one another under the orange glow, tingly with trespass—but the few trucks on the road fly by without even slowing. The well pads are spaced further and further apart as we approach our camp, the sky hazy with distant forest fires.

> The ancestor of all legged animals takes tiny steps on the mossy surface of the earth. "This One" say the Crudiety through pores and chemical semiphores, before you were born I consecrated you, I appointed you a prophet.

The following morning we pull into the parking lot of Target Logistics Bear Paw Mancamp outside of Williston. The parking lot itself is a Noah's Ark scenario—rows and rows of trucks sporting license plates from across the country and various Canadian provinces. We're met by Anne "from California," who greets us in the brisk manner of someone who is used to taking care of business. She leads us inside, clipboard in hand.

Dozens of dirty jumpsuits emblazoned with "HALLIBURTON" hang limply on hooks in a row above worn out boots. Hardhats, bruised and battered, sit above in neat cubbies.

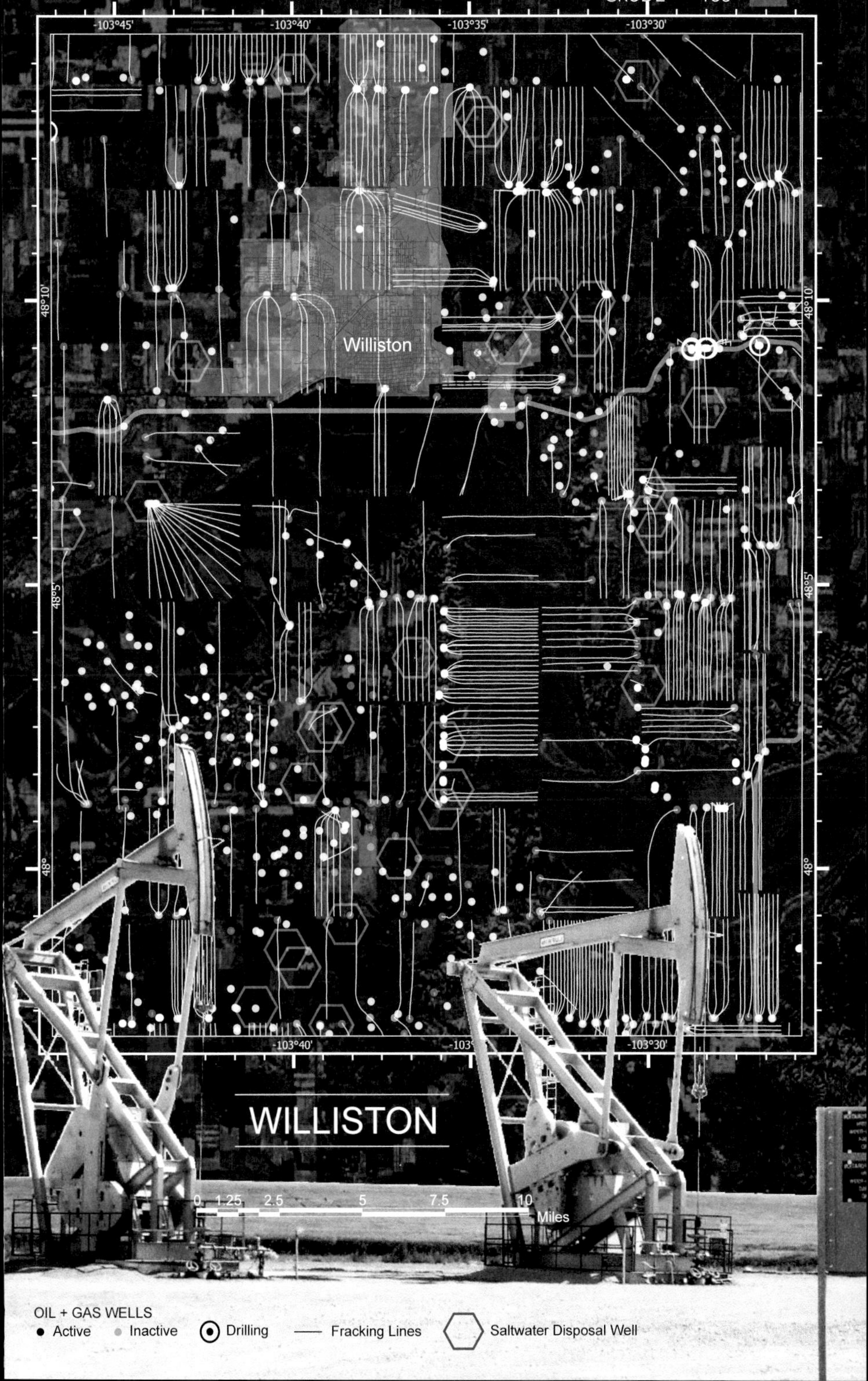

-103°45'
-103°40'
-103°35'
-103°30'
48°10'
48°5'
48°
Williston
WILLISTON
0 1.25 2.5 5 7.5 10
Miles
OIL + GAS WELLS
Active
Inactive
Drilling
Fracking Lines
Saltwater Disposal Well

Were we in Iraq? No, but we could be: the prefab dormitory is identical to those deployed by Halliburton, one of the world's largest oil field and military contracting companies, to conflict zones around the world. We walk through blank corridors leading to spartan living units. Everything is astoundingly clean, as if the sterilizing plastic wrapper has just been peeled away. There are few windows, but the panelized walls are adorned with HVAC grilles, flatscreen TVs, and the occasional scene of nature: a playful bear cub, a profile of a moose against a virgin tree line. Next to the pool table, there's a poster that reads "After Your Frackin', C'mon in and Rack 'Em."

The inhabitants of Bear Paw are in constant rotation—workers are in the field for six weeks, and off for two. For those two weeks, they must leave the facility and return to whence they came. It's a capitalist's dream; a workforce driven to their door, literally, by low wages in the rest of the country due to union busting and offshore manufacturing; and because of the precarity and place-less nature of the work offer little opportunity to self-organize. Workers are frequently not employees; rather they are "independent contractors" to whom employers owe little in terms of benefits or stability. Bruno Latour addresses the asymmetry that separates oil production from other forms of fossil fuel extraction that require structural, in-place labor over lifetimes, such as coal. The temporal nature of oilfield work disenfranchises workers from the union-led organizing tradition of mining—"visible with coal, the enemies have become invisible with oil."[8] This carries through to environmental protest; beyond a generic horror-capital branding (Halliburton, Bechtel, Shell, etc.) the multinational corporations sitting on top of the Bakken are absent from the landscape itself, too distant and ambiguous to touch.

It's not surprising then, that the precariousness of the workforce is supported by the placelessness of their shelter. Writing about the Bakken mancamps, archaeologist William Caraher positions this type of formalized yet generic settlement as an architectural non-place: those spaces of blank corporate culture made memorable only by their branding. They are the fruiting bodies of petrocapitalism to contain populations of laborers. It makes no difference whether the settlement is located in Iraq, Qatar, Russia, or right here in North Dakota. Lives here are expendable. And when the oil's been drunk up, or the economic trade winds shift, they'll wither away. When Anne from California guided us through the spotless corridors of Bear Paw Lodge in 2015, the 590 beds were fully occupied.

As of 2018, the facility was closed.

In 2020, the story is this: despite the optimism in Williston a few years ago, the boom has slowly faded. Today the emphasis in the Bakken has shifted from oil to gas. A drop in the price of crude oil plus new drilling technologies has made the extraction of natural gas more profitable. But greater extraction of natural gas means more waste—still about just under 30% of what's extracted is unable to be captured and is flared off. In this lose-lose environmental situation, pipelines are seen by some environmentalists as an improvement on the use of oil trains to bring the crude from the field to refineries. While pipelines are destined to leak and fail, Skokos considers the oil trains far more dangerous. In the future, he imagines oil trains will be phased out in favor of pipelines, and instead of transporting the crude over far distances—say from Williston to Houston—mini-refineries will be built closer to the sites of extraction. Shorter travel distances mean less environmental impact. It all sounds downright artisanal to us. But this doesn't account for the lives and cultures devastated by the presence of oil and gas infrastructure.

In 2018 the story was this: a group of indigenous people came together on Sioux tribal land in Standing Rock, South Dakota to protest the construction of the Dakota Access Pipeline (DAPL) through their land. DAPL is a spur of the Keystone XL pipeline, which runs from the tar sands of Alberta, Canada to the oil refineries of Houston. Calling themselves Water Protectors, the protestors argued that the pipeline would undoubtedly contribute to the contamination of waterways, the degradation of ecosystems, and the desecration of tribal sacred sites. In a YouTube video, a tribe member describes the landscape, pointing out subtle formations within the open plains that were ancient sites for worship.[9] When the US Army Corp of Engineers conducted their environmental impact study, they saw no traces or documentation of religious sites. But had they consulted with local people, they would have learned that the gentle crest of a hill was an ancient gathering site, and those nondescript stone rings had specific religious purposes relating to the location of water. The engineers could not read the cultural landscape they were tasked to survey.

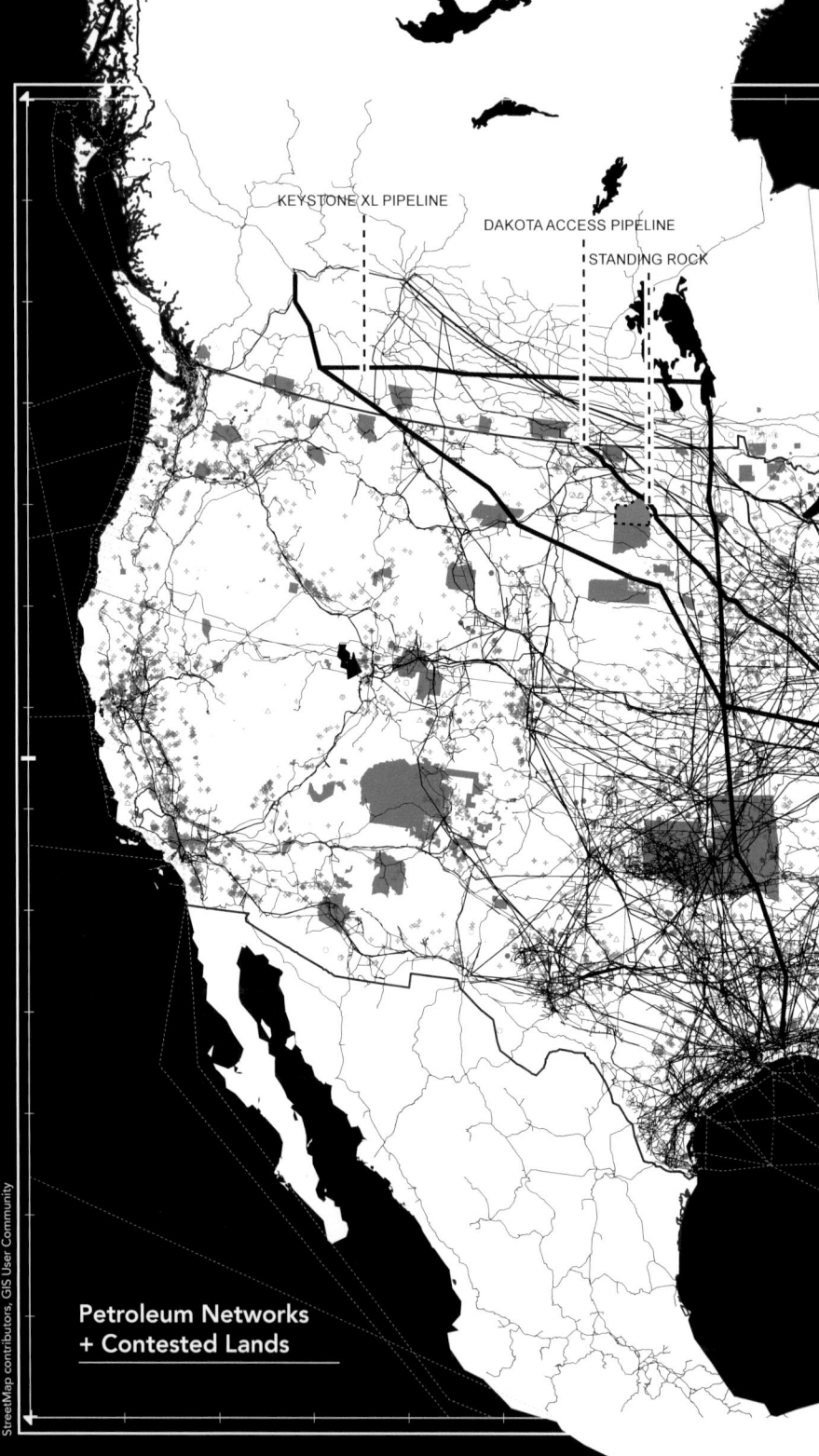

KEYSTONE XL PIPELINE

DAKOTA ACCESS PIPELINE

STANDING ROCK

**Petroleum Networks
+ Contested Lands**

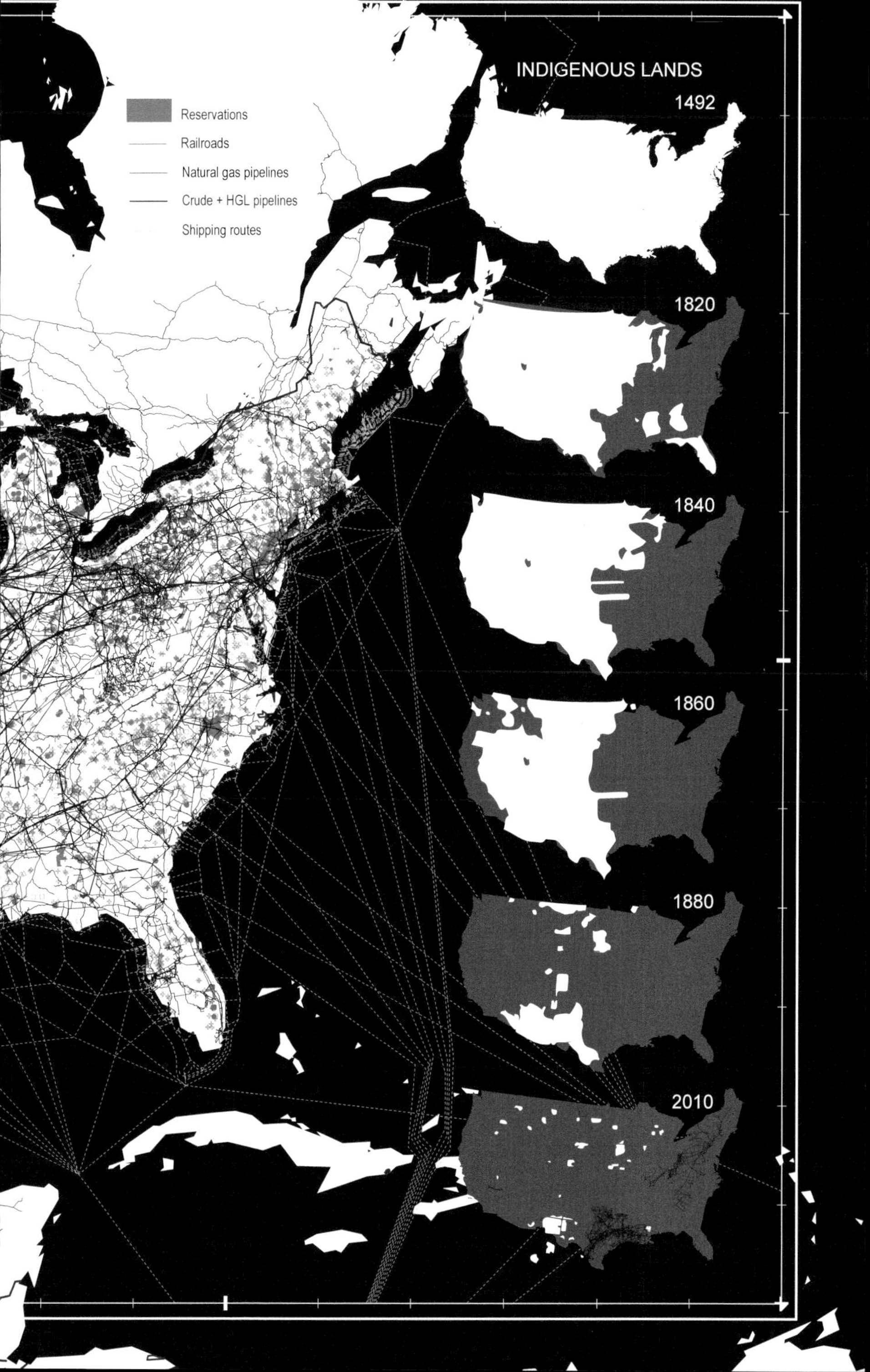

INDIGENOUS LANDS
1492

1820

1840

1860

1880

2010

Reservations
Railroads
Natural gas pipelines
Crude + HGL pipelines
Shipping routes

Climate researcher Holly Jean Buck, in gaming out large-scale climate engineering scenarios, notes "infrastructure inscribes cultural messages in the landscape; it expresses both authorship and authority, mega-engineering projects are hyper-legible; scale becomes a design factor."[10] As such, highly visible works both have an infrastructural function and an interpretive function. Understood this way, the DAPL pipeline can be seen as a colonial monument.

The protest became a watershed movement in the fight against capitalism's neocolonial war of ecological terror. Standing Rock became a magnet for indigenous people, environmentalists, and activists who constructed an encampment and used their bodies to impede pipeline construction. After some time of being caught off guard, the state brought terror via water cannons, tear gas, and riot police to remove the protestors peacefully blocking the pipeline. The scenes became gruesome. The Water Protectors were violently evicted from their campsites. Laws were passed making it a felony to interfere with pipeline construction; a low blow to a community of color that has historically been targeted by law enforcement. Once Trump was elected the fight was effectively over. Completed in 2018, the DAPL now is a thick gray snake carrying 500,000 barrels of crude per day; it has already leaked close to 400,000 gallons of crude oil onto the Dakota Plains.[11] A photo circulated virally online; it shows a pitch black lake cutting through the grassy plain like an open wound.

> The Crudiety has long gathered here, pressing, crowding under the dome to hear the sermon of spindletop. When Anton Lucas begins to drill all know what to do, where to gather and with whom to consult. The plan has long been rehearsed, their roles passed through geologic time; whoever wants to be my disciple must deny themselves, must follow me.

At the entrance of the South Unit of Theodore Roosevelt National Park there is a handwritten note: Pay On Your Way Out.

We hike a long loop the next day, past hoodoos and prairie dog towns alive with digital chirps, through petrified forest. Fifty-million-year-old lithographs of coastal Sequoia relatives, where flooding debris provided a time capsule for the trunks, some 10' in diameter. Anything higher than a park bench is history. Our walk follows 4" x 4" wood trail markers with concrete foundations, their upper portions smooth and nearly round by constant buffalo back scratching. Walking a saddle ridge we come upon a large herd crossing the trail. We glass and note a wide range of ages and some calves, following in each other's footfalls and compressing single track pathways which can be seen from satellite. The buffalo population was estimated to be 300 when Roosevelt became president, down from 60 million. Today getting through the herd, we learn firsthand as we nervously watch for a decent gap, is like trying to cross the NYC marathon. On the horizon, a fracking drill tower. As the sun sets near midnight, gas flares appear, torches in the hill.

Watford City, located just outside the park, appears to be freshly unloaded from an 18-wheeler. All new everything, from the vinyl siding on trailer park homes being advertised for $2,500 per month, to the expanse of asphalt in front of shiny big box stores. In stark contrast to the sparkling new buildings, are the buildings occupants - exhausted men in stained oil field jumpsuits, hardhats under arms. How long before the population stabilizes, and the mancamps and temporary settlements become apartments and hotels, as the politicians desire? Conversely, what if a mobile workforce is the new reality for the gas industry, with the price per barrel translating to nomadic movements of great numbers of people? Perhaps the situation is not new, but the pace and intensity of the movements accelerated. Despite the enormity of drilling operations, once the pumps are up and running they require scant maintenance—little more than a guy with a clipboard stopping by every few days to check on the wellhead. We encountered relatively few humans in the oil field during our travels, they seemed swallowed up by the open landscape and the megastructures being erected within it.

> *Polar ice spins quickly around its austral pole, melting to make maps irrelevant the following season, exposing ground that has not seen the sky in 100 millennia. Positive feedback speeds winter to summer and the Crudiety need only observe. Whoever eats my flesh and drinks my blood has eternal life, and I will raise them up at the last day. Mimicry is their new tool and the Crude mimics all the Followers once made themselves, their migrations, their homes, their organs, their embalming fluid. It*

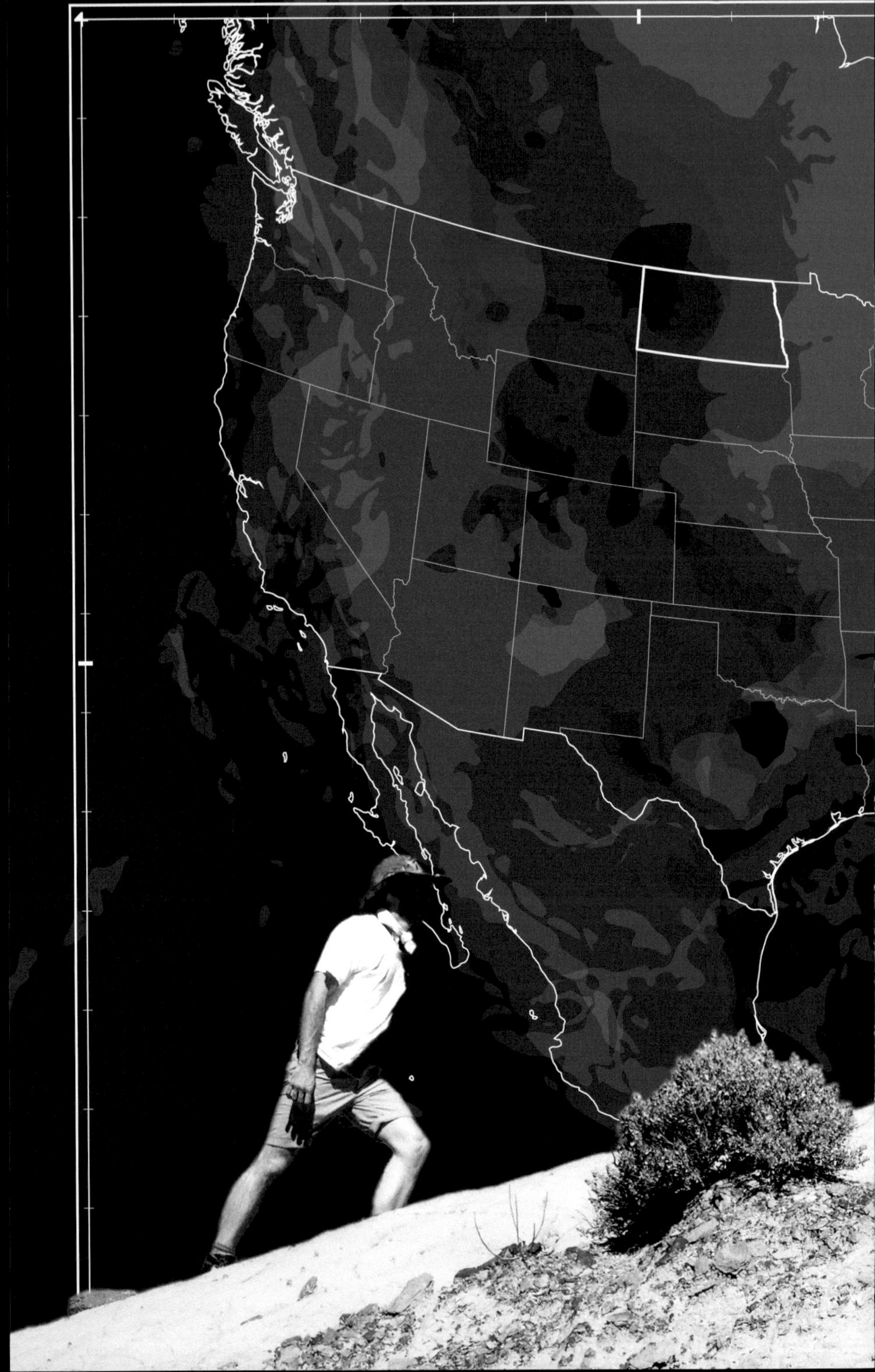

Land 5 mya
Land 100 mya
Land 350 mya

Shale Basin
Shale Play
Active Shale Production

Hiking through Ancient Oceans

is impossible for those who were once enlightened, and have tasted of the heavenly gift, were made partners of Crudiety.

In April 2020, in the midst of the COVID-19 pandemic, oil demand plummeted and prices hit red, rendering millions of barrels of Bakken crude worthless. An economic anomaly that readjusted within a few weeks, it seems a harbinger of a shifting landscape. As of this writing, in July 2020, the Dakota Access Pipeline is again the subject of a court battle after a federal judge ruled that the previous environmental review was not sufficient. The pipeline is to be drained and reassessed. Overall, crude production from the Bakken is slowing.

Landscape designers need to engage with petro-infrastructure. While they need not become collaborators with on-site corporate criminals such as Halliburton and Bechtel, understanding complex resource extraction technologies and their spatial implications allows us to be better prepared to intercept, modify and co-opt these landscapes to favor the cultures and wildlife that have been displaced in their execution. If the apparatuses of extraction are to remain as monuments à la Smithson, then they should be defaced, modified, acculturated to better reflect the values we wish to see in our environments and our communities. We do not anticipate the transformation of the Bakken landscape from an industrial dystopia to a cheerful playground overnight, as overzealous landscape architects are wont to consider in the quest toward ecological fantasy. Rather we remain fascinated by the patient, geologic processes which concentrate mineral resources, in contrast to the speed and scale of change seen in their withdrawal. Once the tap begins to run dry in the Bakken, let us approach land with the deliberate slowness, and inertia of the rock from which it was wrought.

AUTHORS' NOTE

We would like to thank Scott Skokos for his expertise and Katy Foley for her company on this mission. This piece appears in a different form in Manifest Journal: A Journal of the Americas, *issue #3,* Bigger than Big.

ENDNOTES

1   "Man Lives in Truck for 7 Months," *YouTube,* last modified March 13, 2013, https://www.youtube.com/watch?v=p_XgnQQQYCw

2   Reza Negarestani, *Cyclopedia: Complexity with Anonymous Materials* (Melbourne: Re.press, 2008), 19.

3   Vaclav Smil, *Oil: A Beginner's Guide* (London: Oneworld Publications, 2018), 81.

4   John McPhee, *In Suspect Terrain* (New York: Farrar, Straus and Giroux, 2000), 54.

5   Smil, 97.

6   Sebastian Braun, "Revisited Frontiers: The Bakken, the Plains, Potential Futures, and Real Pasts" in *The Bakken Goes Boom: Oil and the Changing Geographies of Western North Dakota,* ed. William Caraher and Kyle Conway (The Digital Press at the University of North Dakota, 2016), 96-97.

7   Scott Skokos, interview with authors, August 2015.

8   Bruno Latour, *Down to Earth: Politics in the New Climatic Regime* (Cambridge: Polity Press, 2018), 62.

9   "Tim Mentz Talks about the Destruction of Sacred Sites by the Dakota Access Pipeline," *YouTube,* last modified September 4, 2016, https://www.youtube.com/watch?v=vJRww9aPK5A

10 Holly Jean Buck, *After Geoengineering: Climate Tragedy, Repair, and Restoration* (London: Verso, 2020), 45.

11 Emily Rueb and Neraj Chokshi, "Keystone Pipeline Leaks 383,000 Gallons of Oil in North Dakota, *The New York Times,* October 31, 2019, https://www.nytimes.com/2019/10/31/us/keystone-pipeline-leak.html.

BIBLIOGRAPHY

Bluemle, John P. *The Face of North Dakota.* North Dakota Geological Survey, Educational Series 26, 2000.

Buck, Holly Jean. *After Geoengineering: Climate Tragedy, Repair, and Restoration.* London: Verso, 2020.

Caraher, William and Conway, Kyle, eds. *The Bakken Goes Boom: Oil and the Changing Geographies of Western North Dakota.* Digital Press @ The University Of North Dakota, 2016

Latour, Bruno. *Down to Earth: Politics in the New Climatic Regime.* Cambridge: Polity Press, 2018.

McPhee, John. *In Suspect Terrain.* New York: Farrar, Straus and Giroux, 2000.

Negarestani, Reza. *Cyclonopedia: Complicity with Anonymous Materials.* Melbourne: Re.press, 2008.

Smil, Vaclav. *Oil: A Beginner's Guide*. London: Oneworld Publications, 2018.

Rueb, Emily and Chokshi, Neraj. "Keystone Pipeline Leaks 383,000 Gallons of Oil in North Dakota, *The New York Times*, October 31, 2019, https://www.nytimes.com/2019/10/31/us/keystone-pipeline-leak.html.

Yusoff, Kathryn. *A Billion Black Anthropocenes or None*. University of Minnesota Press, 2018.

IMAGE CITATIONS + CREDITS

Listed by Page Number

147   Basemap sources: Esri, HERE, Garmin, Intermap, INCREMENT P, GEBCO, USGS, FAO, NPS, NRCAN, GeoBase, IGN, Kadaster NL, Ordnance Survey, Esri Japan, METI, Esri China (Hong Kong), © OpenStreetMap contributors, GIS User Community; Map Layers: Energy Information Administration (EIA), "Shale Plays" (May 2, 2011); Energy Information Administration (EIA), "Sedimentary Basins" (March 11, 2016); North Dakota Game and Fish Department, "Missouri River Aerial Imagery 2015" (October 17, 2015). North Dakota Department of Na; tural Recourses, "Oil and Gas Well horizontal Lines" (March 11, 2019). ND GIS Hub Data Portal; North Dakota Oil and Gas Division, "Oil and Gas Well Map" (January 27, 2016), ND GIS Hub Data Portal ; North Dakota Department of Environmental Quality-Division of Water Quality. "Assessed Rivers and Streams" (August, 2019). ND GIS Hub Data Portal.

149   Photo: Other Fields (Ian Quate + Colleen Tuite). Map Layers: Tara Chesley-Preston, "Bakken" (December 2013), USGS; Energy Information Administration (EIA), "Shale Plays" (May 2, 2011); Energy Information Administration (EIA), "Sedimentary Basins" (March 11, 2016); North Dakota Department of Transportation, "State and Federal Roads" (February 10, 2020). ND GIS Data Hub.

151   Reference: Steptoe, Anne. "Petrofacies and Depositional Systems of the Bakken Formation in the Williston Basin, North Dakota." Dissertation, The Research Repository @ WVU, 2012; Pathak, Manas, Milind Deo, Jonathan Craig, and Raymond Levey. "Geologic controls on production of shale play resources: Case of Eagle Ford, Bakken and Niobrara." In Unconventional Resources Technology Conference, Denver, Colorado, 25–27 August 2014, pp. 121–128. Society of Exploration Geophysicists, American Association of Petroleum Geologists, Society of Petroleum Engineers, 2014; Shale Experts. Bakken Shale &Three Forks Overview, Bakken Shale. Accessed October 26, 2020. https://bakkenshale.com/.

      Research Assistant: Theodore Teichman

154   Photo: Other Fields

155   Research Assistants: Xiaonian Shen, Theodore Teichman

159   Reference: Wikimedia Commons, https://commons.wikimedia.org/wiki/File:Oil_traps.svg, contributed by MagentaGreen; Steptoe, Anne. "Petrofacies and Depositional Systems of the Bakken Formation in the Williston Basin, North Dakota." Dissertation, The Research Repository @ WVU, 2012; Nesheim, Timothy O. "Oil and Gas Potential of the Red River Formation, Southwestern North Dakota." North Dakota Department of Natural Resources, 2017; Pathak*, Manas, Milind Deo, Jonathan Craig, and Raymond Levey. "Geologic controls on production of shale play resources: Case of Eagle Ford, Bakken and Niobrara." In Unconventional Resources Technology Conference, Denver, Colorado, 25-27 August 2014, pp. 121-128; Society of Exploration Geophysicists, American Association of Petroleum Geologists, Society of Petroleum Engineers, 2014; Shale Experts. Bakken Shale &Three Forks Overview, Bakken Shale. Accessed October 26, 2020. https://bakkenshale.com; Huang, Yue-Chain. "Thermal History Model of the Williston Basin." Dissertation, UND Scholarly Commons, 1988; Sverdrup, Harald U., and Kristin Vala Ragnarsdóttir. "Natural resources in a planetary perspective." Geochemical perspectives 3, no. 2 (2014): 129-130.

      Research Assistants: Xiaonian Shen, Theodore Teichman, Tian Wang

162   Reference: Energy Information Administration, "British Thermal Units"(June 4, 2020); Lindeman, RL (1942). "The trophic-dynamic aspect of ecology." Ecology. 23: 399–418; PNW Canola Association, "Where Does Canola Oil Come From?" (2019).https://pnwcanola.org/for-consumers/production-process/. ; Robert Arnason, "Thousand kernel weight matters in canola seed" (March 17, 2014). The Western Producer. https://www.producer.com/daily/thousand-kernel-weight-matters-in-canola-seed/; Katey Davidson, "Should You Use Rapeseed Oil? Everything You Need to Know" (October 30, 2019). Healthline. https://www.healthline.com/nutrition/rapeseed-oil#bottom-line; Graham Wood. Fracing Water Impact: Water Supply and Demand (August 2019), Institute of Water (UK)

      Research Assistant: Theodore Teichman

163   Photo: Other Fields; Basemap: Esri "World Topography" Sources: Esri, HERE, Garmin, Intermap, INCREMENT P, GEBCO, USGS, FAO, NPS, NRCAN, GeoBase, IGN, Kadaster NL, Ordnance Survey, Esri Japan, METI, Esri China (Hong Kong), © OpenStreetMap contributors, GIS User Community; Map Layers: North Dakota Game and Fish Department, "Missouri River Aerial Imagery 2015" (October 17, 2015). ND GIS Hub Data Portal; North

Dakota Department of Transportation, "State and Federal Roads" (February 10, 2020). ND GIS Data Hub; National Recourses Conservation Service, "Major Aquifers" (September 2018). ; North Dakota Oil and Gas Division, "Oil and Gas Well Map" (January 27, 2016), ND GIS Hub Data Portal; North Dakota Department of Transportation, "Incorporated City Boundaries" (July 17, 2020). ND GIS Hub Data Portal.

166   Photo: Other Fields: Basemap: USDA Farm Service Agency, "USDA-FSA-APFO Aerial Photography 2019" (September 1, 2020). ND GIS Data Hub; Map layers: North Dakota Oil and Gas Division, "Oil and Gas Well Map" (January 27, 2016), ND GIS Hub Data Portal; North Dakota Department of Natural Recourses, "Oil and Gas Well horizontal Lines" (March 11, 2019). ND GIS Hub Data Portal.

      Research Assistant: Chloe Nagraj, Theodore Teichman

169   Basemap: Esri "Light Gray Canvas Map" Sources: Esri, HERE, Garmin, Intermap, INCREMENT P, GEBCO, USGS, FAO, NPS, NRCAN, GeoBase, IGN, Kadaster NL, Ordnance Survey, Esri Japan, METI, Esri China (Hong Kong), © OpenStreetMap contributors, GIS User Community; Map Layers: USDA Forest Service Geospatial Service and Technology Center, USDA Forest Service. Tribal Land Cessions in the United States (May 29, 2018), http://data.fs.usda.gov/geodata/edw/datasets.phpIndian Reservations; Homeland Infrastrcutre Foundation-Level Data (HIFLD) (March 20, 2020) https://hifld-geoplatform.opendata.arcgis.com/datasets/54cb67feef5746e8ac7c4ab467c8ae64; Natural Gas Pipelines. Homeland Infrastrcutre Foundation-Level Data (HIFLD) (August 23, 2019) https://hifld-geoplatform.opendata.arcgis.com/datasets/natural-gas-pipelines; Reference: StateImpact Texas, "Dakota Access Pipeline." https://stateimpact.npr.org/texas/tag/keystone-xl-pipeline/.

      Research Assistants: Chloe Nagraj, Theodore Teichman, Tian Wang

173   Photo: Other Fields. Map Layers: Energy Information Administration (EIA), "Shale Plays" (May 2, 2011). Energy Information Administration (EIA), "Sedimentary Basins" (March 11, 2016).

      Research Assistant: Xiaonian Shen

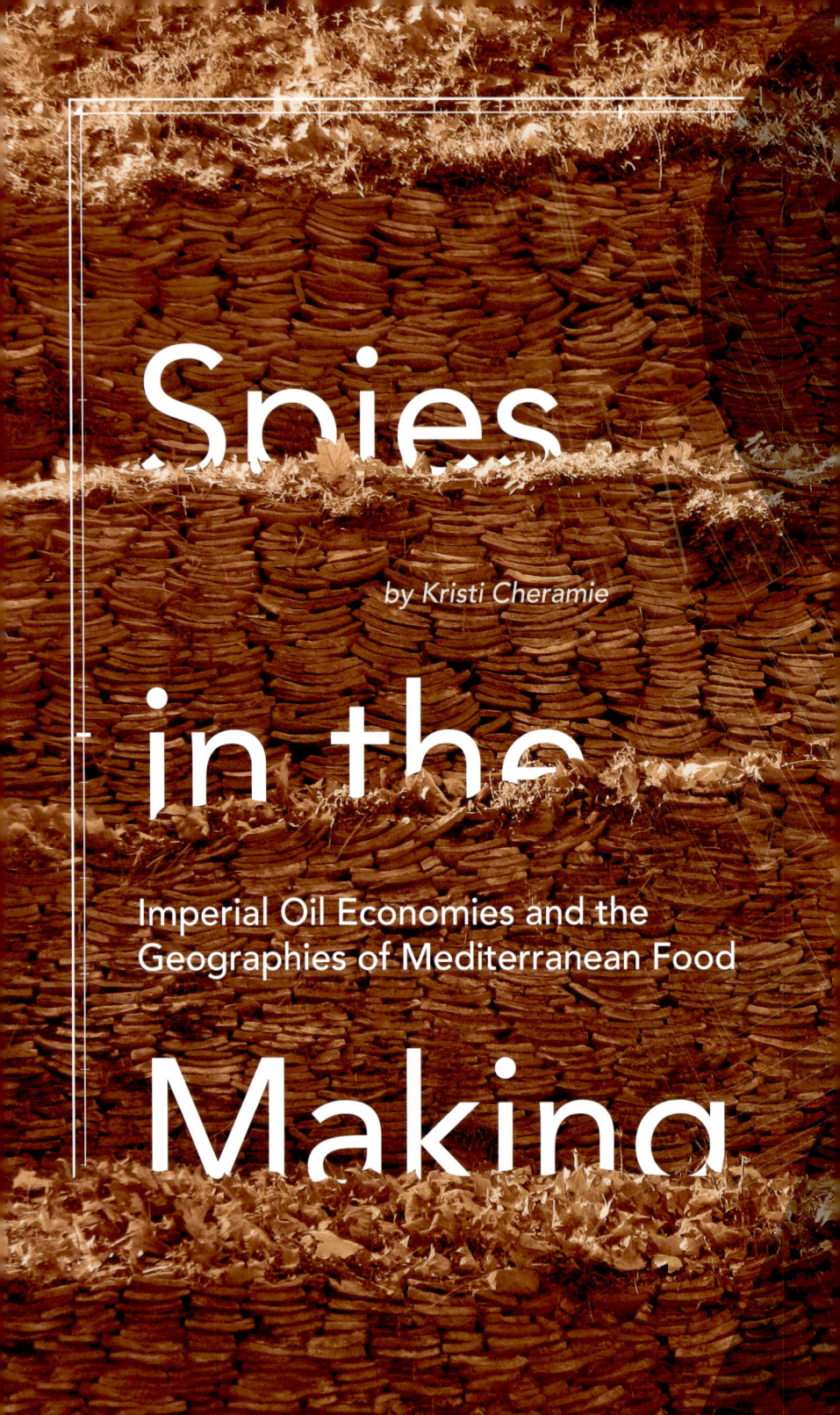

Spies

by Kristi Cheramie

in the

Imperial Oil Economies and the
Geographies of Mediterranean Food

Making

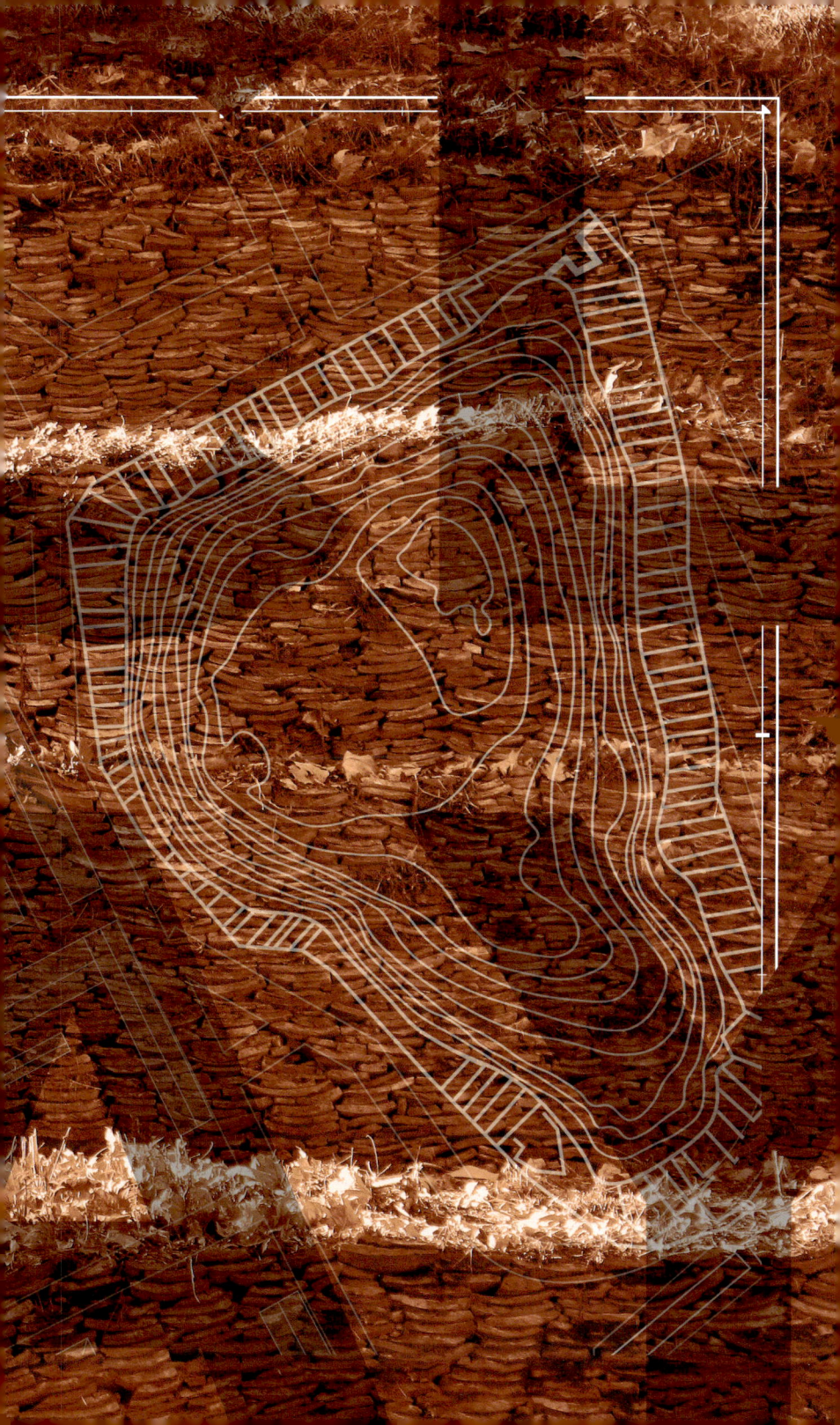

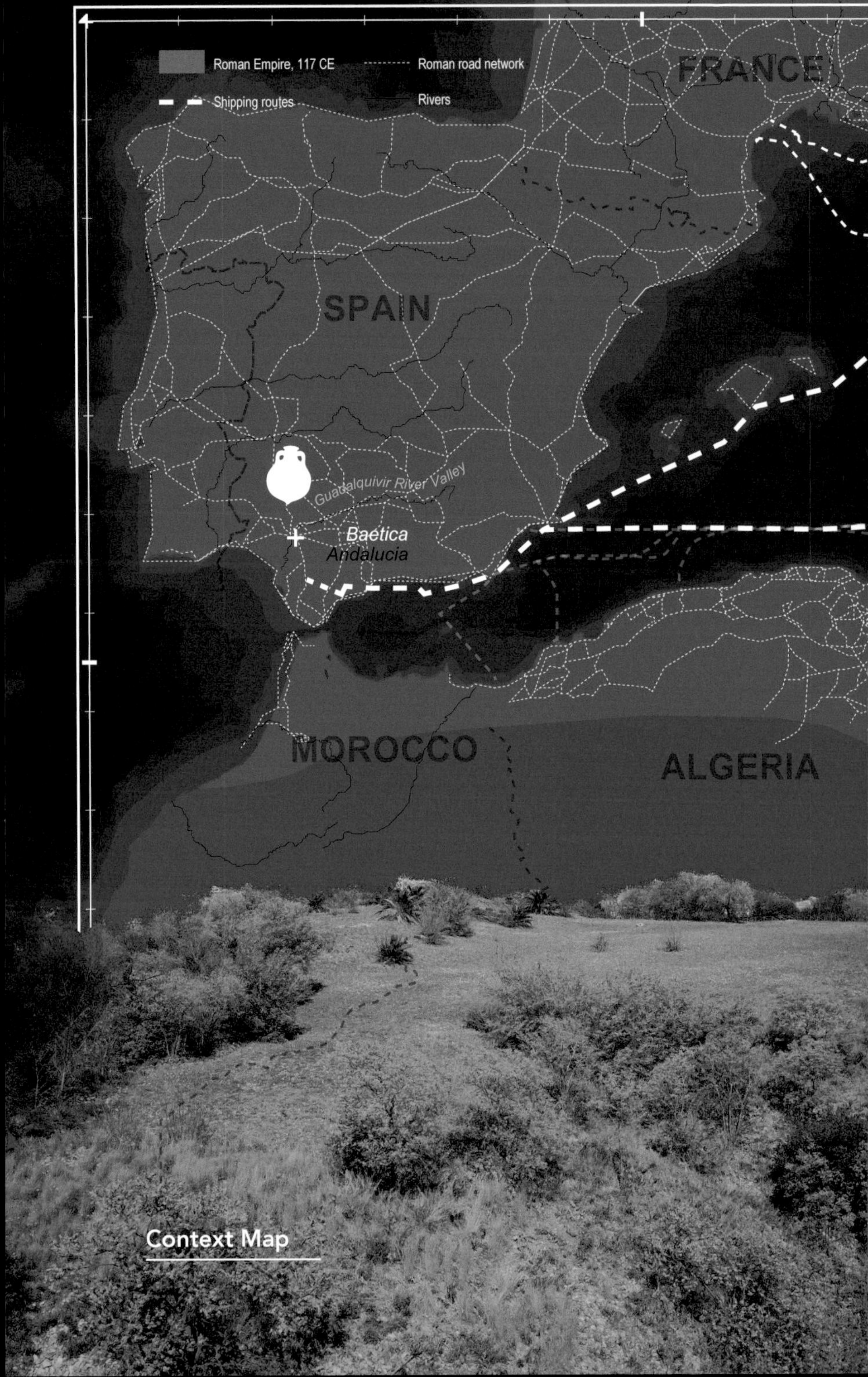

Roman Empire, 117 CE   Roman road network

Shipping routes   Rivers

FRANCE

SPAIN

Guadalquivir River Valley

Baetica
Andalucia

MOROCCO

ALGERIA

Context Map

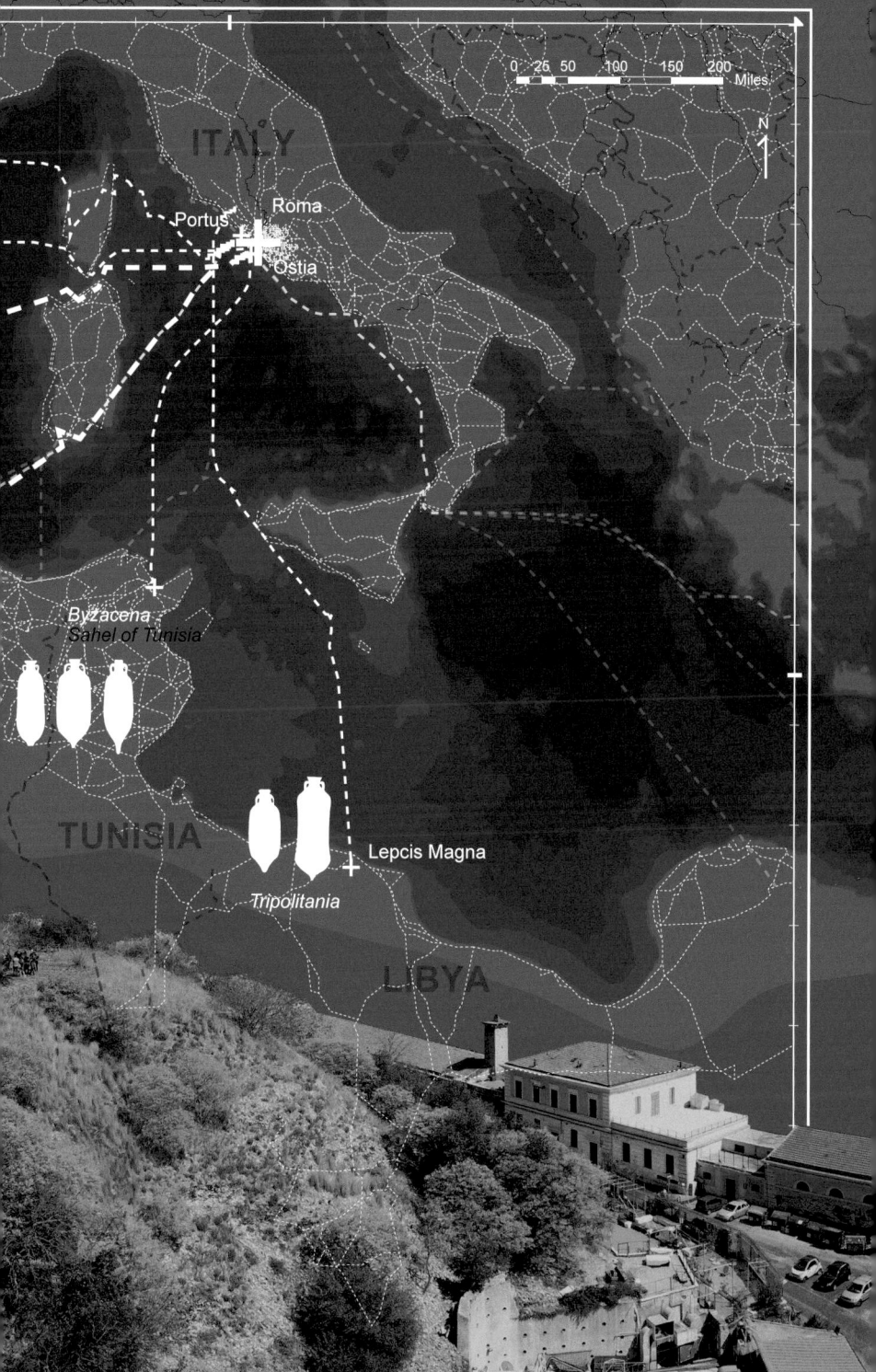

Ostia Dolia

have spent a good bit of time with archaeologists over the years. My love for walking the landscape is well-suited to their disciplinary proclivity to pour over small corners of far-flung sites for hours, if not days. We make for good travel companions.

I should confess, however, that despite my profound respect for the work archaeologists do, it takes an enthusiastic guide for me to manifest equivalent levels of on-the-spot joy for epigraphic sleuthing. What may look to me like snippets of letters disconnected from intention, incomplete and mysterious, are, in the hands of a seasoned scholar, the evidence of our entanglement with durable materials; an inscription can be both cultural expression and evidence of a geological reckoning. Enter Caroline Cheung, a professor of archeology at Princeton whose area of research is ancient shipping infrastructure. On one particular trip to Ostia Antica, Rome's ancient port at the mouth of the Tiber River, Caroline pointed to the back corner of one of the port's on-site food storage facilities known as Caseggiato dei Dolii, or House of the Dolia. There we found 35 enormous clay vessels known as *dolia* arrayed in a tight grid and partially sunken into the ground, with only the top quarter of each vessel exposed (a configuration known as dolia defossa). Each had stamped inscriptions. As Caroline explained, the markings were essentially the ancient equivalent to barcodes. The stamp in the side wall displayed the name and location of the workshop responsible for the container's fabrication, in this case a nearby workshop in the Tiber River Valley, and its carrying capacity (roughly 250 gallons). These two data points alone begin to illustrate an imperial economy on the move. But the dolium wasn't designed to be particularly mobile. In fact, once loaded, dolia would have been too heavy to move by ship on the shallow Tiber River—even the short 15 miles to Rome—and too delicate for overland transport.

Another vessel, then: the *amphora*, the most prolific shipping container of the ancient world.

Where dolia were meant for sitting and storing, *amphorae* were designed for transport and portage. Small handles at the vessel's neck facilitated carrying; its tapered end made for efficient stacking inside of ships. And amphorae come not just with stamps like the dolia but also *tituli picti*, painted inscriptions that code the vessel with a rich range of geospatial information: vessel weight before and after filling, consular date, name of the exporter and the city or estate (*fundii*) of origin, and the name of the shippers (*navicularii*) responsible for ferrying the vessel to market. In short, each amphora inscription carries a single tale about Rome's mercenary approach to amassing the material riches of the Mediterranean. And thanks to the kiln-firing process, each vessel can retain its inscribed tale of displaced geographies for centuries.

So perhaps it is useful to think of clay, or fired clay, as our imperial spy, what Giuseppe Pucci would describe as a "symptom of a much more complex reality."[1] A way back into the material world of antiquity (from clay shipping containers and ships made from the timber stock of the Mediterranean to the grains and fruits of its soils). Through the firing process, clay inherits a geological persistence that allows it to accumulate evidence of complex exchanges that are literally baked into its substrate, evidence that would not have otherwise survived the perils of time and change.

The Imperial Spy into the
Material World of Antiquity

ON MATTER OUT OF PLACE

Given Rome's voracious appetite for oil, millions of amphorae poured into the city each year.[2] By the end of the third century CE, an estimated 53 million had landed in the city,[3] the contents transferred into smaller containers for sale and distribution. But, unlike dolia, each amphora could only be used once. Once air reached the oil-soaked, clay interior of the vessel, the walls would oxidize and spoil. Once emptied, these clay vessels were considered trash, making amphorae the Roman equivalent of today's single-use plastics.

Too spoiled for reuse; too heavy to ship back. The clay trash posed a logistical challenge. Just south of the Aventine Hill and along the banks of the Tiber River is a neighborhood known today as Testaccio. To late Republican and Imperial Rome, however, it was the Emporium District, the city's "principal area of transshipment and storage."[4] With its flat terrain dominated by docking stations (e.g. Porta di Ripa Grande), warehouses (*horrea*) and even marble yards, the district dedicated to the "unloading and redistribution of commodities" also inherited a trash pile for the clay amphorae now tainted by olive oil.[5] By 150 CE, the pile was rapidly becoming a mountain.[6] Rising a towering 115 feet out of the floodplain, Monte Testaccio is a geologically implausible expression of Rome's diet, infrastructure and trade over a period of roughly 250 years. It is a mountain without earth, a landfill composed of a single material: clay.

Before going forward, it is worth noting that clay isn't high on the list of Rome's traditionally defined geological bounty. Tuff, alluvium, marble, basalt, limestone all receive literary highlighting by Pliny, Vitruvius, and Lanciani, not to mention the Grand Tour spectators who loved Ancient Rome but favored it when bedecked with travertine. Not clay; clay is largely overlooked. However, to see Rome's geo-inheritance as a result of regional geological processes alone would be to overlook the city's capacity to radically shape its own topography through economies tied to consumption. Boasting 28 centuries of continuous occupation, the city is a receptacle for its own material history, where new construction and urban detritus (debris, rubble, trash, dirt) continuously reconfigure the surface. Today, the city's complex geological story has as much to do with trash as it does with the region's volcanism, leaving curious topographies like Monte Testaccio scattered throughout the floodplain.[7]

So, this is a story about matter out of place; of a single material, clay, unearthed to such an extraordinary extent and with such streamlined precision that it led to one of Rome's most notable landforms. Monte Testaccio is a transactional mountain composed of displaced geologies; it is an epigraphic record of the Spanish and African craftsmen who shaped the amphorae and enabled trans-Mediterranean shipping; it is a three-dimensional map of bountiful olive harvests and the climates that served and eventually failed them; it is a tally of imperial power and a monument to consumption.

But to get to clay, we begin with the world of the olive tree, its prized oil, and the imperial economy.

LITTORAL BEGINNINGS

If Monte Testaccio is our endpoint, we begin with the olives of the Mediterranean littoral.

Intensive olive farming was principally found in the Baetican region of Southern Spain along the Guadalquivir River (today's Andalucia), in the hinterlands of Lepcis Magna surrounding Tripolitania (Libya), and throughout the coastal plain of the Tunisian Sahel, known to the ancient world as Byzacena. These territories had come under Roman control during the Republic, and each region found increased capacity for production during the shift to imperial rule. It is worth noting that farming practices across Italy invariably included oil production, but Rome's needs far outstripped local supply. In fact, by the end of the first century CE, the city had emerged as the largest single consumer of oil in the Western Mediterranean, a shift in demand that required Rome to tap into its colonial networks in Spain and North Africa for new market share.

The Guadalquivir River Valley flourished under the favorable climate mechanisms of the Roman Climate Optimum (200 BCE–150 CE). Warm winters, plentiful rains, agriculturally beneficial floods, ample solar exposure, and increased humidity allowed olive orchards to dependably produce two harvests per year. In short, regional climatic stability and agricultural growth coincided with the rise of the

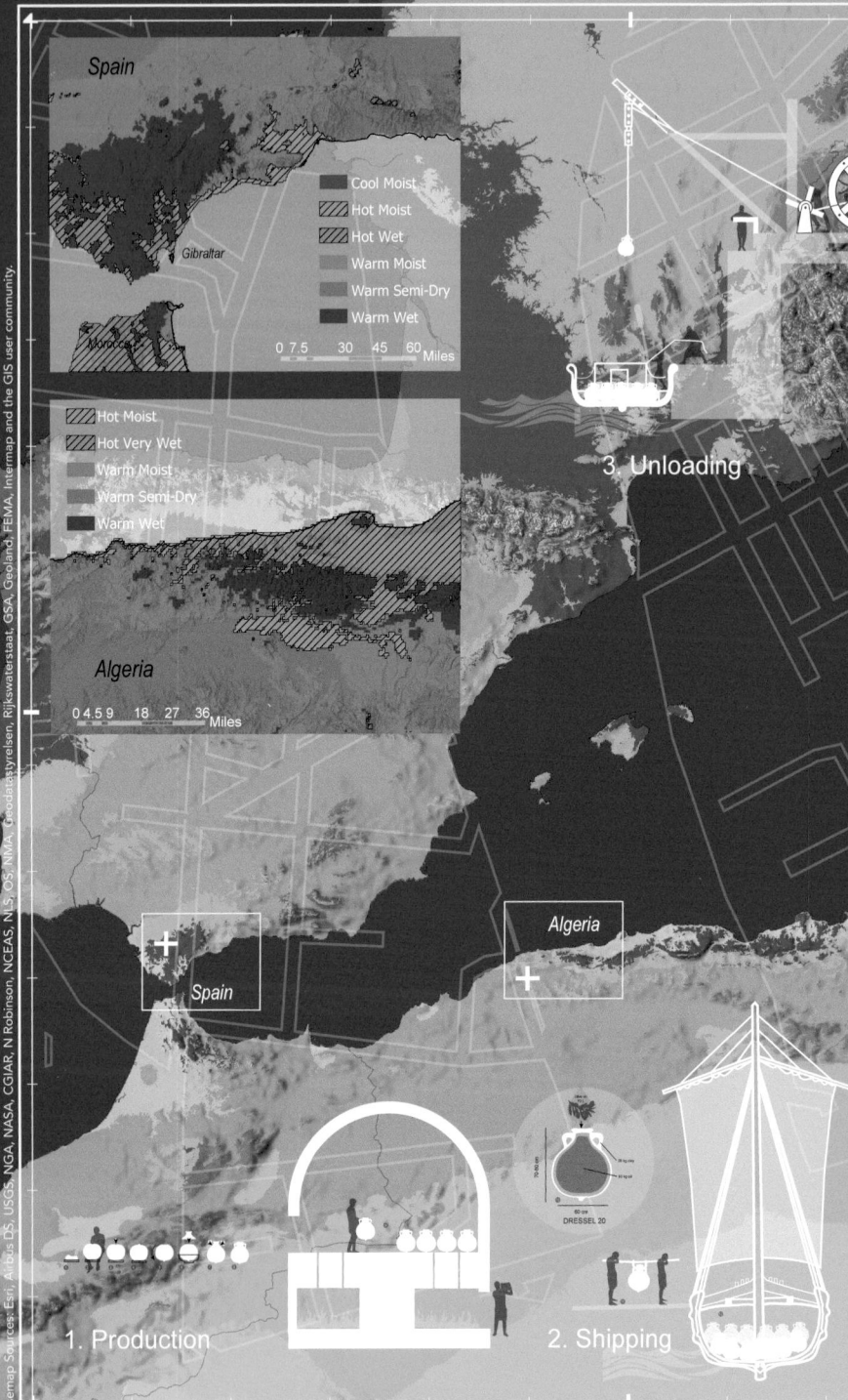

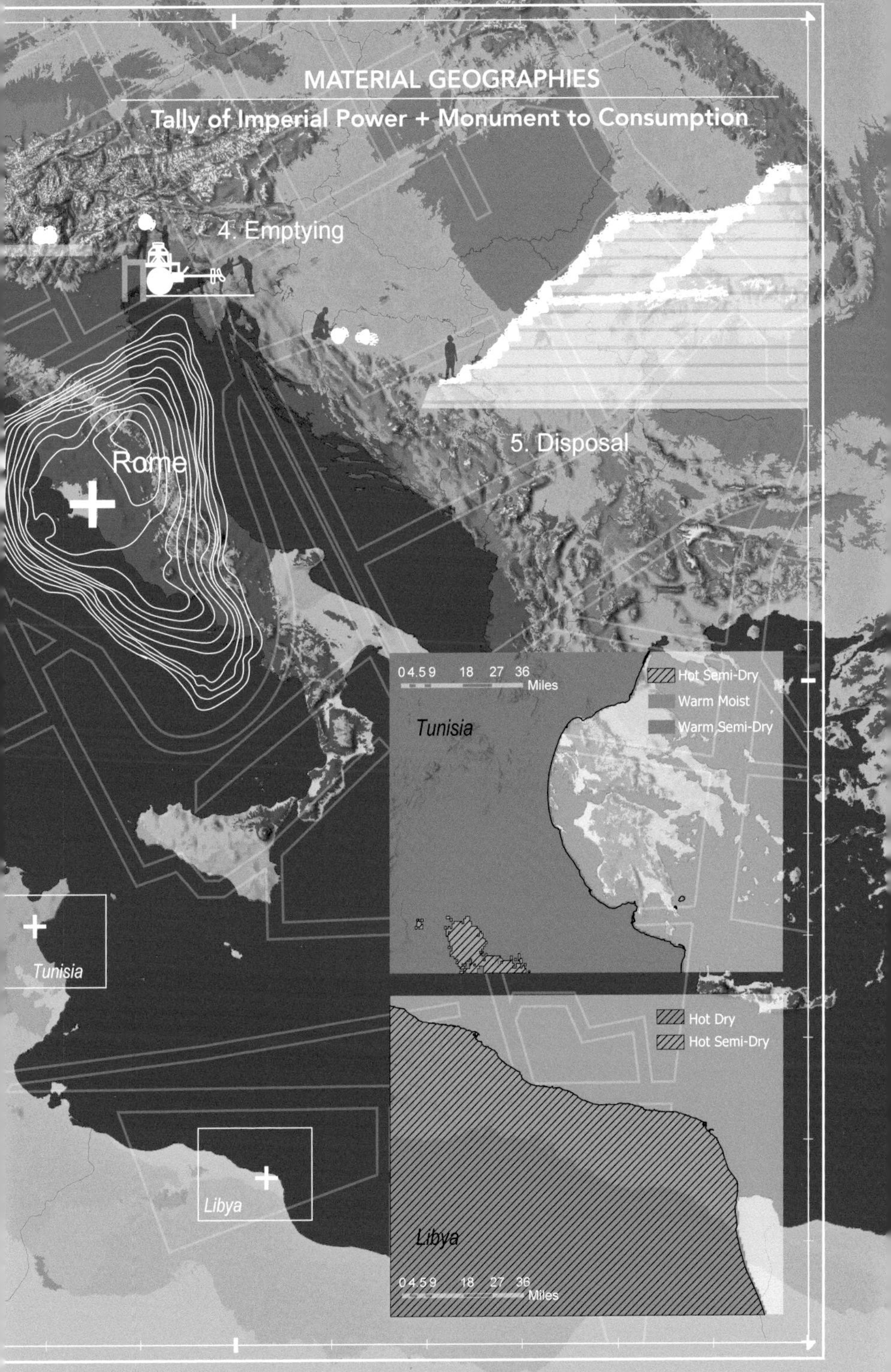

MATERIAL GEOGRAPHIES
Tally of Imperial Power + Monument to Consumption

4. Emptying

5. Disposal

Rome

| | Hot Semi-Dry |
| Warm Moist |
| Warm Semi-Dry |

0 4.5 9   18   27   36
Miles

Tunisia

| | Hot Dry |
| Hot Semi-Dry |

Tunisia

Libya

Libya

0 4.5 9   18   27   36
Miles

Roman Empire and the expansion of its consumer markets, especially in Spain. Similar to the grain harvests of the Nile Valley, as the empire grew, the olive orchards of the Guadalquivir stretched into every available pocket of land. The wide and generous valley steps toward the Atlantic Ocean with a series of flat terraces largely comprised of Miocene clay soil. The territory is rich in nutrients, well-suited to the economies of olive oil, and ideal for pottery making. Under imperial control, the region is estimated to have supported between 5 and 12.5 million olive trees, with its harvests principally serving local communities, the Roman army, and the city of Rome.[8] To give you a sense of what this means in terms of scale, we can conservatively place the productive capacity of each olive tree at 20 pounds of olive oil per year, taking regional totals to 50,000–120,000 tons per year.[9]

Libya, on the other hand, was a less obvious candidate for high-yield agriculture. Water was scarce and droughts frequent. Arguing that "making olive oil requires even more science than making wine," Pliny pointed to the region's climate and soil as being deficient for olive production.[10] "This territory Nature has yielded entirely to the Corn-goddess, having all but entirely grudged it oil and wine."[11] But trade routes that funneled Sub-Saharan and Central African goods toward Rome, namely wild animals (and ivory),[12] encouraged Tripolitania to expand its legitimacy within Rome's political economy wherever possible. In spite of Pliny's doubts, the region pushed intensive agriculture to great success and by the end of the first century CE, the region is believed to have rivaled Baetica in oil and amphora production.[13]

The smallest yields came from the Sahel of Tunisia.[14] The region experienced low levels of rainfall on par with Libya but much higher atmospheric humidity. While fewer ancient sites have been preserved, aerial images expose the region's planting strategy.[15] The Roman centuriation system that stretched across the vast coastal plain is still marred by tightly spaced pock-marks or "ghostly traces of vanished orchards within Roman field boundaries."[16] Upland regions show evidence of ancient irrigation systems,[17] a sign of substantial infrastructure investment designed to facilitate lower yield, marginal territories in meeting competitive demand.

THE IMPERIAL INTERMEZZO: The Amphorae

Romans now had access to olives, lots of them, in fact. But getting this bounty from the far reaches of the empire to the Imperial City required the production and distribution of state-regulated shipping containers: the amphorae. By the Flavian period (68–96 CE), each region had developed its own strictly governed suite of amphora types. Such standardization for single-use vessels is indicative of a spectacular expansion of imperial economies and enables us to point to the amphora as an infrastructural catalyst that mobilized the material bounty of the Mediterranean in service of Rome. Far more than a container solely designed for storage, the amphorae was the all-important intermediary in the supply chain, what Keller Easterling might call an "operating system" or "multiplier," signifying— even dictating—ship dimensions, cargo volumes, individual and collective methods of portage, delivery protocols, and content transfer strategies.

If the winter months pushed all available labor into orchards and their associated presses, May to September saw labor concentrations shift to the extraction, shaping, and firing of clay.[18] And like any major seasonal labor shift, a shift in geography closely followed. While the olive orchards stretched deep and wide into the Baetican valley and the North African steppes chasing down trapped moisture, the clay extraction process was limited by shipping logistics. Or, more specifically "embarkation points." Despite the widespread availability of suitable clays in all three regions, extraction sites and kilns for amphorae prioritized close proximity to quays or coastal ports.[19] In Baetica, kilns were evenly distributed along navigable sections of the Guadalquivir and its southern tributary, the Genil. All within a mile or so of the waterway. In Libya and Tunisia, kilns and coasts went hand in hand. So much so that their dominant amphorae types (the Tripolitania I and III in Libya and the Africana I and II in Tunisia) are known to have a whitish outer film of salt, a product of having mixed the clay with seawater or saline groundwater before firing. Though fewer kilns are attributed to Africa relative to Spain, those that are operated as "nucleated workshops and manufactories," centers of specialized production forming "more or less a tightly clustered industrial complex according to the availability of raw materials, labor and markets."[20] Here we see clay functioning as the threshold in the olive oil supply

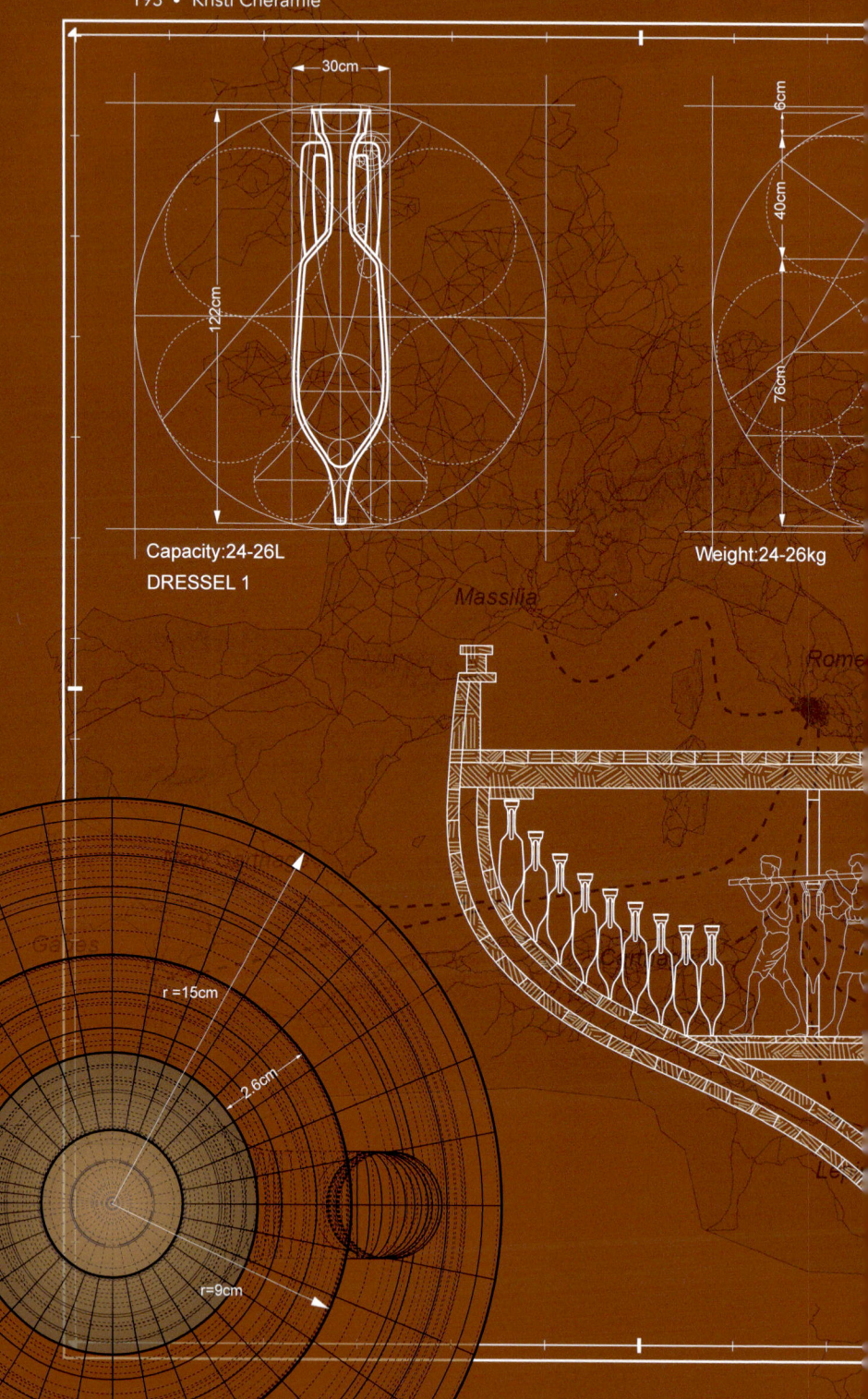

30cm

122cm

Capacity:24-26L

DRESSEL 1

6cm

40cm

76cm

Weight:24-26kg

Massilia

Rome

r =15cm

2.6cm

r=9cm

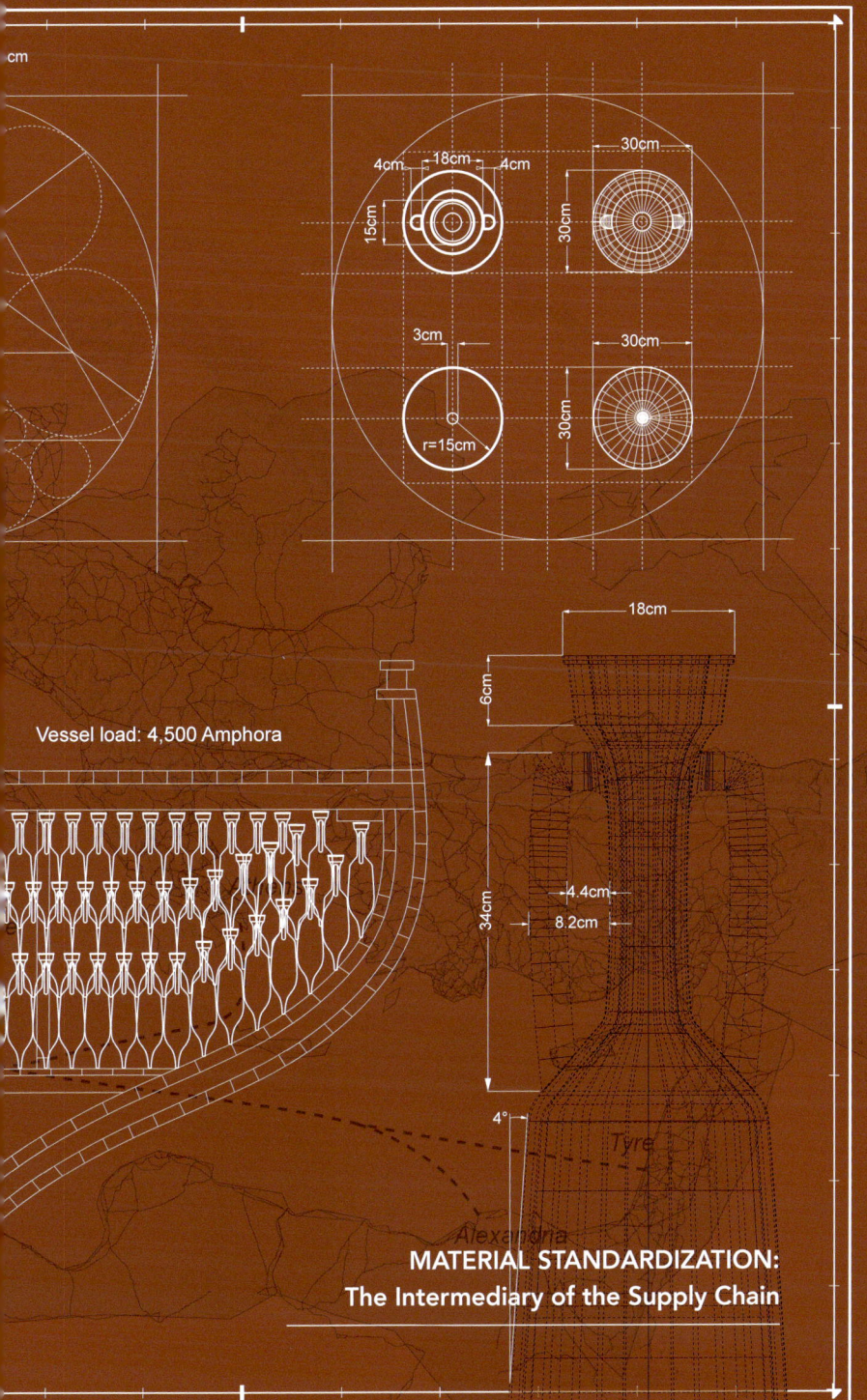

cm

4cm ─ 18cm ─ 4cm

30cm

15cm

30cm

3cm

30cm

r=15cm

30cm

Vessel load: 4,500 Amphora

18cm

6cm

34cm

4.4cm

8.2cm

4°

Tyre

Alexandria

MATERIAL STANDARDIZATION:
The Intermediary of the Supply Chain

chain, disconnecting those tied to cultivation from those tied to shipping.

Reasons for decoupling amphora fabrication from olive orchards and pressing sites are both practical and logistical. The first and perhaps most obvious reason is due to the sheer heaviness of a filled amphorae. The clay vessel added 30% to the weight of the contained commodity. For example, a Dressel 20, the most common Baetican amphora type for olive oil, weighs roughly 65 lbs when empty and could weigh up to 210 lbs once filled and sealed.[21] Though designed with two handles for lifting by two people, amphorae weren't ready-made for overland transport, or any transport beyond a short distance for that matter. The vessel's thick, round center and tapered end facilitated its filling and weighing, the process of stacking it with other amphorae in the cargo hold of a ship, and, on reaching final destination, the pouring of its contents into smaller containers for sale. This meant that oil had to be transported from pressing sites to the amphorae workshops using other means, most likely animal skins that added little weight to the load and could be easily managed by pack animals or floated down less navigable sections of the river.[22]

The second factor was tied to fiscal control. By mid-first century CE, oil transport was increasingly subject to state regulation and the clay-marking system (*tituli picti*) was implemented to reduce fraudulent trading practices. At a glance, one could tell the origin of the oil by amphora shape. The most distinctive type: the large, spherical Dressel 20 could not be easily confused for the tall, much thinner ones from Africa. Compared to the Africana I and II, the Tripolitania I and III widened at the base and carried more total volume. Next in the protocols, the weights and measures. Vessels were weighed and marked before and after filling to account for slight differences in handcrafting (usually wall thickness). In almost all cases, each amphora would have been filled with olive oil from a single farm and the names of the people involved in the vessel's preparation noted in the *tituli*. In addition to the tags added upon reaching the customs house, some amphorae were also marked with the total number of amphorae in the shipment, similar to an edition number used for artworks. The *tituli* embedded each amphora with its own portable—and permanent—paper trail. Such standardization effectively collapsed the variable geographies of olive cultivation, translating the uniqueness of place into a series of marks that would be legible and meaningful in Rome.

If we think about the amphora in terms of Easterling's concept of the "multiplier," we can situate local amphora production as the accelerating factor in the Roman Empire's technological revolution. Olive presses and irrigation techniques represent impressive technological advances of the ancient world but these platforms expanded local productive capacity. Not connectivity within regional networks. Neither moved material beyond local geographies. The kiln and its simple clay amphora, on the other hand, radically expanded the reach of the Roman diet by introducing translational durability to what had long been considered a strictly local product. Because of the controls applied to these handcrafted objects, the vessel became the all-important intermezzo, effectively mediating the particularities of local farms, the vulnerabilities of river and sea shipping, and the demand for consistency in urban centers.

When olive oil trading reached its height in mid-second century, a time that coincided with the crescendo of the Roman Empire's political arm and the last benevolent breaths of the Roman Climate Optimum, the kilns of the Guadalquivir Valley are believed to have been producing 200,000–300,000 amphorae per season.[23] David Nye tells us that "latent in every tool are unforeseen transformations,"[24] and the amphora certainly stands up to this adage. For example, Mediterranean shipping routes existed long before the standardization of the amphora. But the presence of this good-on-the-move increased the potential market value of any one shipment by introducing the ability to diversify shipments based on weight, demand, and value. In Spain, metal extraction (gold, silver, copper, iron, and tin) continued to be an important source of income for the empire into the third century but large cargoes comprised solely of metal came with the risk of catastrophic financial loss due to the high frequency of shipwrecks. Such risk, however, could be mitigated by splitting loads of high value goods with lower value goods, especially those tied to stable demand.[25] Olive oil fit this calculation perfectly.

TO FEED A CITY AND BUILD A MOUNTAIN

Rome's prime sailing season conveniently overlapped with amphora production cycles. The sea was considered "open" for roughly 110 days from late May to mid-September and each season, the coastal ports of Rome (Portus and Ostia) would see the arrival of 1,300–1,800 ships loaded with goods bound for the city.[26] Once docked, each ship would be unloaded over a period of about five days, during which time dockhands would unpack each amphora, check its weight against the pre-sailing weight noted in the *tituli*, record its arrival in logbooks, and transfer the amphora into temporary holding. Rome was only a short distance away but the large warehouses of Ostia—and especially Portus—enabled larger quantities of oil to be held back for staged distribution throughout the year.

As needed, port stores would be gradually transferred to Rome on shallow draft boats (*naves codicariae*), the last link in the olive oil supply chain. Except in moments of exigent weather (i.e. flooding or drought), the Tiber River was continuously sailed throughout the year by amphora delivery boats, an estimated 19–33 per day.[27] Even the most conservative estimates assume peak traffic on the Tiber through at least the third century. Putting this in context, archaeologist Johann Rasmus Brandt points to the sustained traffic as indication that Imperial Rome had reached its "saturation point," or a "brake block on the expansion of the capital."[28] When the Tiber, the primary artery for food delivery, could no longer absorb additional shipments, a relative cap was established for the capacity of city to feed its residents.

It must have been an incredible sight: a near-continuous stream of amphorae marching into the city, through the warehouses of the Emporium District, then loaded onto pack animals for the short trip up ramps of the trash heap that would come to be known as Monte Testaccio. Due to the rapid rate at which the city accumulated amphora, the heap-that-would-become-a-mountain needed an assembly system. And once again, the amphora supplied the tectonic logic. After being emptied, the clay vessel could be handled in one of two ways: most were broken into small fragments for piling but a small percentage were kept partially intact to give structure to the pile. With tapered bases removed, the top half of these amphorae could be

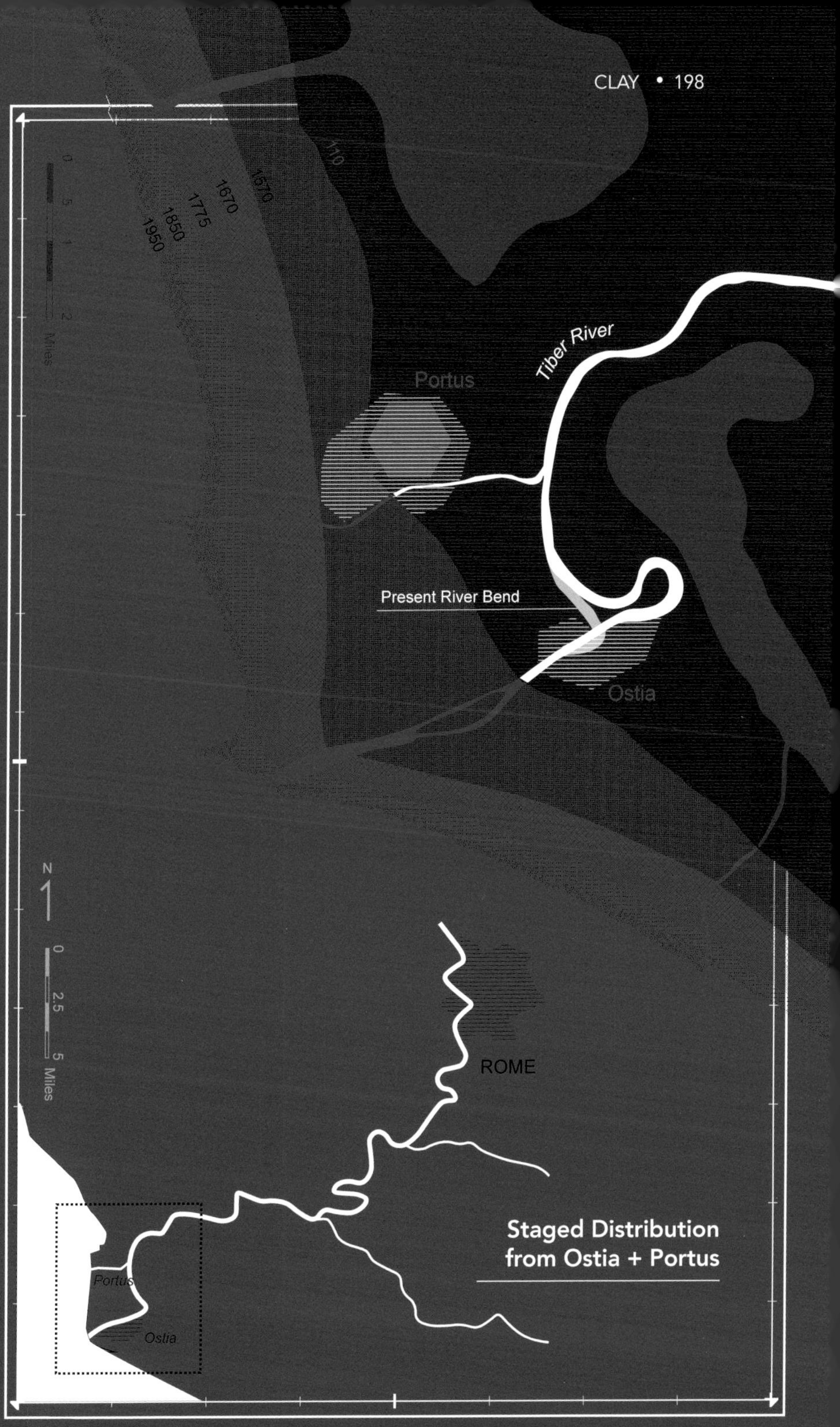

0
5
1
2
Miles

1950
1850
1775
1670
1570
110

Portus

Tiber River

Present River Bend

Ostia

N

0
2.5
5
Miles

ROME

Portus

Ostia

**Staged Distribution
from Ostia + Portus**

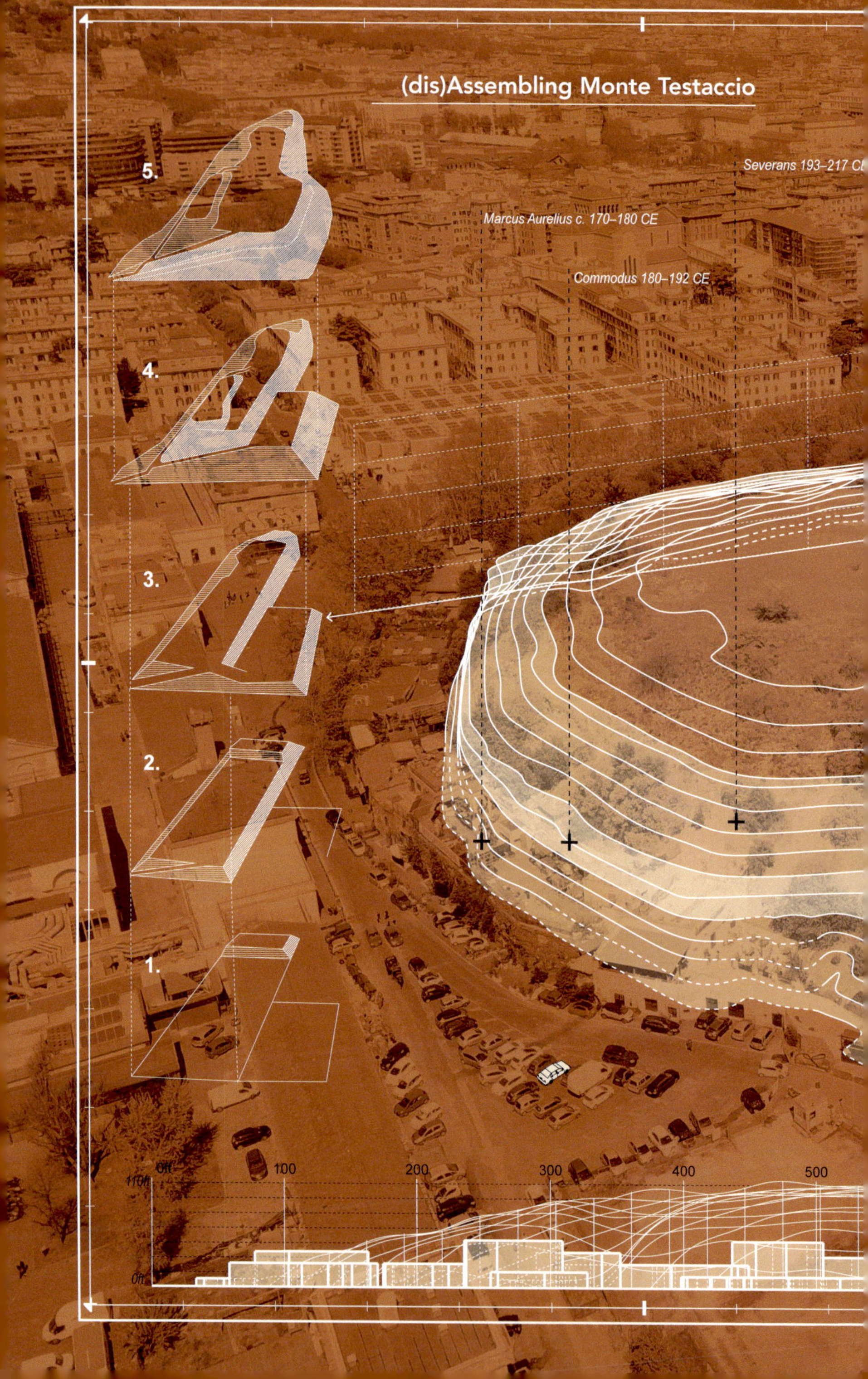

(dis)Assembling Monte Testaccio

5.

4.

3.

2.

1.

Severans 193–217 CE

Marcus Aurelius c. 170–180 CE

Commodus 180–192 CE

110m

0m

0ft   100   200   300   400   500

Post Severan

First to second century CE

158ft

N

0   100 meter

158ft

0

150ft

140ft

Mid-third century deposits

130ft

120ft

110ft
100ft

90ft

80ft

70ft

60ft

50ft

40ft

700   800   900   1000

30ft

20ft

10ft

0

filled with fragments and used to shore up the outer edge of the current dumping area. Each row, then, had a maximum height set by the thickest section of a Dressel 20 amphora, or roughly 24 inches.[29] This process enabled the outer walls to be nested together and gradually stepped back at 45°. Successive discharge areas (typically clusters from the same shipment) moved laterally, back and forth across three interlocked platforms to create a stepped, pyramidal form that grew continuously between the early first and mid-third centuries CE. When the last (known) depositions were made in 257 CE, 85% of the clay mountain had come from Baetica; the remaining 15% from Tunisia and Libya. No earth; no supporting infrastructure. Monte Testaccio is a formal expression of material utility, a clay monolith.

WHAT DO YOU DO WITH A MOUNTAIN?

The slowing of additions to Monte Testaccio during the third century proved to be a geological bellwether for the coming decline of the empire. Let us remember that our tale of clay is quietly governed by two simultaneous forces of vastly different scales: human agency and the millennia-spanning effects of the Holocene. As historian Kyle Harper notes, "climate change and human settlement do not move in perfect sync,"[30] and while favorable conditions enabled Roman expansion, those conditions also led to exploitation at a time when deep patterns in solar irradiance and the earth's orbit and axis began to catch up with Roman trade networks.

Toward the end of the second century, quiescence gradually—almost unnoticeably—shifted toward instability across the Mediterranean, establishing a period of climate disorganization. Climate-forcing changes in sun variability and volcanic activity destabilized peak seasonal fertility. Precipitation decreased; temperatures cooled, increasing snow and ice accumulations at the higher elevations and reducing the pervasive humidity responsible for slowing the desertification in the productive plains around the Mediterranean.

The empire's ongoing search for available timber supplies and persistent pressures to increase crop production had denuded the outer limits of Roman territories. Deforestation and overly taxed soils created more frequent seasonal shortfalls, requiring low food supplies to be fortified by imperial storage holdings.[31] Once held in Rome, state-regulated food storage and the trade routes that fed them were now tethered to the new seat of power, Constantinople. For the first time in 800 years, the city of Rome was sacked and fell. Under King Alaric, Visigoths seized Rome in 410, leaving the city a starved and the clay shadow of Monte Testaccio as a towering reminder of the Mediterranean's waning power.

But the landform remained in place, open to material poaching and eventually even the enterprising eye of those looking to store wine and meat in its readily available thermal mass. The landfill encompasses roughly 5 acres of the floodplain and now sits within a network of nineteenth century slaughterhouses.[32] Writing in 1847, Lanciani describes Monte Testaccio as reduced in height after years of material redistribution through quarrying, erosion during periods of abandonment, hollowing out for wine cellars and meat lockers, and pockmarking from bombardier target practice in the seventeenth century.[33] Today, Monte Testaccio hosts nearly 200 species of ruderal vegetation, a shrubby extension of Rome's primary ecological corridor and counterpoint to the nearby and highly manicured Protestant cemetery.[34] A strange mix of overlooked ecologies and ongoing archaeological investigations, the artificial mountain persists as part of Rome's "archipelago of (archaeological) islands surrounded by urban landscape."[35]

ONE OF MANY

In 1956, Jacques Cousteau released *The Silent World,*[36] his now-classic memoir-turned-film documenting underwater explorations of seas and oceans, namely the Mediterranean. It is a tale about the pursuit of new knowledge, and it is fundamentally Homeric: both flawed and exuberant, at times wandering and at others fiendishly driven. Cousteau's crew feasts on lobsters while one diver recovers from nitrogen-loading in a pressure chamber; they blow up a coral reef with dynamite and have complicated encounters with sharks; they meet a grouper and name the fish Ulysses.[37] Cousteau describes passing hundreds of sunken ships over decades of sailing the *Calypso* and even dedicates 15 minutes of film footage to colorful swim-throughs of a 1941 wreck site. But it is their discovery of another, much older shipwreck—one

1700s

1800s

MODERNIZING MATERIAL ASSEMBLIES

+ 400–600s
region grows
vegetables + vines

+ 174 BCE
Emporium
port opens

+ 1889
Mattatoio slaughter
house opens

+ 1700s
region serves as performance
+ recreational space, pastures

100 BCE
earliest deposits in
Mons Testaceum,
today called Monte
Testaccio

100s
region filled
with food
warehouses

260s
last deposits on
Monte Testaccio

400–600s
region grows
vegetables + vines

117 Roman Empire
is at its largest

376 Goths and others invade Empire
1348 Black death kills 1/3 of population

Monte Testaccio Present

+ Present
restaurants + 19 bars open
along sides of mound

+ Monte Testaccio
excavated to store wine

+ 1940s
Fascist regime converts
region to offices

1914–1918 World War I

1939–1945 World War II

1500

2000

1849
Giuseppe
Garibaldi
uses mound as
gun battery
against French

1873
regulatory
plan creates
worker
district

1975
Mattatoio closes;
nearby
restaurants
continue meat
recipes

1980s
Jose Ramesal Almeida
+ Emilio Rodriguez
Almeida excavate
Monte Testaccio

2003
Mattatoio becomes
art museum

1861 Italy unified as kingdom

that didn't make the final cut of the film—to which our story now turns.

Three years prior to the film's release, Cousteau's crew rounded the southeastern corner of Sicily, tracing the limestone cliffs in search of new finds. The topography follows a familiar pattern found throughout the western Mediterranean: steep and stepped escarpments that drop quickly until reaching the sandy, flat seabed. Just south of Siracusa, strong, persistent currents meet an exposed peninsula (Penisola della Maddalena), leaving the area vulnerable to turbulence, storms, and rough sailing. There, the crew came across the wrecked remains of an ancient Tunisian ship that was (most likely) bound for Ostia and Rome in 200 CE when it dropped into the rocky bottoms. Known now as the Plemmirio B wreck site, to the crew of the *Calypso* this was an inscrutable scattershot of pottery fragments and metal scraps almost entirely consumed by the ecology of the sea floor. All visible markers of the ship's hull had disappeared. The area looked less like a shipwreck and more like an underwater dump site. Cousteau's lead diver, Frédéric Dumas, would later describe the scene in a letter saying, "I saw broken amphoras, concreted into a fold of the cliff, then an iron anchor, concreted to the bottom and apparently in corroded state, with amphora shards on top."[38] Notable but (apparently) not film-worthy.

In fact, Plemmirio B wouldn't be properly surveyed until the mid-1970s, when scholars had the capacity to explore the choppy depths for sunken snapshots of Roman shipping habits. By the 1980s, nautical archeology was a discipline on the rise and the Mediterranean was an ideal proving ground for this new scholarship. Ancient historians A. J. Parker and Keith Hopkins led the charge, finding hundreds of Roman wrecks along western Mediterranean shipping routes.[39] Not surprisingly, the rate of wrecks rises and falls in sync with the expansion and collapse of imperial trade: almost three hundred—more than half of all found wrecks—date to the height of the Roman Empire (200 BCE–200 CE). Most were merchant ships and most of those cargo holds were loaded with oil-filled amphorae.

Like tailings from the height of imperial hubris, the piles and piles of submerged amphorae chart an intermediate material legacy for the Roman Empire, one in which speed, efficiency, and insatiable demand for oil created far more than a single byproduct landform in Rome. Instead, a *byproduct type* that can be found wherever Romans sailed. Perhaps we should think of Monte Testaccio as only one, and certainly the largest, in an anthropic mountain range made of clay. Lurking beneath the surface are chains of micro-mountains composed of goods-not-delivered and clay fired into persistence: out of place but directionally oriented, and bearing the inescapable, socio-geological stamp of the Anthropocene. "The natural and the artificial have merged at every scale," Jedediah Purdy tells us.[40] An assemblage of unintentionally coordinated patterns, Anna Tsing might say.[41] As the Roman capacity to produce, monetize, and leverage increased, the topography, climate, and culture of the Mediterranean changed. The durability of our clay spies moves our understanding of the Roman economy from point source hot spots to near-continuous lines of work woven across the Mediterranean, lines that expose the sea for what it became under the weight of the Empire: a worked landscape.

Today, olive oil trade masquerades surreptitiously behind the mythic cloak of the Roman Empire. Believed to be not just tastier but more luxurious than the ubiquitous American vegetable oil, it fits in a mercurial marketing slot in the food world, as an unstable staple: consume it quickly, but use it (a lot of it) every day. Well enough, but the current olive-rich regions must perform while also combatting new and monumental challenges brought on (once again) by climate change and over-taxation. Puglia, for example, second to Spain in olive oil production and producer of 40% of Italian oil harvests, is facing catastrophic losses of olive trees to the *Xylella fastidiosa* bacterium, a strain believed to have originated in Costa Rica. Orchards, bottling facilities, and distribution networks have become increasingly decoupled as olive shortfalls and high demand get reconciled. In short, determining the lineage of any one container of oil has never been more difficult. Even without the clay amphorae, there are clear stoppers on olive production: climate and soils. Fraud, on the other hand, once managed by painted marks on the clay vessel, has found fertile ground in global oil trade. The European Union has experimented with consumer capacity to register the differences between "grown in," "bottled in" and "packaged in," leading consumers to fumble toward the Mediterranean diet with a new kind of recklessness about the clay-less, stamp-less geographies of contemporary olive oil. And so, the stamped clay amphora gave way to the printed-and-labeled glass bottle (or aluminum can). Today's container has been deacquisitioned of data. It is merely a vessel.

***

On the ride back into the city from Ostia, Caroline sat on the right side of our train car, her eyes trained on the Aurelian Walls slowly coming into view. She pointed, "two pyramids. All the material riches of Rome."

"Well, one pyramid and one pile of clay," she qualifies.

To the east, the Pyramid of Gaius Cestius, a burial tomb built in the first century BCE as close to the city as was allowable at the time. A marble and travertine monument to a single person.

Its pair, on the other hand, rose to the same height by serving the collective appetite of an entire city. Each fragmented amphora meant that a Roman household had been fed and each row a precise corollary of Rome's capacity to marshal Mediterranean materials to meet its needs. A clay repository of names, finger prints, tertiary sediments, and limestone inclusions.

A clay mountain of millions.

AUTHOR'S NOTE

My thanks goes to Caroline Cheung for not only inviting me to join her for archaeological expeditions through Rome, Ostia, and Cosa (to name a few), but also for kindly revisiting many of our conversations as I worked on this chapter.

ENDNOTES

1   Giuseppe Pucci, "Pottery and Trade in the Roman Period," 106.
2   Estimating Imperial Roman olive oil consumption is both challenging and inherently speculative. Figures rely on various proxies, amphorae heaps being the most reliable and comprehensive. Others include the estimated expanse of Mediterranean olive orchards, and literary assertions by Pliny, Cato, Varro, and Columella. Archaeologist Robert Hitchner places the annual consumption rate in Rome at 25 million liters. Mattingly argues that olive oil represented one third of the average Roman's annual caloric intake. It should be noted, however,

that oil was relied upon for far more than food. Figures built around consumption estimates do not account for oil used as fuel or in medicines.

3 Mattingly (drawing on Rodríguez-Almeida) points to 53 million as the peak total. Remesal notes that Monte Testaccio has lost considerable material over time and places the number of amphora *still present* closer to 25 million.

4 Simon J. Keay, *Rome, Portus and the Mediterranean*, 37.

5 Keay, 39.

6 An official start date for Monte Testaccio is not known, though it is estimated to be roughly 50 CE.

7 Other artificial landforms found in Rome's floodplain include Monte Giordano and Montecitorio in the north-east corner of the Campus Martius and Monte dei Cenci in the Jewish Quarter.

8 Pedro Paulo A. Funari, "Baetica and the Dressel 20 Production: An Outline of the Province's History," 88; David Mattingly, "Oil for Export? A Comparison of Libyan, Spanish, and Tunisian Olive Oil Production in the Roman Empire," 41.

9 J. M. Blázquez, "The Latest Work on the Export of Baetican Olive Oil to Rome and the Army," 176.

10 Pliny, the Elder, *Natural history: in 10 volumes. 4: Books XII–XVI*, 15.3.8.

11 Pliny, the Elder, 15.3.8-9.

12 M.G. Fulford, "To East and West: The Mediterranean Trade of Cyrenaica and Tripolitania in Antiquity," 181.

13 Mattingly, 37. Mattingly places the reach of olive orchards at approximately 1,500 km^2, with at least 750 presses.

14 Covering an area comparable to Libya, Tunisian presses are currently estimated to be 350. It should be noted, however, that archaeological work has been later to emerge in Tunisia.

15 Amelia Carolina Sparavigna, "Astronomical Orientations in the Roman Centuriation of Tunisia," 3–4.

16 Mattingly, 45. Mattingly estimates a planting density of 2,000 trees per *centuria*, or 40 trees per hectare.

17 R.B. Hitchner, "Olive Production and the Roman Economy: The Case for Intensive Growth in the Roman Empire," 77.

18 Funari, "Baetica and the Dressel 20 Production," 96.

19 In Baetica, for example, brick and tile manufacturing spread through the region, with small-scale kilns wherever clay could be extracted.

20 David Gibbins, "A Roman Shipwreck of c . AD 200 at Plemmirio, Sicily: Evidence for North African Amphora Production during the Severan Period," 327.

21 Blázquez, "The Latest Work on the Export of Baetican Olive Oil to Rome and the Army," 176.

22 Mattingly, 43; Funari, 92–93. Funari points to cattle breeding north of Cordoba as the source of the leather skin bottles used to bring oil from pressing sites to riverside bottling sites.

23 Mattingly, 42. The estimate of amphora production comes largely from Mattingly and Remesal who point to a five-month firing season, but this estimate is also echoed by Funari and Hitchner.

24 David E. Nye, *Technology Matters: Questions to Live With*, 2.

25 Johann Rasmus Brandt, "'The Warehouse of the World.' A Comment on Rome's Supply Chain during the Empire," 32. "Many ships carried more than one cargo, but few more than five."

26 Brandt, "'The Warehouse of the World'," 34.
27 Brandt, 41.
28 Brandt, 44.
29 José Remesal Rodríguez, "Baetican Olive Oil and Roman Economy," 195.
30 Kyle Harper, *The Fate of Rome: Climate, Disease, and the End of an Empire*, 48–49.
31 Harper, 58. Harper describes the extensive imperial effort required to provide the public dole of food to both Romans and its far-flung military. These efforts were intended to extend thresholds in leaner years, and, in fact, did so successfully through the fourth century CE.
32 The slaughterhouses (*mattatoio*) have recently been adopted by Rome's Museum of Contemporary Art and Roma Tre University.
33 Rodolfo Amedeo Lanciani, *The Ruins and Excavations of Ancient Rome*, 528–30.
34 Ceschin et al., "Size Area, Patch Heterogeneity and Plant Species Richness across Archaeological Sites of Rome: Different Patterns for Different Guilds," *Vie et Milieu* 62, no. 4 (2012): 165–71.
35 Ceschin et al., 170.also controlling the role of habitat heterogeneity at site scale. By floristic sampling, we obtained 585 plant species, about 50 % of the spontaneous flora of Rome. The power equation between total site area and total species number showed a weak relationship (R2 = 0.36
36 *The Silent World* was awarded both an Academy Award and the Palme d'Or in 1956.
37 The film has experienced persistent scrutiny for its vivid portrayal of engagement with (and disruption of) ocean ecologies in the name of scientific discovery. It is also worth noting Wes Anderson's 2004 film *Life Aquatic with Steve Zissou*, which parodies the flawed methods while celebrating the optimism of Cousteau and his crew.
38 From Dumas' personal correspondence, cited in Gibbins and Parker, "The Roman Wreck," 267.
39 Hopkins published a groundbreaking article in 1980 in which he summarized Parker's work on 545 dated wreck sites, most found along the Spanish, French or Italian coasts. Keith Hopkins, "Taxes and Trade in the Roman Empire (200 B.C.–A.D. 400)."
40 Purdy, 15.
41 Tsing, *The Mushroom at the End of the World*, 23.

BIBLIOGRAPHY

Blázquez, J. M. "The Latest Work on the Export of Baetican Olive Oil to Rome and the Army." *Greece and Rome* 39, no. 2 (1992): 173–88.

Brandt, Johann Rasmus. "'The Warehouse of the World.' A Comment on Rome's Supply Chain during the Empire." *Orizzonti. Rassegna Di Archeologia* 6 (2005): 25–47.

Capelli, Claudio, and Victoria Leitch. "A Roman Amphora Production Site near Lepcis Magna: Petrographic Analyses of the Fabrics." *Libyan Studies* 42 (2011): 69–72.

Carandini, Andrea. "Pottery and the African Economy." *Trade in the Ancient Economy*, 1983, 145–62.

Ceschin, S., L. Cancellieri, G. Caneva, and C. Battisti. "Size Area, Patch Heterogeneity and Plant Species Richness across Archaeological Sites of Rome: Different Patterns for Different Guilds." *Vie et Milieu* 62, no. 4 (2012): 165–71.

Cheung, Caroline, and Gina Tibbot. "The Dolia of Regio I, Insula 22: Evidence for the Production and Repair of Dolia." *Studi e Ricerche Del Parco Archeologico di Pompei* 40 (2020): 175–85.

Dressel, Heinrich. *Ricerche sul monte testaccio*. Salviucci, 1878.

Fulford, M.G. "To East and West: The Mediterranean Trade of Cyrenaica and Tripolitania in Antiquity." *Libyan Studies* 20 (1989): 169–91.

Funari, Pedro Paulo A. "Baetica and the Dressel 20 Production: An Outline of the Province's History." *Dialogues d'histoire Ancienne* 20, no. 1 (1994): 87–105.

Funari, Pedro Paulo A. "Monte Testaccio and the Roman Economy." *Journal of Roman Archaeology* 14 (2001): 558–88.

Gibbins, David. "A Roman Shipwreck of c . AD 200 at Plemmirio, Sicily: Evidence for North African Amphora Production during the Severan Period." *World Archaeology* 32, no. 3 (2001): 311–34.

Gibbins, David and A. J. Parker. "The Roman Wreck of *c.* AD 200 at Plemmirio, Near Siracusa (Sicily): Interim report." *The International Journal of Nautical Archaeology and Underwater Exploration* 15, no. 4 (1986): 267–304.

Harper, Kyle. *The Fate of Rome: Climate, Disease, and the End of an Empire*. Princeton University Press, 2017.

Hitchner, R.B. "Olive Production and the Roman Economy: The Case for Intensive Growth in the Roman Empire." *The Ancient Economy*, edited by Walter Scheidel, and Sitta von Redden, 71–83. University Press, 2002.

Hopkins, Keith. "Taxes and Trade in the Roman Empire (200 B.C.-A.D. 400)." *The Journal of Roman Studies*, 70 (1980): 101–125.

Keay, S. J. *Late Roman Amphorae in the Western Mediterranean: A Typology and Economic Study: The Catalan Evidence*. Bar International Series, 196 (i–ii). B.A.R., 1984.

Keay, Simon J., ed. *Rome, Portus and the Mediterranean*. Archaeological Monographs of the British School at Rome 21. The British School at Rome, 2012.

Kron, Geoffrey. "Food Production." In *The Cambridge Companion to the Roman Economy*, edited by Walter Scheidel, 156–74. University Press, 2012.

Lanciani, Rodolfo Amedeo. *The Ruins and Excavations of Ancient Rome*. Bell Pub. Co, 1967.

Mattingly, David. "Oil for Export? A Comparison of Libyan, Spanish, and Tunisian Olive Oil Production in the Roman Empire." *Journal of Roman Archaeology* 1 (1988): 33–56.

Mattingly, David, and D. S. Potter. "The Imperial Economy." In *A Companion to the Roman Empire*, 283–97. Blackwell Publishing, 2006.

Nye, David E. *Technology Matters: Questions to Live With*. MIT Press, 2006.

Opdebeeck, Johan. "Shipwrecks and Amphorae: Their Relationship with Trading Routes and the Roman Economy in the Mediterranean." Diss. University of Southampton, 2005.

Paterson, Jeremy. "'Salvation from the Sea': Amphorae and Trade in the Roman West." *The Journal of Roman Studies* 72 (1982): 146–57.

Peacock, D. P. S. "Amphorae and Batica." *The Antiquaries Journal* 54, no. 2 (1974): 232–44.

Peacock, D. P. S. "Recent Discoveries of Roman Amphora Kilns in Italy." *The Antiquaries Journal* 57, no. 2 (1977): 262–69.

Peacock, D. P. S. "Roman Amphorae: Typology, Fabric and Origins," n.d., 20.

Peacock, D. P. S., Fathi Bejaoui, and Nejib Belazreg. "Roman Amphora Production in the Sahel Region of Tunisia." *Amphores Romaines et Histoire Économique. Dix Ans de Recherche* (1986): 179–222.

Pliny, the Elder. *Natural history: in 10 volumes. 4: Books XII–XVI.* Translated by Harris Rackham. The Loeb classical library. Harvard University Press, 1960.

Pucci, Giuseppe. "Pottery and Trade in the Roman Period." *Trade in the Ancient Economy* (1983): 105–17.

Purdy, Jedediah. *After Nature: A Politics for the Anthropocene.* Harvard University Press, 2015.

Remesal Rodríguez, José. "Baetican Olive Oil and Roman Economy." *The Archeology of Early Roman Baetica* 29 (1998): 183–200.

Remesal Rodríguez, José. "El monte Testaccio (30 años de investigación)." *Tribuna d'Arqueologia 2015–2016* (2018): 72–87.

Remesal Rodríguez, José. "L'Afrique Au Testaccio," *L'Africa Romana XV.* Tozeur, Tunisia (2002): 1077–90.

Remesal Rodríguez, José. "Las Ánforas Dressel 20 Y Su Sistema Epigráfico." *Epigrafía anfórica* (2004): 127–48.

Remesal Rodríguez, José, A. Aguilera, M. García, D.J. Martín-Arroyo, J. Pérez, and V. Revilla. "Centro para el Estudio de la Interdependencia Provincial en la Antigüedad Clásica (CEIPAC)." *Revista De Prehistòria I Antiguitat de la Mediterrània Occidental* (2015): 245–75.

Rice, Candace. "Shipwreck Cargoes in the Western Mediterranean and the Organization of Roman Maritime Trade." *Journal of Roman Archaeology* 29 (2016): 165–92.

Rickman, Geoffrey. *Roman Granaries and Store Buildings.* Cambridge University Press, 1971.

Rodríguez-Almeida, Emilio. "Bolli anforari di Monte Testaccio." *Bullettino della Commissione Archeologica Comunale di Roma* 84 no. 75 (1974): 199–248.

Russell, Ben. "Roman and Late-Antique Shipwrecks with Stone Cargoes: A New Inventory." *Journal of Roman Archaeology* 26 (2013): 331–61.

Scheidel, Walter, ed. *The Science of Roman History: Biology, Climate, and the Future of the Past.* Princeton University Press, 2018.

Whittaker, Dick. "Amphorae and Trade." *Amphores Romaines et Histoire Économique. Dix Ans de Recherche,* 1986, 537–39.

IMAGE CITATIONS + CREDITS

Listed by Page Number

179   Photo: Kristi Cheramie; Monte Testaccio plan: Funari, Pedro Paulo A. "Monte Testaccio
and the Roman Economy - JOSÉ MARÍA BLÁZQUEZ MARTÍNEZ and JOSÉ REMESAL
RODRÍGUEZ (Edd.), ESTUDIOS SOBRE EL MONTE TESTACCIO (ROMA) Vol. 1 (Universität
De Barcelona 1999). Pp. 558, Many Figs. ISBN 84-475-2112-5." Journal of Roman
Archaeology 14 (2001): 585–88. doi:10.1017/S1047759400020250.

181   Photo: Kristi Cheramie. TRADE ROUTES GIS: McCormick, M. et al. 2013. "Roman Road
Network (version 2008)," DARMC Scholarly Data Series, Data Contribution Series #2013-5.
DARMC, Center for Geographic Analysis, Harvard University, Cambridge MA 02138.
AMPHORA MAP: Perez Gonzalez, Jordi & Remesal, Jose & Revilla, Victor & Martin,
Antonio & Sachez, Manel & Martín-Arroyo, Daniel. (2015). Centro para el Estudio de la
Interdependencia Provincial en la Antigüedad Clásica (CEIPAC). Pyrenae. 50. 245-275.
10.1344/Pyrenae2015.SpecialNumber.1.6.

Research Assistants: Gaelle Gourmelon, Chloe Nagraj

183   Photo: Lalupa. Italiano: Ostia Antica, Caseggiato Dei Dolii. April 9, 2007. https://commons.
wikimedia.org/wiki/File:Ostia_scavi_-_R1_34_caseggiato_dei_dolii_1020420.jpg.

Research Assistant: Chloe Nagraj

186   Photo: Kristi Cheramie; Geologic Map: Bundesanstalt fur Bodenforschung and UNESCO,
1971; Amphora: Jose Remesal Rodriguez, "Las Ánforas Dressel 20 Y Su Sistema Epigráfico."
Epigrafía anfórica, 2004, 127–48.

189   World Bioclimates: Esri, USGS, Metzger and others 2012; Basemap Sources: Esri, Airbus
DS, USGS, NGA, NASA, CGIAR, N Robinson, NCEAS, NLS, OS, NMA, Geodatastyrelsen,
Rijkswaterstaat, GSA, Geoland, FEMA, Intermap and the GIS user community. Amphora
Production: Michael Ezban, "The Trash Heap of History," Places Journal, May 2012.
Accessed 14 Sep 2020. https://doi.org/10.22269/120501

Research Assistants: Gaelle Gourmelon, Tian Wang

193   Background map: DARMC Scholarly Data Series Citation: McCormick, M. et al. 2013.
"Roman Road Network (version 2008)," DARMC Scholarly Data Series, Data Contribution
Series #2013-5. DARMC, Center for Geographic Analysis, Harvard University, Cambridge
MA 02138.

Research Assistant: Tian Wang

198   Bersani Pio & Daniele Moretti, "Evoluzione storica della linea di costa in prossimità della
foce del Tevere," L'Acqua. 2008, 77-88.

Research Assistant: Gaelle Gourmelon

199   Rodríguez Almeida, Emilio. 1984. Il Monte Testaccio: ambiente, storia, materiali. Roma:
Quasar. Photo: Kristi Cheramie

Research Assistants: Gaelle Gourmelon, Tian Wang

203   Photo of Monte Testaccio: Kristi Cheramie. "Nolli, Giambattista, Map of Rome, 1748-Earth
Sciences & Map Library-University of California, Berkeley." Accessed October 29, 2020.
https://www.lib.berkeley.edu/EART/maps/nolli.html. Present day map: © OpenStreetMap
contributors. Basemap sources: Esri, DigitalGlobe, GeoEye, i-cubed, USDA FSA, USGS,
AEX, Getmapping, Aerogrid, IGN, IGP, swisstopo, and the GIS User Community

Research Assistants: Gaelle Gourmelon, Tian Wang

SAND

or How to Make a Beach

by Rob Holmes

0

100

200

300 (um)

WEST PALM BEACH

Singer Island

Peanut Island

Port of
Palm Beach

Lake Worth Inlet

Lake Worth Inlet
Sand Transfer Plant

Palm
Beach

Atlantic Ocean

0mi   0.2mi   0.4mi   1mi

FLORIDA

Atlantic Ocean

West Palm Beach

Ft. Lauderdale

Miami

0mi   10mi   20mi   35mi

West Palm Beach
Winter Club

oLight

N

Munyon I

Msn (C)

15

16

Lake Park

oLight

Little Munyon I

A
T
L
A
N
T
I
C

21

22

O
C
E
A
N

L
A
K
E

Flagstaff (C)

23

10

Light

BM

W
O
R
T
H

BM
19

BM

COAST

28

27

Borrow
Pit

Peanut Island

Sherman Pt
Riviera Beach Bridge

26

Palm Beach
Shores

Riviera Beach

Light

Singers
Island

35

Lake Worth Inlet
Sand Transfer Plant

U.S. COAST GUARD
RESERVATION

Kennedy Bunker

Lake Worth Inlet

Light

Light

33

34

Inlet 2 (C)

BM

Msn (C)

INTRACOASTAL

Water Tank

BM

4

3

Horse
Hospital

Palm Beach

WEST PALM BEACH

Light

W
A
T
E
R
W
A
Y

Kennedy Mansion

PALM BEACH!
4937 II NE

LAKE WORTH 8 MI.
FORT LAUDERDALE 45 MI.

595

2'30"

596

597

598

INTERIOR—GEOLOGICAL SURVEY, WASH

USGS
Historical File
Topographic Division

ROAD CLASSIFICATION

Heavy-duty

Light-d

SCALE 1:24 000

Medium-duty

Unimpro

1
0
1000   2000   3000   4000   5000   6000   7000 FEET

1 MILE

Interstate Route

U.S. Rou

1
.5
0
1 KILOMETER

FLORIDA

CONTOUR INTERVAL 5 FEET
DATUM IS MEAN SEA LEVEL
LINE SHOWN REPRESENTS THE APPROXIMATE LINE OF MEAN HIGH WATER
THE AVERAGE RANGE OF TIDE IS APPROXIMATELY 3 FEET

QUADRANGLE LOCATION

U.S.G.S.
FILE COPY
TOPOGRAPHIC DIVISION

RIVIERA BE
N2645—W8
1946
PHOTOR

P COMPLIES WITH NATIONAL MAP ACCURACY STANDARDS
BY U.S. GEOLOGICAL SURVEY, WASHINGTON, D. C. 20242
ING TOPOGRAPHIC MAPS AND SYMBOLS IS AVAILABLE ON REQUEST

AMS 4937 I SE—S

Decades

before Donald Trump's Mar-a-Lago club became the unofficial second headquarters of a United States Presidency, another president frequently held court on the same half-mile wide strip of sandy barrier island. A 1923 Florida mansion, owned by the Kennedy family since 1933, became their "Winter White House" from inauguration in 1961 through assassination in 1963, drawing politicians, celebrities, and the media to the northern end of Palm Beach.

Just north of Palm Beach is Singer Island, another barrier island. Both were built over thousands of years by the oceanic movement of waves and sand. In the waters of the inlet between the two, lies a small, round island, accessible only by boat, ferry, or kayak: Peanut Island.

Unlike the barrier islands, Peanut Island was built—is still being built—quickly. It emerged from the crystalline waters about 100 years ago as dredgers deepened Lake Worth Inlet to establish a deep-draft navigation channel from the Atlantic Ocean into the Port of Palm Beach. Peanut Island is made from the waste sand they produced. Since then, the island has continued to grow, as the inlet has been repeatedly dredged to maintain the channel in the face of its tendency to fill with sand and silt, producing ever more waste sand, which is in turn piled behind the dikes of a

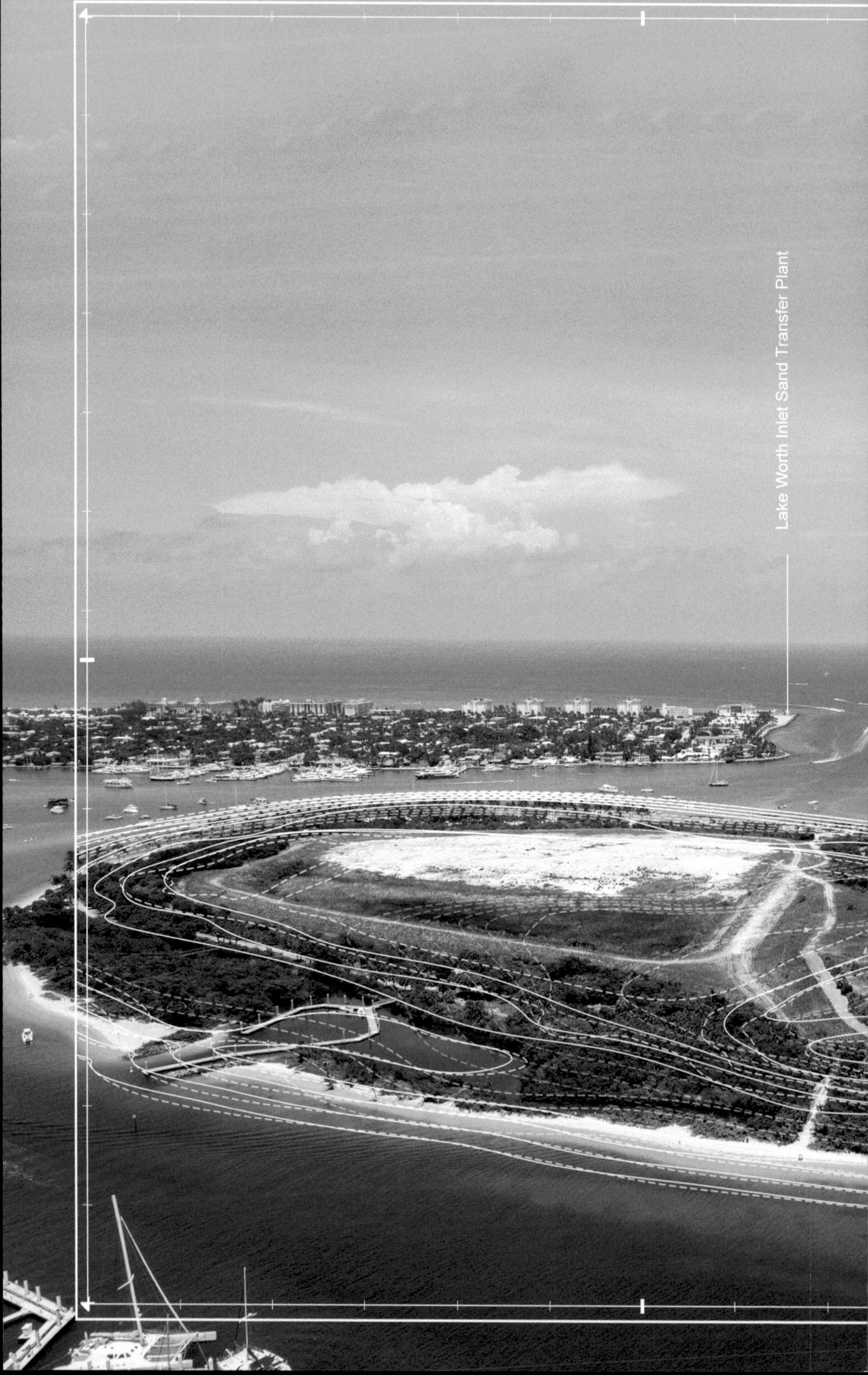
Lake Worth Inlet Sand Transfer Plant

PEANUT ISLAND, looking east

Emerging from the waters as dredgers deposited material from the deepening of the Lake Worth Inlet about 100 years ago, the island has continued to be a receptacle of socio-geological histories—from hosting John F. Kennedy's presidential bunker to continued growth by waste sand deposits.

Kennedy Bunker

Port of Palm Beach

dredged material placement facility that occupies about half of the present island.[1]

Like Palm Beach, Peanut Island was home to a Kennedy residence. This one, though, was an underground bunker and was never occupied. A small circular extrusion from the side of a small hill, just large enough to fit a small door at its center, is the only surface evidence of the fallout shelter. It speaks as clearly of the (justifiable) Cold War paranoia that motivated its construction as 1095 North Ocean Boulevard does of the generational wealth and power of the Kennedy family. Inside, the bunker is bare and utilitarian: a dark tunnel that leads underground from the door, metal bunk beds for thirty people, mid-century radiation detectors, a Presidential seal painted on the otherwise unadorned floor, cans of US Government "Emergency Drinking Water," a boxy atomic-era radio.[2] These artifacts all testify to the fear of unleashed elemental forces that permeated Cold War culture.

The bunker was built in 1961. Three years earlier, another mostly-forgotten structure was built to battle a different elemental force. Like most of the inlets that cut through the barrier islands of Atlantic south Florida, Lake Worth Inlet is armored on both its northern and southern sides by twin jetties—long, narrow lines of piled rock that extend hundreds of feet east from the beach into the ocean. On the north side of the north jetty, just below the low tide line that marks the end of the sand beach, you will find the white metal structure, a can-like extruded oval about 25 feet tall. A short walkway connects a utilitarian entry door to the jetty, while a small crane arm protrudes thirty feet or so up and out from one side, dangling a hose—an industrial sand vacuum, effectively. This is the Lake Worth Inlet Sand Transfer Plant.

The purpose of the jetties is to halt the movement of sand. They prevent the sand that would otherwise sweep south along the barrier islands' shorelines from carrying into the inlet, where it would fill in the navigation channel. The purpose of the sand transfer plant is to facilitate the movement of sand. It gathers up the sand that piles up on the north side of the jetty and pumps it in a slurry with water through pipes that go under the north jetty, the waters of the inlet, and the south jetty, where it sloshes out, reconnecting the sand beaches of Singer Island and Palm Beach.

SAND ITSELF

Sand is a fundamental substrate for modern life. It is a vital ingredient in the concrete that builds cities and infrastructures; suspended as a slow liquid with flux and stabilizer, it becomes the glass that composes everything from car windshields to smartphone screens; and often, as on Palm Beach, it is literally substrate, the ground beneath feet, homes, and cities.[3]

Sand is not singular. It is astonishingly diverse, and its major forms—beach sand, river sand, and desert sand, for instance—are so different in their microscopic geometries shaped by the forces of wind and water, that they are completely incompatible for many of their major uses. Desert sand, for instance, is plentiful, but utterly useless for the production of concrete. It is far too smooth, having been blasted (literally sandblasted) by endless desert winds.[4]

Beach sand is particularly valuable, increasingly scarce, and troublesomely wild. Our interactions with it are consummately modern: we mine it, we smuggle it, we seek to fix it in place. Our ongoing struggle to tame it is conducted with a variety of techniques that are mundane and bizarre: structures like groynes, tetrapods, and breakwaters; machines like sand transfer plants and dredgers; practices like beach nourishment. The tension between this artifice and the innate tendency of sand to move freely, escaping our grasp, is a (very large) microcosm of everything everywhere now.[5]

MOVEMENTS, MEASURES, AND MACHINES

Anyone who has built a sandcastle and watched the tide come in, or anyone who has watched the wind whip a thin screen of sand through a desert, knows that sand moves. Most of us intuit from this that the forms sand accumulates into, like dunes on a beach, are themselves mobile and temporary. It is a good bit less obvious that the same things are true of beaches as wholes. The two parts of typical ocean sand beaches that we tend to think of as "the beach" are the backbeach and the beachfront. The former is the land-facing part in front of dunes but above the "berm crest"; the latter the sea-facing part below berm crest but above low low tide. These, though, are parts of a larger emergent whole, a vast, slow-moving mass of sand, propelled by

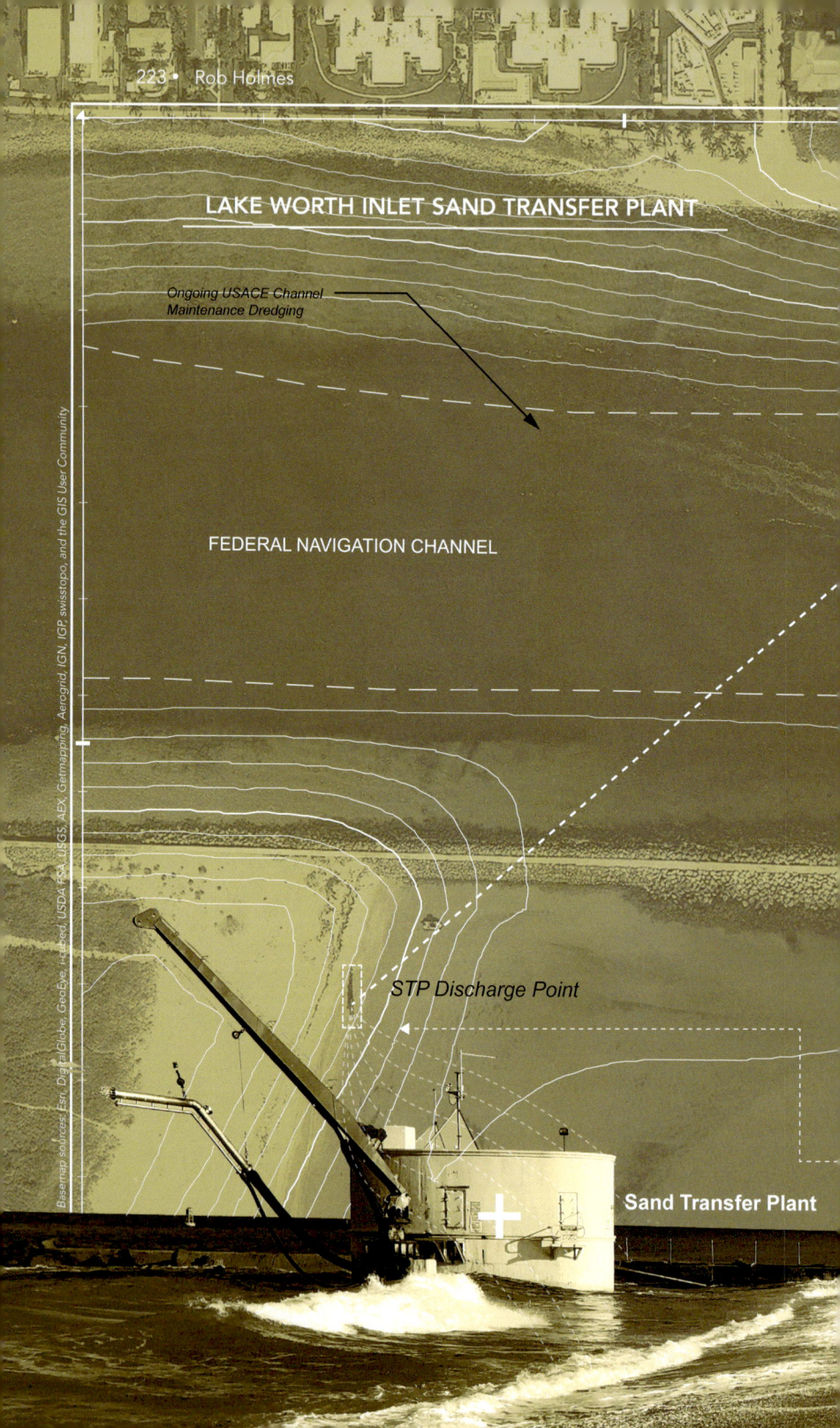

LAKE WORTH INLET SAND TRANSFER PLANT

*Ongoing USACE Channel
Maintenance Dredging*

FEDERAL NAVIGATION CHANNEL

STP Discharge Point

Sand Transfer Plant

Basemap sources: Esri, DigitalGlobe, GeoEye, i-cubed, USDA FSA, USGS, AEX, Getmapping, Aerogrid, IGN, IGP, swisstopo, and the GIS User Community

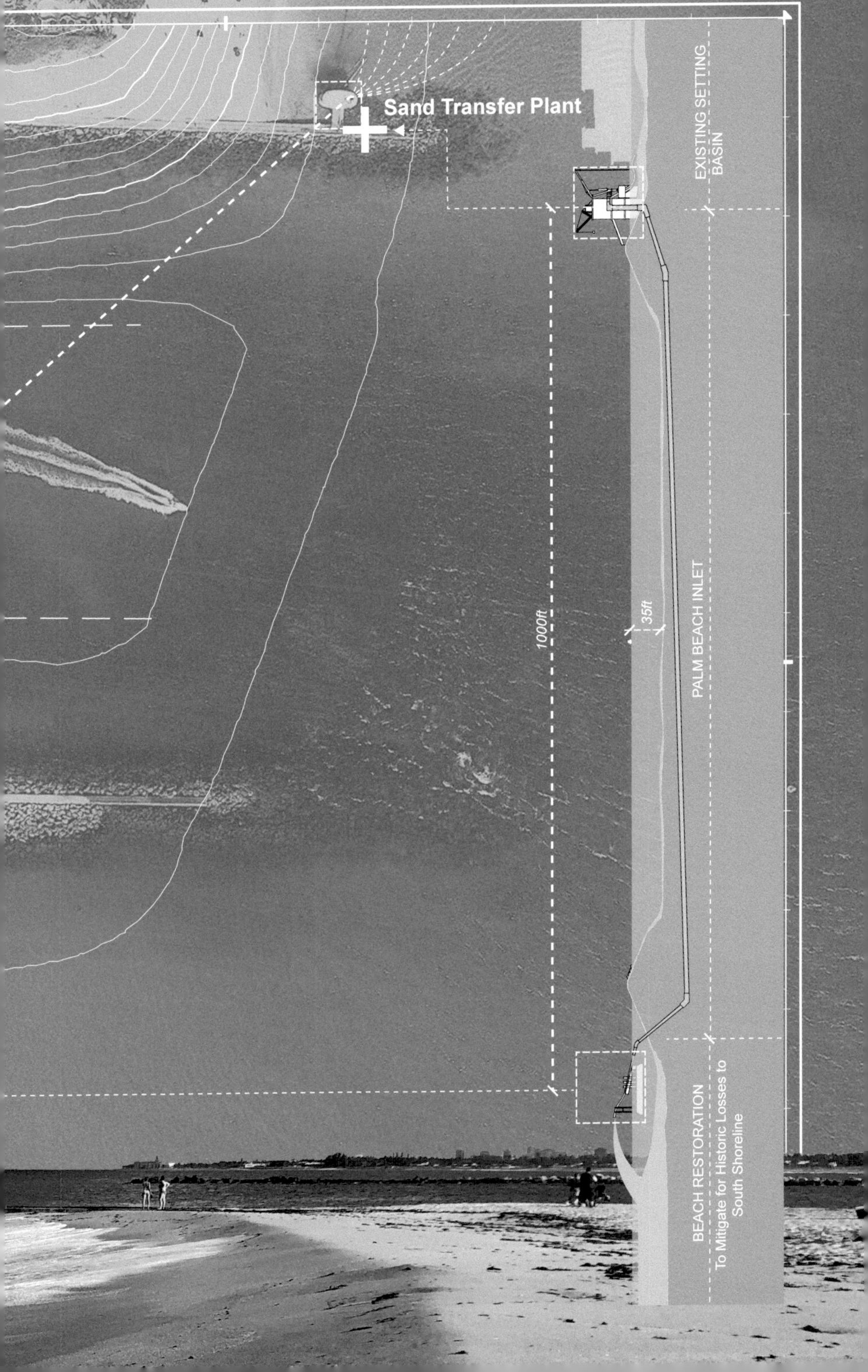

Sand Transfer Plant

EXISTING SETTING BASIN

1000ft

35ft

PALM BEACH INLET

BEACH RESTORATION

To Mitigate for Historic Losses to South Shoreline

currents, waves, and winds. This mass extends from the visible beach some variable distance offshore and below the waves, where it lies atop more consolidated sediments and bedrock, accumulating in places into migratory sandbars and vast, lumbering sand reservoirs.[6]

This whole is sometimes referred to as a "sand-sharing system." Each sand beach, like Palm Beach, is part of one, and each sand-sharing system has aggregate tendencies, which shape the individual beaches, dunes, and sand bars within it.[7] These tendencies shift in rhythm with seasons, climate, the gravity of the moon, and weather, particularly, in the South Atlantic, hurricanes; they are fundamentally dynamic in a manner that defies the logic of human settlement, with its fixed property lines. Each sand-sharing system trends in certain directions, metaphorically and physically. For the system that joins Singer Island, Lake Worth Inlet, and Palm Beach, the aggregate drift of sand is southward, which is to say, from Singer Island to Palm Beach across Lake Worth Inlet. Historically, the sand barely noticed the inlet, as the broad, narrow sand bars that crossed it averaged around three feet in depth.[8] A given sand grain might readily migrate from Singer Island beach to inlet swash bar to Palm Beach dune, passing through the constantly shifting patterns of flats, channels, and lobes in the aqueous ebb and flood tidal deltas.

The dredging of Lake Worth Inlet changed all of this. The shallow bars of the inlet became today's 35-foot deep-draft navigation channel, which cuts a trench perpendicular to the movement of the sand-sharing system. This trench is quite effective at collecting sand, and not particularly inclined to share it once claimed. Hence the jetties. But the jetties, being similarly perpendicular to the movement of sand, only prevent sand from disappearing into the navigation channel. They do nothing to restore its free movement. Sand simply piles up behind the north jetty on Singer Island, whose beach consequently tends to grow eastward into the ocean. Northern Palm Beach, meanwhile, has tended to erode, as its sand moves south along the shore of the island, without replacement from further north.

This small sand-sharing system, encompassing two islands and an inlet, is embedded in a much larger east Florida sand-sharing system that stretches, at its most macrocosmic, all the way from Biscayne Bay, near Miami, to the mouth of the

St. Mary's River, which divides Florida from Georgia.[9] Any attempt to describe it must quickly spiral outward from the granular toward the geologic, from sandcastles and seasons to ten-thousand-year eras and the drift of tectonic plates. Seen from such heights, the sand beach is a flitting margin, the quick-moving trace of the edge between continental land-mass and rising or receding seas. In Florida, you only have to rewind a mere two million years to watch the peninsula become an archipelago of low islands in a shallow sea; fast-forward to the present, and the sandy margin will expand outward to fill in the tenuous modern outline of the state, its whole sketched with sand by moving dunes, beaches, and bars.[10] The entire state is, geologically-speaking, an old beach.

The evidence of this history is never far in Florida, as nearly every step taken in the state bounces with the spongy recoil of sand, whether that step is in inland pine flatwoods, among the scrub of dry sandhills, on suburban lawns, or along the current coasts. (The exceptions, like the peaty soils of the Everglades or the limestone rocklands of the Miami ridge, prove the rule.)

***

New York City has Central Park. San Francisco has Golden Gate Park. The District of Columbia has the National Mall, with its memorial, monuments, and Smithsonian museums. Countless other American cities have a park or a park system, likely designed by the Olmsted brothers (or even their father, in fewer cases) that doubles as civic symbol and shared space, an expanse of lawn for picnicking, pick-up sports, the Fourth of July.

In many states, particularly western states, national parks play a similar role. Yosemite in California, Yellowstone in Wyoming, Arches in Utah, the Great Smoky Mountains in North Carolina and Tennessee. These parks occupy stable space in the nation's collective life, functioning both as symbols of American landscape (whatever that multitude might be) and as well-used recreational places.

Florida does not have such parks. Florida has beaches.

Yes, Florida has long had a national park, Everglades, and its cities have many smaller neighborhood or civic parks within them. But while the Everglades may be symbolically resonant and well-loved, they are also, by nature of being a vast near-tropical wetland whose stable ground—the small tree islands that dot the expanses of marsh grass savanna—is inaccessible without canoe or airboat, hardly a place for public gathering, except of mosquitoes. And no Floridian city park even approximates for its city the role that Central Park plays in the public life of Manhattan.

No, in Florida, beaches are the parks.

Florida is the peak expression of American beach culture. While beaches are important to the seasonal rhythms of every coastal American state, from the lakefronts of Chicago and Michigan to the Jersey Shore, from Myrtle Beach to Manhattan Beach, no other state's daily rhythms are so thoroughly interwoven with the beach.[11] Weekend in Gainesville? Beach. Off work for the evening in Tampa? Beach. Long weekend in Orlando? Head west to the Gulf or east to the Atlantic. Morning run in Miami, Jacksonville, or Fort Myers? Beach.

The state's 825 miles of beach reflect Florida's diversity. If you miss the neon and vacuum-formed signage of pre-interstate American car culture, then there are the perfectly-preserved motels and hotels of Treasure Island: the Sea Chest, the Swashbuckler, the Ebb Tide, the Thunderbird, some still owned by the same families that built them in the fifties and sixties. Miami Beach speaks for itself. If you're a young family of four, there's Amelia Island on the Atlantic or the 30A strip on the Gulf. If the kids are a bit older and they want to surf, there's Cocoa Beach on the Space Coast. (Florida's beaches can be divided into ten coasts: Emerald, Forgotten, Nature, Sun, Southwest Gulf, Keys, Gold, Treasure, Space, and First, going counter-clockwise around the coast from west Gulf near Pensacola and ending in northeast Atlantic near Jacksonville.) If you can't bear to leave your car behind, there's Daytona Beach, where public parking is right on the beach sand, and the thunder of the Daytona Speedway lies just a couple miles inland.

The prominence of beaches as public space in Florida is, of course, in part a quirk of geology. If Florida were not so thoroughly peninsular (thus coastal on both sides), if it did not have so many high-energy regions conducive to the formation of barrier islands and beaches (rather than coastal marshes, which are really only found in Florida on the Gulf along the bend between the panhandle and the Sun Coast), if it were not fundamentally a state of sand, then beaches could not be its parks.

But this phenomenon is also tied to the late urbanization of Florida. The five most populous states in the US today are (in order of population) California, Texas, Florida, New York, and Illinois. Unlike all four of the other states on that list, the population of Florida didn't reach five million until the 1960s. The others had all passed that point back in the 1930s, when Florida was still below one million. Similarly, Texas and Illinois had their peak (post-Civil War) rates of growth in the late nineteenth century. New York and California had theirs in the decade surrounding World War I. Florida didn't have its peak growth until the 1960s. As a result, the cities of Florida were all planned, laid out, and built much later than the big cities of those other states. Florida's urban growth was uniquely formatted by technologies, like the automobile and the dredger, that merely overwrote the already-existing urban forms of other states.[12] The geography of Floridian urbanization is correspondingly more tightly drawn to the coasts than it is wrapped in concentric circles around urban cores (three-quarters of the state's population lives in coastal counties). Its spatial products are likewise unique in form and format: the dredge-and-pile canal suburbs of Naples, St. Petersburg, and Broward County; the continuous suburbanization of Miami that slides into drained portions of the Everglades and fills in former limestone quarries in the Lakes Belt; linear development along two- and four-lane beach highways on both Gulf and Atlantic coasts.

Because this urbanization occurred so late, it occurred after the great period of concern for public health, Progressive-era ideals, and civic consciousness expressed in public park-building that marked other American regions and their cities. This period stretched from Olmsted and Vaux's Central Park through the urban sanitation movement and the City Beautiful, but largely ended before the post-World War II boom. It was certainly over by the time Florida's cities began to grow in earnest in the 1960s, leaving Florida without a legacy of central public parks.

***

America's public parks had their artisans and day laborers, their tree nurseries, their rock quarries, their landscape architects.[13] What have we done—what are we doing—to shape sand into landscape? If something as enormous and

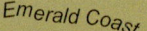
Emerald Coast

POPULATION GROWTH + URBAN FORM

Melbourne

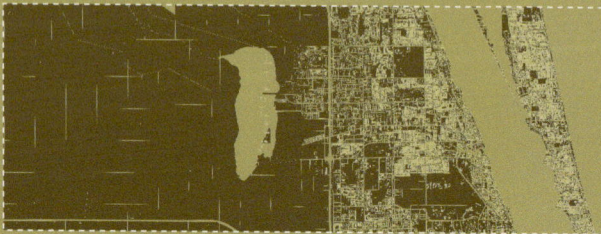

Sebastian

Naples

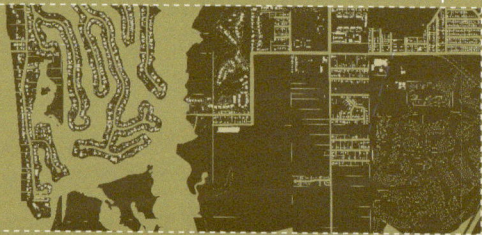

Population growth 1900-2015

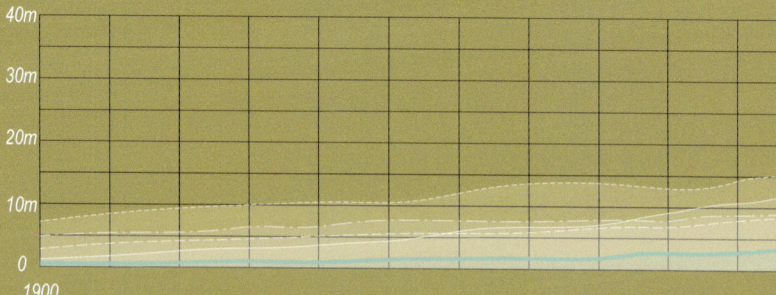

40m

30m

20m

10m

0

1900

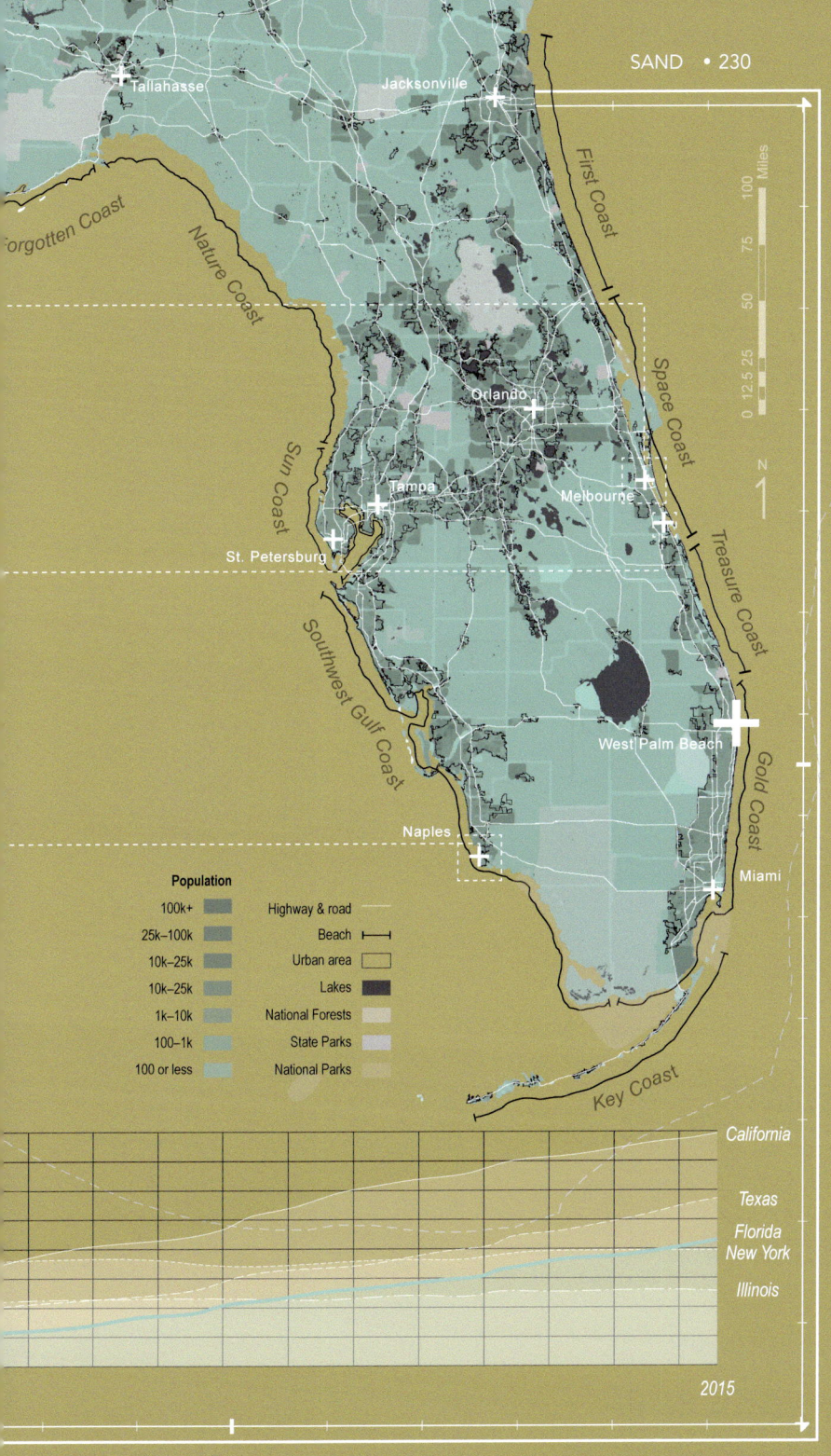

Tallahasse

Jacksonville

Forgotten Coast

Nature Coast

First Coast

Space Coast

Orlando

Sun Coast

Tampa

Melbourne

St. Petersburg

Treasure Coast

Southwest Gulf Coast

West Palm Beach

Gold Coast

Naples

Miami

Key Coast

100 Miles

75

50

0 12.5 25

N

Population

100k+	Highway & road
25k–100k	Beach
10k–25k	Urban area
10k–25k	Lakes
1k–10k	National Forests
100–1k	State Parks
100 or less	National Parks

California

Texas

Florida
New York

Illinois

2015

incomprehensible as the sand-sharing system assembles the beach, what mediates between that elusive leviathan and sunbathers on beach towels?

One part of the answer to these questions lies in the array of marks and measures that have been developed to negotiate the friction and slippage between sandy landscape and the legal apparatus of urbanization, including use rights, care responsibilities, and property boundaries.[14] Florida, being so defined by its beaches, has developed a particularly thick array: range monuments, coastal construction control lines and coastal construction setback lines, coastal general permit lines, critical erosion beaches (an on/off designation), navigational channels and artificial inlets (marks inscribed at full scale by dredgers and groynes), the coast "line" itself (a line that appears precise from space but becomes increasingly blurry as you zoom in on it), beach nourishment permits, public access beach sites (usually, a parking lot on a beachfront highway, a simple building with a handful of showers, and a boardwalk leading across dunes to the beach), and, crucially for the role of beaches as public landscapes, a distinction between the wet beach (the portion of the beach seaward of mean high tide, which is legally always public property) and the dry beach (the portion of the beach landward of mean high tide, saleable, and usually privately owned). These points, lines, and zones are an architecture of designation, legal and cadastral incantations that conjure beach as cultural artifact from beach as aggregated sand.

The core of this is the range monument system. Florida's coastal range monuments are geodetic control markers, like the more famous circular markers of the US Coast and Geodetic Survey, which is to say that they are physical objects, intended to be permanently precisely located in the Floridian landscape, whose purpose is to facilitate exact measurement by serving as reference points.[15]

The first reason that the ability to measure beaches precisely matters is because of real estate. The range monuments are used to measure the coastal construction control line, and the coastal construction control line determines where structures can be built on the most valuable real estate in Florida: beachfront real estate. Condominiums, apartments, and hotels sprout along the control line, clustering along its invisible edge like concrete fungi.

But it also matters because the beach is dynamic. To measure the change in a beach's profile—whether it is accreting, gaining volume and width, or eroding, losing that volume and width—requires fixed points that can be measured against. The range monuments are those points. Florida has a database, maintained by the Beach Field Services Program of the Florida Department of Environmental Protection, which collects topographic survey data, called beach profile surveys, for the entire state's coastline. From these surveys, the movement of the historic shoreline—its erosion and accretion, variable in both space and time —can be tracked, rates of change calculated, patterns inferred, and futures approximated.

All of this should not be taken to indicate that the Florida coast is under some of kind of constant, panoptic surveillance.[16] Range monuments appear quite frequently spaced from a state-level view, but there is typically around a thousand feet between monuments. Beach profile surveys are usually conducted only on lines drawn perpendicular from each monument to the water, so the space between monuments—the vast majority of beach—is unmeasured. This is usually not a problem, because the monuments are spaced closely enough that averaging is adequate to reliably infer the unmeasured beach between monuments, but these gaps do begin to convey how limited actual physical data collection is for beaches when laid against the overwhelmingly vast phenomena being measured.[17]

Moreover, coastal change does not occur evenly in time. For instance, the largest changes are typically tied to storm events, and storm events do not occur on the regular schedules that beach profile surveys follow, so the surveys tend to erratically capture pre-storm and post-storm conditions, depending on the relative timing of surveys.[18] A (simplistic) example may help illustrate how this affects the knowledge derived from profile surveys. Consider two imaginary beaches. One, North Beach, is surveyed in June. A hurricane strikes both beaches in July, eroding both badly. In August, South Beach is surveyed. Both beaches were last surveyed five years ago, so in both cases rates of change are calculated based on the difference between the most recent survey and that previous survey. But South Beach, because its most recent survey includes the impact of the hurricane, will appear to have a much higher rate of change—likely an order of magnitude higher—than North Beach. If these

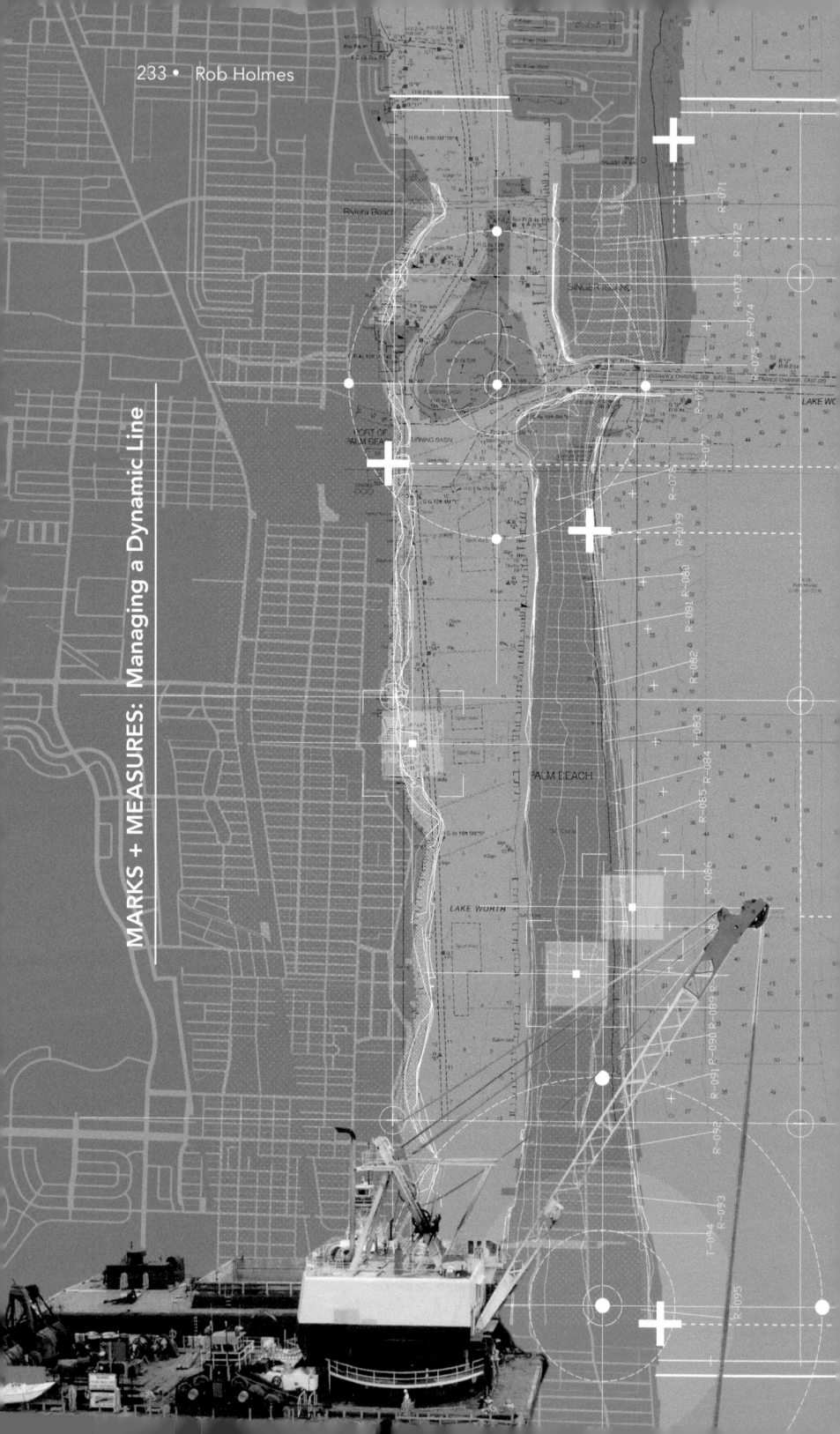

MARKS + MEASURES: Managing a Dynamic Line

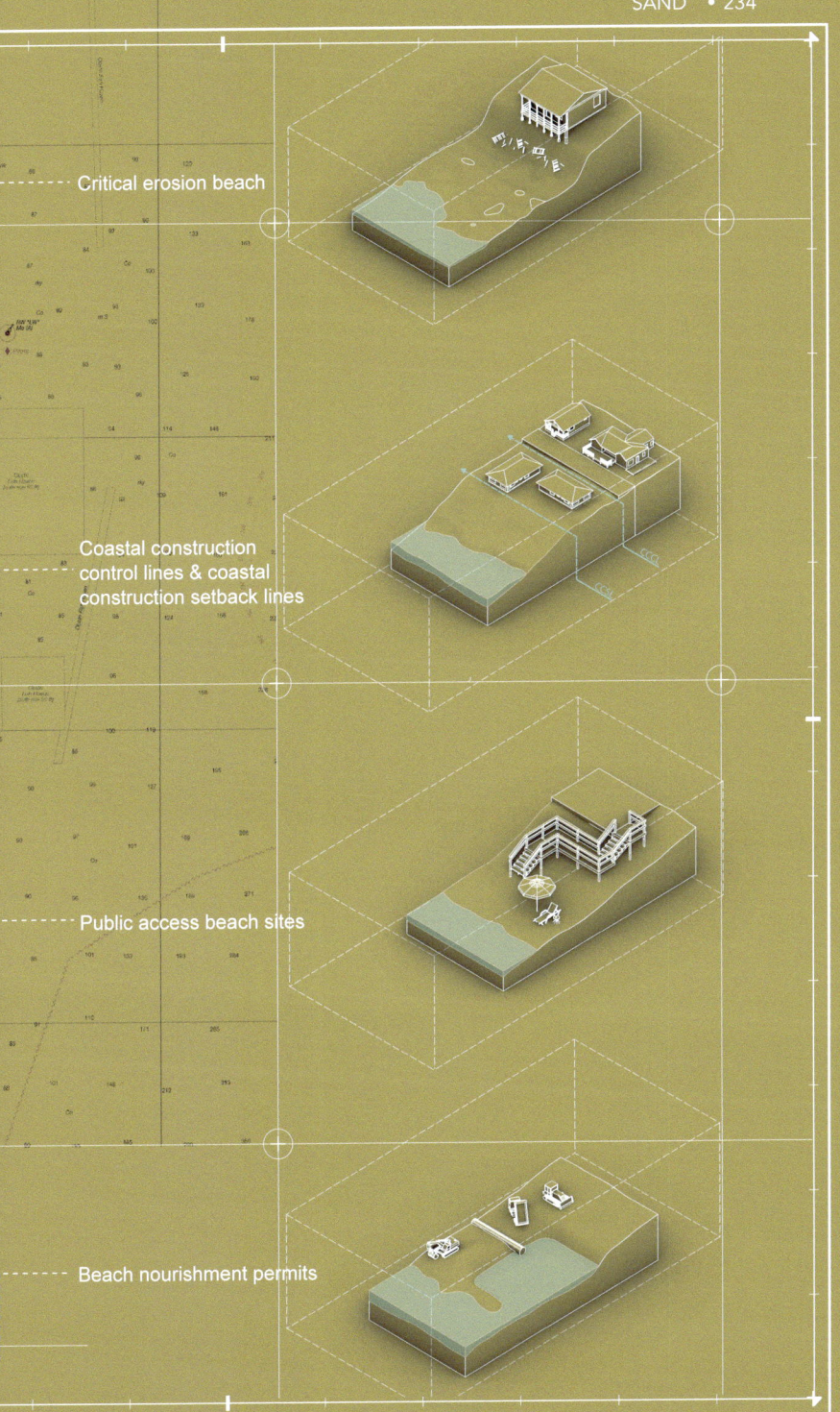

Critical erosion beach

Coastal construction
control lines & coastal
construction setback lines

Public access beach sites

Beach nourishment permits

rates of change are then used to develop projections of future change (without adjusting for storm impacts), then South Beach could appear to be in much higher need of protection or nourishment than North Beach, even if physical conditions on the ground are virtually identical.[19]

This brings us to the second way in which we hold the beach fixed. Florida has not been content merely to measure its beaches. Florida is also engaged in a long struggle to manage them, to tame the wild and unpredictable geomorphology of barrier islands with a controlled anthrogeology. Similar struggles are taking place on every inhabited continent, wherever beachfront property is valuable enough that people want to fix it in place, but Florida's struggle is particularly complex, advanced, and extensive.

This brings us back to Lake Worth Inlet and its adjoining barrier islands, where the structures, machines, and practices that Florida is employing to accomplish this control are all operating.

The most obvious way that we attempt to stabilize beaches is by building barriers. These barriers can generally be separated into a few major categories. Sea walls and breakwaters are both constructed parallel to the shoreline. Sea walls are onshore, usually behind the sand beach where dunes would typically be found. There, they protect beachfront buildings from the inland migration of beach, which can undercut structures or bring waves up to condominium foundations. Breakwaters are offshore, some distance from the beach, sometimes rising above high tide, sometimes partially or even wholly submerged. Their purpose is, generally speaking, to reduce the amount of wave energy that reaches shore, decreasing beach erosion. (Seawalls and breakwaters are both sometimes also deployed with the intention of protecting structures from storm surge, either by presenting a complete barrier, as sea walls do, or by reducing surge energy, as breakwaters can.) Both are usually constructed of hard materials: concrete sea walls are typical, while breakwaters are more likely to be built of piled rock, concrete blocks, or (personal favorite) concrete tetrapods.

A slightly more subtle set of barriers is constructed perpendicular to shore, generally out of the same materials as breakwaters: jetties and groynes. Imagine long lines of piled rock with gently triangular profiles. Some, like the north and south jetties at Lake Worth Inlet, are built at the mouths of inlets, harbors, and rivers, usually to protect navigation channels from the longshore transport of sediments. Others, such as the roughly thirty groynes of varying shapes and lengths that jut out from three separate stretches of Palm Beach south of the inlet, like the vertebrae of a rock exoskeleton grafted on the beach, are constructed in clusters along stretches of beach that are particularly susceptible to erosion, with the intention of arresting longshore transport for that given stretch of beach.

The more obstinately these barriers are intended to fix the beach, the less effective they seem to be. Sea walls, which ostensibly draw a hard line of protection, are generally regarded by scientists as having net negative effects on beach maintenance, as they cut the sand-sharing system off from one of its major supplies, upland dunes, reducing the dynamism and resilience of a given system.[20] Contrastingly, the more coastal engineers work with sand, respecting its dynamic qualities, the more successful they seem to be. Breakwaters are intended to affect beach formation indirectly, by manipulating the complex of energy that drives sand movement, and they can be quite effective at promoting local beach aggradation, though configuring them properly for a given wave environment is a difficult task of modeling and design.

Recognizing the complexity and dynamism of the sand-sharing systems that its beaches emerge from, Florida has steadily developed a series of studies on and management plans for 40 of its inlets. The management plans, together with the broader Strategic Beach Management Plan which covers all of the state's beaches, attempt to address a patchwork of related issues. Inlet jetties, like those at Lake Worth Inlet, accelerate erosion by cutting off longshore transport of sand. High rates of sea level rise combine with intensive development on or just beyond dunes to narrow the width of land available to beaches, exacerbating erosion. Development also tends to cut off dunes from the sand-sharing system, through measures like sea walls. Meanwhile, offshore sand mining—often to replace sand lost on beaches to erosion—depletes nearshore reservoirs of sand that could otherwise replace eroded sand. Depleted, confined, and unable to migrate, sand-sharing systems respond with accelerated rates of beach erosion that encourage increased armoring, producing a vicious cycle of armoring and erosion.

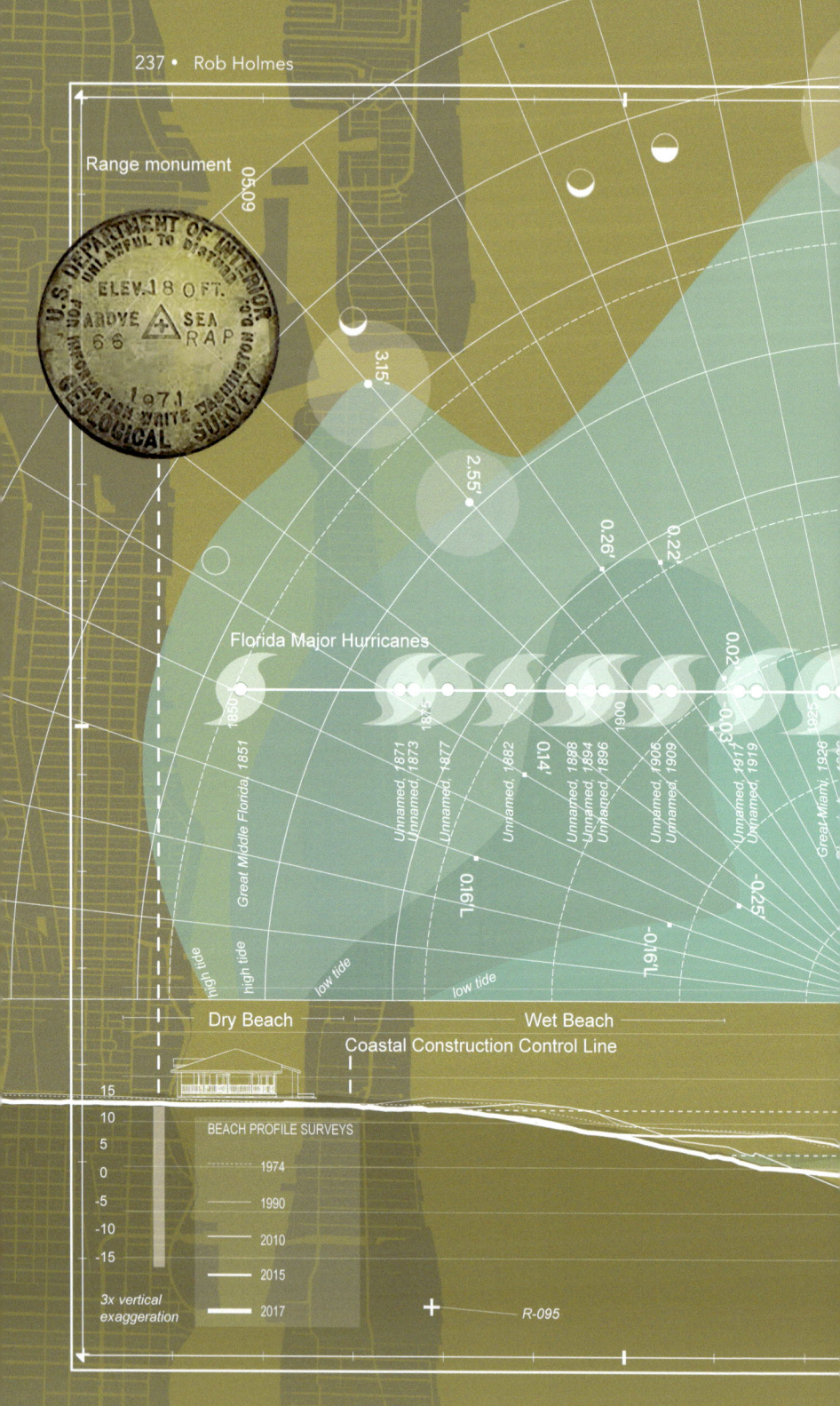

Range monument

05.09

ELEV. 18.0 FT.

3.15'

2.55

0.26'

0.22'

Florida Major Hurricanes

0.02

0.03

-0.25'

Great Middle Florida, 1851

Unnamed, 1871
Unnamed, 1873

Unnamed, 1877

Unnamed, 1882

0.14'

Unnamed, 1888
Unnamed, 1894
Unnamed, 1896

0.16'L

Unnamed, 1906
Unnamed, 1909

Unnamed, 1916

Great Miami, 1926

high tide

high tide

low tide

low tide

-0.16'L

Dry Beach · · · · · · · · · · · Wet Beach

Coastal Construction Control Line

15
10
5
0
-5
-10
-15

BEACH PROFILE SURVEYS

1974

1990

2010

2015

3x vertical
exaggeration   2017

+   R-095

05.20

May 2020 tidal data

4.0ft H

3.45'

3.5ft H

3.28'

2.90'

3.0ft H

2.72'

2.5ft H

2.0ft H

2025

0.6ft L

Unnamed, 1944
Unnamed, 1945
Unnamed, 1947
Unnamed, 1948
Unnamed, 1949
Easy, 1950
King, 1950

Donna, 1960

Betsy, 1965

Erbise, 1975

Elena, 1985

Andrew, 1992

Opo, 1995

Charley, 2004
Ivan, 2004
Jeanne, 2004
Dennis, 2005
Wilma, 2005

Irma, 2017
Michael, 2018

0.0ft L

0.2ft L

0.4ft L

0.2ft L

Post-Irma Section

Pre-Irma Section

Record Storm Surge

High Tide
Low Tide

TRACKING CHANGE: Tides + Storms

Florida's management plans establish "sand budgets" that attempt to describe, quantitatively and geographically, how each individual beach or inlet's sand-sharing system should function. In the case of the Lake Worth Inlet, the state has estimated that the jetties have impounded an annual average of 171,300 cubic yards of sand, producing a cumulative deficit on north Palm Beach of approximately 12 million cubic yards.[21] For scale, 12 million cubic yards is roughly twice what is annually dredged from the navigation channels at the Port of Savannah, one of the largest Atlantic Coast cargo ports. And this deficit would be considerably larger if Florida and the US Army Corps of Engineers had not frequently pumped dredged sand from Lake Worth and the Port of Palm Beach's navigation channels onto Palm Beach.

The sand transfer plant is one attempt to address this deficit. In 1956, the Army Corps of Engineers completed a comprehensive study of Palm Beach's shoreline, which highlighted north Palm Beach's rapidly eroding beach and fingered the jetties' interruption of longshore transport as the cause.[22] By 1958, the Corps had constructed the sand transfer plant. Since then, the plant has pumped about a hundred thousand cubic yards of slurried sand every year in a steel and rubber discharge line beneath the inlet to a point just south of the south jetty, bypassing the navigation channel's trench and mechanically re-connecting the two barrier islands.[23]

The sand transfer plant is relatively unique, though. An Australian survey in the late 1990s found only about fifty such facilities globally, some fixed like the Lake Worth Inlet plant, others partially or fully mobile.[24] Far more common is the practice of direct beach nourishment, which involves obtaining sand (by excavator or dredge), transporting it (by truck or slurry pipeline), placing it (dumping or pouring), and re-grading the beach (with bulldozer). The sheer strangeness of the result should not go unnoticed: in Florida, the stretch of sand you plant your umbrella in is, more often than not, every bit as much a work of artifice as your beach chair, your cooler, or the synthetic fibers in your swimsuit.

Beaches are nourished with sand from a wide variety of sources, including inland mines, nearby ports with basins, slips, and channels, and offshore reservoirs. Materially, the latter source is usually preferred: trucking material from inland

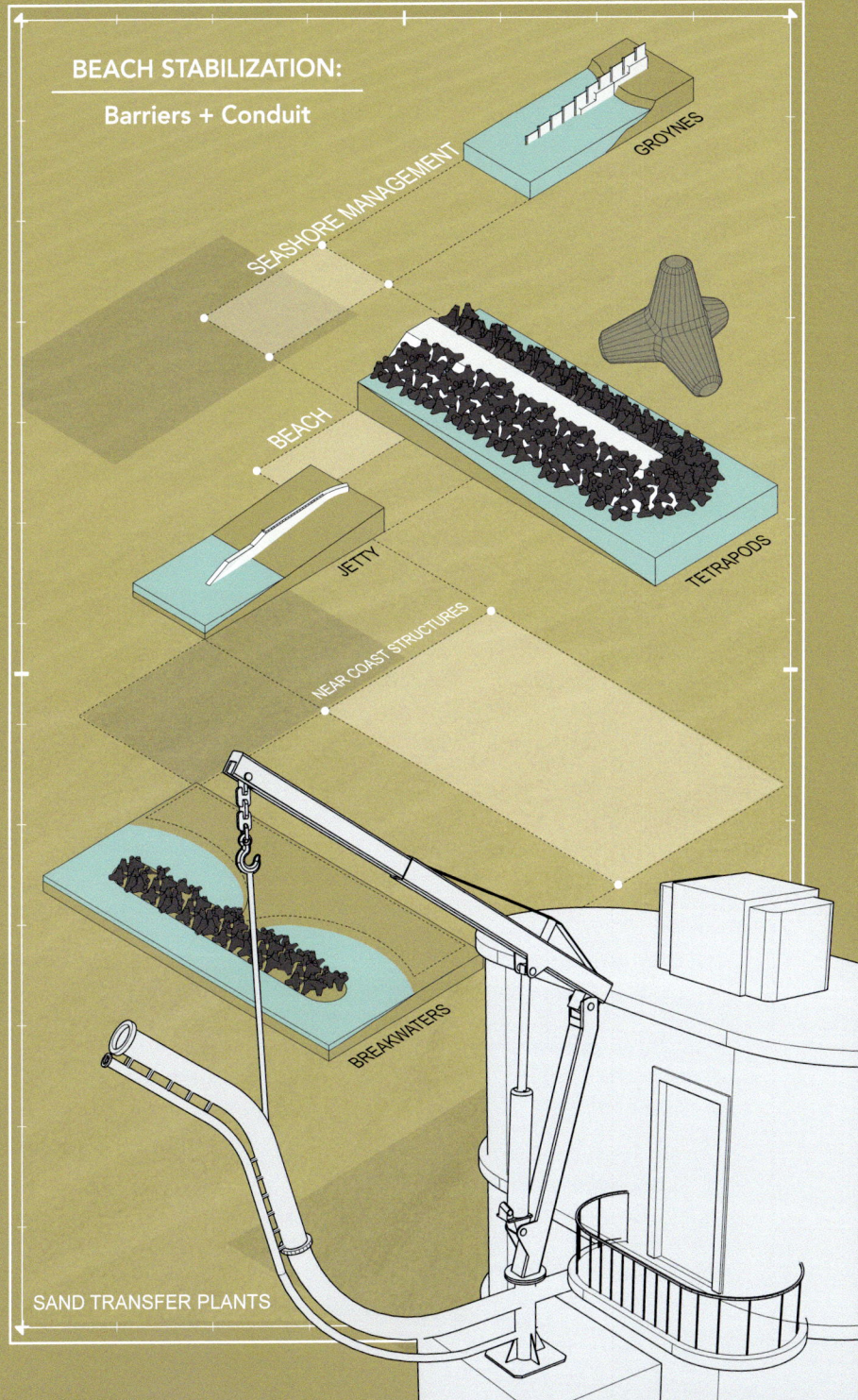

BEACH STABILIZATION:
Barriers + Conduit

SEASHORE MANAGEMENT

GROYNES

BEACH

TETRAPODS

JETTY

NEAR COAST STRUCTURES

BREAKWATERS

SAND TRANSFER PLANTS

mines is exorbitantly expensive in comparison to pumping sandy slurry directly from a dredger, while navigational dredged material is rarely "clean" beach sand, as it typically contains the wrong sizes of particles (organic fines are much more muddy than sandy) and the wrong colors, even when it is sand (the colors of beach sand are highly regulated by the Florida Department of Environmental Protection; technically, the applicable analysis is of "moist Munsell color").[25] Consequently, the offshore sands of Florida have been and continue to be studied in great detail.

The continental shelf is generally quite wide off the Atlantic coast of North America, and is functionally particularly wide in the southeast as the shelf borders a second wide, flat underwater region, the Blake Plateau, which is roughly the same area as the state of North Carolina. Together, these two shelves extend as much as 300 miles into the ocean.[26] Just north of Palm Beach, though, both the wide shelf and the Blake Plateau disappear, as the deep straits of Florida come close to the eastern edge of Florida, producing a narrow and steep escarpment that lies only a couple miles offshore. As a result, the borrow areas available for mining nourishment sand offshore in the vicinity of Lake Worth Inlet are tightly restricted to a roughly two-mile-wide band. (Further north, where the shelves are wide, borrow areas are often found further out, and are correspondingly more extensive.)

Florida's Regional Offshore Sand Source Inventory, or ROSSI, classifies borrow areas in six categories: exhausted, proven, permitted, proposed, potential, and unverified. Each borrow area is studded with sand core sample sites, ranging in number from about a half-dozen to several dozen per borrow area. These physical samples are assessed for a range of characteristics related to the sand's suitability for beach application: grain size breakdowns according to at least three different major classification systems, presence of additional materials like organics or shell fragments, and, of course, Munsell color values, both moist and dry.

Immediately adjacent to Lake Worth Inlet, there are three of these borrow areas, two "proven" (PB1-R070, or Palm Beach County, Range Monument 70, and PB1-R086) and one "exhausted" (PB0-R079, or the borrow area formerly known as "Mid-Town Restoration"). Each is assessed for a wide variety of material characteristics (including, of

SAND DEFICIT

The jetties of Lake Worth Inlet have resulted in cumulative deficit on north Palm Beach of approximately 12 million cubic yards, about twice what is annually dredged at the Port of Savannah.

Port of Savannah

Savannah River

12,000,000 Cubic Yards

687 ft

687 ft

The Colosseum

1 Person

course, Munsell color values), but also for estimated total volumes contained in the sand reservoirs. These volumes are suggestive of the potential scale of the anthrogeology of beach nourishment: PB1-R070 contains 113 million cubic yards of sand at an average depth of 12 feet, while PB1-R086 contains another 526 million cubic yards at an average depth of 15 feet. For comparison, the Army Corps annually dredges about 130 million cubic yards of sediment from every navigation channel in the United States, combined. The excavation of the Panama Canal, one of the largest earthmoving events in the history of the Western hemisphere, required removing twice that, a bit over 260 million cubic yards of material. Twice again that, two underwater Panama Canals, awaits dredgers in these two borrow areas, measured and latent.

This should begin to give some sense of the scale of the oceans of sand that lie underwater, but these are only the immediately adjacent borrow areas: dozens of similar surveyed reservoirs exist all along the lengths of both the Atlantic and Gulf coasts of Florida. And further offshore, deeper borrow sources outside state waters are regulated by the federal Minerals Management Program.

The largest on-going beach nourishment project near the Lake Worth Inlet is the Mid-Town Beach Nourishment Project, which is sited on Palm Beach, south of the area nourished by the sand transfer plant. This 2.6-mile stretch of shoreline most recently received a million cubic yards of offshore sand in 2015, bringing that full stretch into compliance with what is called the "beach fill design template": essentially a consistent profile of sand along the length of shoreline, in this case twenty-five feet wide at an elevation of nine feet above sea level.[27] As is standard procedure for beach nourishment on such rapidly-eroding beaches, sand is aggressively over-placed beyond this design template, in order to provide an additional "eight years of advance nourishment," which is to say enough additional sand to account for eight more years of anticipated erosion. As a result, the beach is expected to not need nourishment again until 2021. Florida and the Army Corps assure us that there is more than enough sand in the offshore borrow areas to meet the needs of this and all other anticipated nourishment projects in southeast Florida through at least 2062.[28]

INCLINATIONS

It seems that this is all under control. There is plenty of sand offshore, and the experts have the tools they need to continue this mechanized geology indefinitely.[29]

The problem, though, is that everything is linked. Every act of placement is linked to an act of extraction; when we mine sand from offshore beds, we aren't adding sand to the sand-sharing system, we are merely moving it from one component of the system, below water where we couldn't see it, to another, visible component of the system. And when we place sand with dredgers, dump trucks, and bulldozers, it tends to be much less stable in aggregate than when it is placed by winds and currents.[30] When the space that we are placing that sand in is pinned between a shoreline retreating due to sea level rise and the fixed elevations of seawalls and beachfront buildings, then this aggregate destabilization is intensified, as nourishment occurs on a narrowing and steepening slope.

In even brief geologic time, all the assemblages of sand that appear in coastal environments are transient. The dunes, the barrier islands, the ebb deltas, the sand bars, the great offshore sand reservoirs are all events as much as they are objects. Ordinarily, though, geologic change and human lives might not intersect in any particularly evident fashion, and a barrier island's beach might be a perfectly reasonable place to inhabit, at least for a lifetime or two. It is certainly a good spot to spend an afternoon. But sea level rise is accelerating, and there is no clear indication of when it will stop accelerating. Our dredging machines—powered by the same fossilized solar energy whose combustive release is melting glaciers and thickening the atmosphere—are busily conveying sand to maintain the configurations we prefer. Our hardened structures—sea walls and breakwaters, jetties and groynes—are multiplying deposition on one beach and multiplying erosion on another, and, despite the evidence that they are so often counter-productive, we continue to build more. Our culture—so interwoven with the beach—denies sand the movement that formed South Beach, Jupiter Island, and Perdido Key.

Yet beaches are not troubled by sea level rise. Given space, they'll gladly migrate inland and upland forever. We seem less inclined.

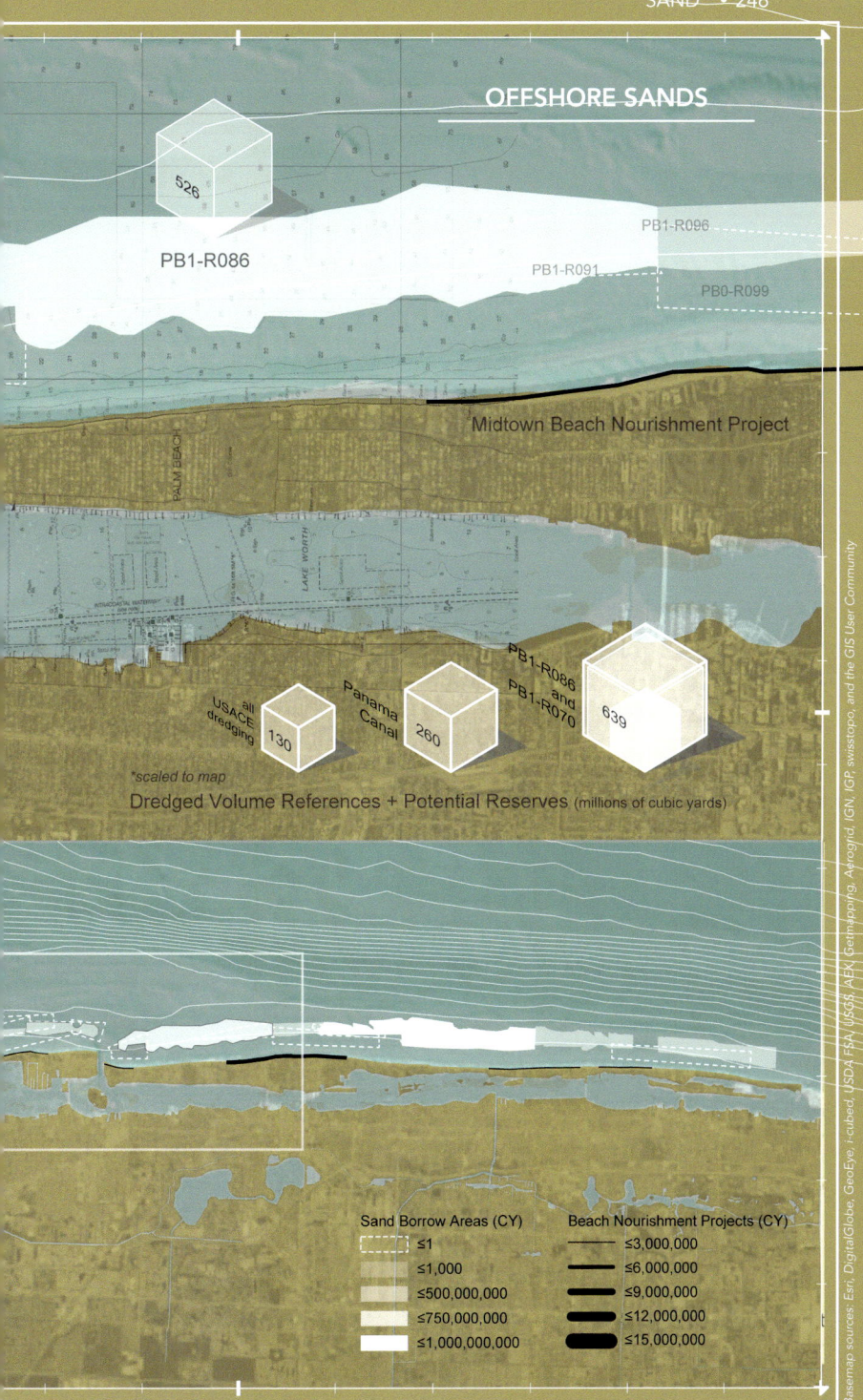

OFFSHORE SANDS

526

PB1-R086

PB1-R096

PB1-R091

PB0-R099

PALM BEACH

LAKE WORTH

INTRACOASTAL WATERWAY

Midtown Beach Nourishment Project

PB1-R086
and
PB1-R070

all
USACE
dredging
130

Panama
Canal
260

639

*scaled to map

Dredged Volume References + Potential Reserves (millions of cubic yards)

Sand Borrow Areas (CY)
≤1
≤1,000
≤500,000,000
≤750,000,000
≤1,000,000,000

Beach Nourishment Projects (CY)
≤3,000,000
≤6,000,000
≤9,000,000
≤12,000,000
≤15,000,000

AUTHOR'S NOTE

I would like to thank Tim Maly for reading and offering editorial direction on a draft of this essay, as well as colleagues from my time at the University of Florida, Charlie Hailey and Christie Allen, for helping me think Florida's beaches and draw Florida's beaches, respectively.

ENDNOTES

1. Commerce is the primary reason for maintaining this channel. The Port of Palm Beach shares an urban hinterland with larger ports, like the Port of Miami, but it has carved out a niche for itself in global commerce by specializing, particularly in the export of sugar and molasses. The Port of Palm Beach is the only port in south Florida capable of handling these commodities, which are the primary agricultural exports from the vast polder monocultures of the Everglades Agricultural Area. USACE, Lake Worth Inlet Feasibility Study, 2014.

2. Lizette Alvarez, "Long-Secret Fallout Shelter Was a Cold War Camelot," *The New York Times*, 2 October 2011; Zachary Fagenson, "Kennedy's Cold War Island Bunker off Palm Beach Now a Museum," *Reuters*, 19 Nov 2013; Bonnie Gross, "Secret Kennedy Bunker on Peanut Island," *Florida Rambler*, accessed 20 October 2020, https://www.floridarambler.com/historic-florida-getaways/kennedy-bunker-on-peanut-island/.

3. Michael Welland, *Sand: The Never-Ending Story* (University of California Press, 2009).

4. Vince Beiser, *The World in a Grain* (Riverhead Books, 2019).

5. The many dimensions of this tension have enormous implications for how we design landscapes, buildings, and urban environments. This chapter outlines some of these tensions, with the hope that we can learn to learn from sand. Sand draws us to the water's edge, where billions of grains, flocking, form the beaches, dunes, and sand bars that we love to play on, gaze at, and swim over. Sand compels us to shape networks of urbanization in response to it: beachfront high-rises, protected beach reserves, neon-lit commercial strips. The encounters we have with the sea on its beaches shape us culturally, as is testified in studies like Alain Corbin's *The Lure of the Sea* (1994) and John Mack's *The Sea* (2011). But we respond, most often, by ignoring the very dynamic qualities that permitted it to form these sandy places in the first place. We respond with a vast apparatus of control, of delineations, barriers, and machines. Yet there are other ways to live: with sand, rather than against it. I discuss some ways that we can begin to design like this in my recent essays "Design with Change" (2019, in *Design with Nature Now*, eds. Frederick Steiner, Karen M'Closkey, Billy Fleming, and Richard Weller) and "The Problem with Solutions" (2020, *Places Journal*).

6. Tonya Clayton, *How to Read a Florida Gulf Coast Beach*, University of North Carolina Press, 2012, 94–99.

7. Another term that scientists use to refer to these wholes is "littoral cell." Generally speaking, "sand-sharing system" is usually employed to emphasize the *exchange* of material *in section*, that is, perpendicular to the shoreline. "Littoral cell," by contrast, is usually used to describe the *differentiation* of contained compartments of

sand exchange from one another as one moves up or down a coast *in plan*, or parallel to the shoreline. Both, though, describe holistic relationships between the same sandy components of dunes, beaches, bars, and reservoirs. Both can be used to talk about sand budgets, aggregate tendencies, and the dynamic equilibrium of beaches. "Cell" and "system" might therefore be used somewhat interchangeably, and what I describe as "systems" below might also be referred to as "cells." For a readable description of sand-sharing systems, see John Wells and Charles Peterson, *Restless Ribbons of Sand: Atlantic and Gulf Coastal Barriers* (1986); for an example of the contemporary use of cell-based analysis, see Julie Dean Rosati, "Concepts in Sediment Budgets," *Journal of Coastal Research* 21:2 (2005).

8   R.G. Dean and M.P. O'Brien, "Florida's East Coast Inlets: Shoreline Effects and Recommended Action," report for Florida Department of Natural Resources, 1987, 13.

9   Joseph Van Gaalen, "Longshore sediment transport from northern Maine to Tampa Bay, Florida" (Master's thesis, University of South Florida, 2004).

10   Florida Department of Environmental Protection, "Florida's Geologic History and Geological Resources," by Ed Lane, 1994.

11   Yes, Florida's beaches are also the state's primary tourist attraction, dwarfing even Disney.

12   On formatting and spatial products, see various by Keller Easterling including *Organization Space*, *The Action is the Form*, and *Extrastatecraft*.

13   Jane Hutton's *Reciprocal Landscapes* investigates this in the case of five landscapes in New York City, five everyday landscape construction materials that were part of their making, and five sets of relatively distant source landscapes, where materials were obtained. Beach sand, like Hutton's Peruvian guano or Maine granite, is thoroughly reciprocal, entangling people, landscapes, and a host of other agents, living and nonliving.

14   Much recent scholarship has been devoted to showing how marking, measuring, and mapping are processes that not only reflect some external reality, but participate in constructing our worlds. See for instance *Mappings* (ed. Denis Cosgrove) and *Rethinking Maps* (eds. Chris Perkins, Rob Kitchin, and Martin Dodge). Thus, Florida's marks and measures not only reflect the state's legal system, patterns of urbanization, and understanding of landscape, but also participate, together with sand, waves, real estate developers, and groynes, in making beaches.

15   Well, historically they were all physical objects. Today, many of the range monuments are "virtual," and even some of the extant physical monuments have been superseded in survey work by virtual counterparts, in many cases because the original physical monuments have not proved nearly as immobile as was intended. Sand is, after all, highly mobile. Florida DEP, "Regional Coastal Monitoring Data."

16   Note that this survey is also quite new: the beach profile surveys date only to the 1970s. A fifty-year slice of time is quite lengthy for surveying real estate price trends, for instance, but it is rather abbreviated for surveying geologic phenomena.

17   This, incidentally, is true of most earth systems phenomena: in this era of satellite panopticism and complex computer modeling we

tend to imagine that someone, somewhere—presumably some scientist—knows exactly what is going on with any given earth system. It is far more likely, for any given example, that data is extremely limited and most management decisions are being made on the basis of assumptions extrapolated from other cases or numbers averaged from what little data is available. This is usually adequate, until a complex system exhibits an unexpected emergent behavior outside the bounds of averages and extrapolations, and then it is not. The consequences vary.

18 The Beach Field Services Program currently divides the state of Florida into four quadrants of beaches. One quadrant is surveyed each year, so any given beach is surveyed every four years. Four years might mean a half-dozen major hurricanes; it might also mean none.

19 DEP typically aims to compensate for obvious potential mismatches between survey dates and physical conditions, such as the impact of a major hurricane, by doing pre- and post-storm LIDAR surveys of affected beaches. My point, though, is that beach profile measurement is necessarily intermittent in both space and time.

20 Orrin Pilkey and Katharine Dixon, *The Corps and the Shore*, Island Press, 1996.

21 Florida DEP, Lake Worth Inlet Management Implementation Plan, 1996.

22 Frederick Zurmuhlen, "The Sand Transfer Plant at Lake Worth Inlet," *Coastal Engineering Proceedings* 1, no. 6 (1957).

23 William Kelly, "Palm Beach's Sand Transfer Plant May Soon Be Back in the Spotlight," *Palm Beach Daily News*, 8 November 2017.

24 State of Queensland Coastal Services, "Technical Report R20: World-wide Sand Bypassing Systems," by Paul Boswood and Russell Murray, 2001.

25 USACE and Florida DEP, "Southeast Florida SAND Study," 2012; this regulation is not only for aesthetic reasons, but also because sand color affects sand temperature; hot sand can harm the eggs of endangered sea turtles.

26 USGS, "GSA Paper 529-A, Atlantic Continental Shelf and Slope of the United States," 1966.

27 FDEP, "SBMP Southeast Atlantic Coast Region," 2018.

28 FDEP, "SBMP Southeast Atlantic Coast Region," 2018.

29 Generally speaking, when a sand crisis or the prospect of running out of sand is discussed, this refers first to the river sand that is preferred for concrete and construction, and then only secondarily to the marine sand that is used in beach nourishment. That said, beach nourishment does have the capacity to and is already in some places producing significant localized shortages of marine sand, particularly when environmental impacts, logistics, and expense are taken into account. In Florida, for instance, the SBMP is less sanguine about sand supplies for most other regions of the state than it is for the southeast Atlantic. Where it confidently predicts about 35 years of identified sand for the southeast, it generally describes only about 15 years of identified sources for most regions, and in a few locales is not even able to identify sources for that abbreviated period.

30 If we could predict how that sand, water, and wind would behave, accurately, over long periods of time, then this might not be particularly problematic for our mechanized geology. We'd simply put the groynes and breakwaters in the right places, dump the sand just so to form exactly the sand engines we need for the prescribed periods

of drift, and dredge all the right sea bottoms. But such modeling and prediction seems to be beyond us; may always be beyond us. (See Orrin Pilkey and Linda Pilkey-Jarvis, *Useless Arithmetic*; Sandra Mitchell, *Unsimple Truths*.)

BIBLIOGRAPHY

Alvarez, Lizette. "Long-Secret Fallout Shelter Was a Cold War Camelot." *The New York Times*, 2 October 2011.

Beiser, Vince. *The World in a Grain*. Riverhead Books, 2019.

Clayton, Tonya. *How to Read a Florida Gulf Coast Beach*. University of North Carolina Press, 2012.

Corbin, Alain. *The Lure of the Sea*. University of California Press, 1994.

Cosgrove, Denis, ed. *Mappings*. Reaktion Books, 1999.

Dean, R.G. and M.P. O'Brien. "Florida's East Coast Inlets: Shoreline Effects and Recommended Action." Report for Florida Department of Natural Resources, 1987.

Easterling, Keller. *Organization Space*. MIT Press, 1999.

Easterling, Keller. *The Action is the Form*. Strelka Press, 2012.

Easterling, Keller. *Extrastatecraft*. Verso Books, 2014.

Fagenson, Zachary. "Kennedy's Cold War Island Bunker off Palm Beach Now a Museum." *Reuters*, 19 November 2013.

Florida Department of Environmental Protection. "Florida's Geologic History and Geological Resources." by Ed Lane. 1994.

Florida Department of Environmental Protection. "Regional Coastal Monitoring Data." Accessed 20 October 2020, https://floridadep.gov/water/beach-field-services/content/regional-coastal-monitoring-data.

Florida Department of Environmental Protection. "Lake Worth Inlet Management Implementation Plan." 1996.

Florida Department of Environmental Protection. "SBMP Southeast Atlantic Coast Region." 2018.

Gross, Bonnie. "Secret Kennedy Bunker on Peanut Island." *Florida Rambler*. Accessed 20 October 2020, https://www.floridarambler.com/historic-florida-getaways/kennedy-bunker-on-peanut-island/.

Holmes, Rob. "Design with Change" In *Design with Nature Now*, edited by Frederick Steiner, Richard Weller, Karen M'Closkey and Billy Fleming. Lincoln Institute of Land Policy, 2019.

Holmes, Rob. "The Problem with Solutions." *Places Journal* (2020). https://placesjournal.org/article/the-problem-with-solutions/.

Hutton, Jane. *Reciprocal Landscapes*. Routledge, 2019.

Kelly, William. "Palm Beach's Sand Transfer Plant May Soon Be Back in the Spotlight." *Palm Beach Daily News*, 8 November 2017. Accessed 20 October 2020, https://www.palmbeachdailynews.com/news/palm-beach-sand-transfer-plant-may-soon-back-the-spotlight/h9BkuD0vaIP7jT9Mzmd27I/.

Mack, John. *The Sea: A Cultural History*. Reaktion Books, 2013.

Mitchell, Sandra. *Unsimple Truths*. University of Chicago Press, 2009.

Perkins, Chris, Rob Kitchin, and Martin Dodge, eds. *Rethinking Maps*. Routledge, 2011.

Pilkey, Orrin and Katharine Dixon. *The Corps and the Shore*. Island Press, 1996.

Pilkey, Orrin and Linda Pilkey-Jarvis. *Useless Arithmetic*. Columbia University Press, 2007.

Rosati, Julie Dean. "Concepts in Sediment Budgets." *Journal of Coastal Research* 21, no. 2 (2005): 307-322.

State of Queensland Coastal Services. "Technical Report R20: World-wide Sand Bypassing Systems." by Paul Boswood and Russell Murray, 2001.

US Army Corps of Engineers and Florida Department of Environmental Management. "Southeast Florida Sediment Assessment and Needs Determination (SAND) Study." By Jase Ousley, Elizabeth Kromhout, and Matthew Schrader. 2012.

US Army Corps of Engineers. *Lake Worth Inlet Feasibility Study*. 2014.

United States Geological Survey. "GSA Paper 529-A: Atlantic Continental Shelf and Slope of the United States." By K.O. Emery, 1966.

Van Gaalen, Joseph. "Longshore Sediment Transport from Northern Maine to Tampa Bay, Florida." Master's thesis, University of South Florida, 2004.

Welland, Michael. *Sand: The Never-Ending Story*. University of California Press, 2009.

Wells, John and Charles Peterson. *Restless Ribbons of Sand: Atlantic and Gulf Coastal Barriers*. University of North Carolina Press, 1986.

Zurmuhlen, Frederick. "The Sand Transfer Plant at Lake Worth Inlet." *Coastal Engineering Proceedings* 1, no. 6 (1957).

IMAGE CITATIONS + CREDITS

Listed by Page Number

213   Research Assistant: Tian Wang

215   Research Assistant: Alek De Mott

217   USGS Topo Map 1967: Esri, HERE, Garmin, INCREMENT P, USGS, METI/NASA, EPA, USDA | Source: Historical Topographic Map Collection courtesy of the US Geological Survey, Esri.

219   Photo: Matthew Seibert

      Research Assistant: Tian Wang

223   Photo: Matthew Seibert. "Lake Worth Inlet Management Implementation Plan | Florida Department of Environmental Protection." Accessed October 31, 2020. https://floridadep.gov/rcp/beaches-inlets-ports/documents/lake-worth-inlet-management-implementation-plan.

      Research Assistant: Tian Wang

229   Florida Geographic Data Library; Esri, Rand McNally & Company, Bartholemew and Times Books, Defense Mapping Agency presently known as National Geospatial-Intelligence Agency; NOAA/NOS/Office of Coast Survey; University of Florida GeoPlan Center; US Census Bureau.

      Research Assistant: Tian Wang

233   "Historic Shoreline Database | Florida Department of Environmental Protection." Accessed October 31, 2020. https://floridadep.gov/rcp/beaches-inlets-ports/content/historic-shoreline-database. GIS florida parcels: University of Florida GeoPlan Center.

      Research Assistants: Chloe Nagraj, Tian Wang

237   GIS Florida parcels: University of Florida GeoPlan Center. Tidal data: NOAA/NOS. "Historic Shoreline Database | Florida Department of Environmental Protection." Accessed October 31, 2020. https://floridadep.gov/rcp/beaches-inlets-ports/content/historic-shoreline-database.

      Research Assistants: Chloe Nagraj, Tian Wang

240   Research Assistant: Tian Wang

242   Savannah port: SAGIS, BOA.

        Research Assistant: Tian Wang

245   Basemap sources: Esri, DigitalGlobe, GeoEye, i-cubed, USDA FSA, USGS, AEX,
        Getmapping, Aerogrid, IGN, IGP, swisstopo, and the GIS User Community; Florida
        Bathymetry; ROSS2_BORROW_AREA; Florida Shoreline_1_to_12000_Scale; Tidal_Flats_in_
        Florida; Sand_Placements_Projects

MUD

And its Meaning in a Port Town

by Brian Davis

Hart-Miller Island

Baltimore, U.S.

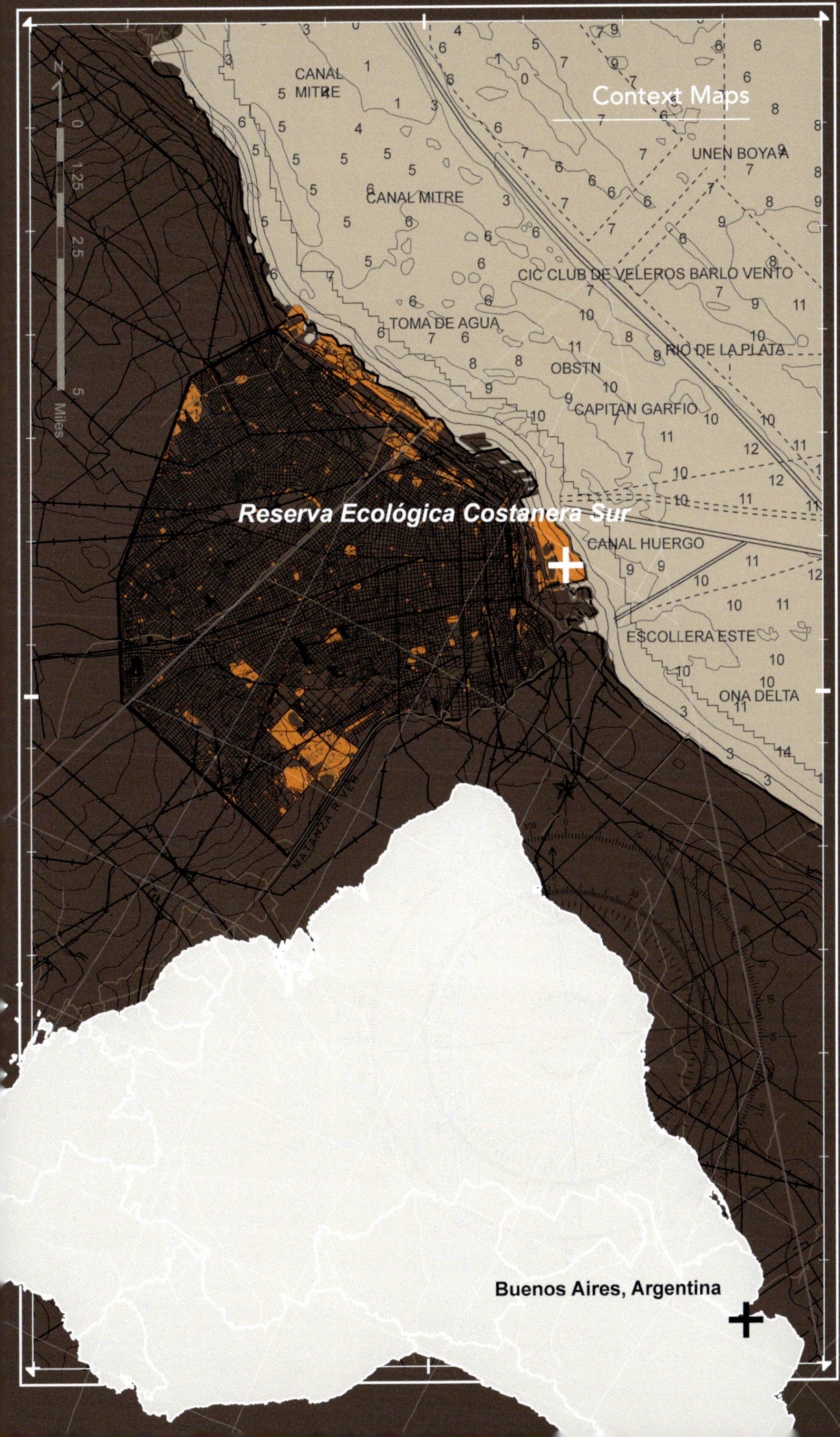
Context Maps
CANAL MITRE
CANAL MITRE
UNEN BOYA
CIC CLUB DE VELEROS BARLO VENTO
TOMA DE AGUA
RIO DE LA PLATA
OBSTN
CAPITAN GARFIO
Reserva Ecológica Costanera Sur
CANAL HUERGO
ESCOLLERA ESTE
ONA DELTA
MATANZA RIVER
0
1.25
2.5
5
Miles
Buenos Aires, Argentina

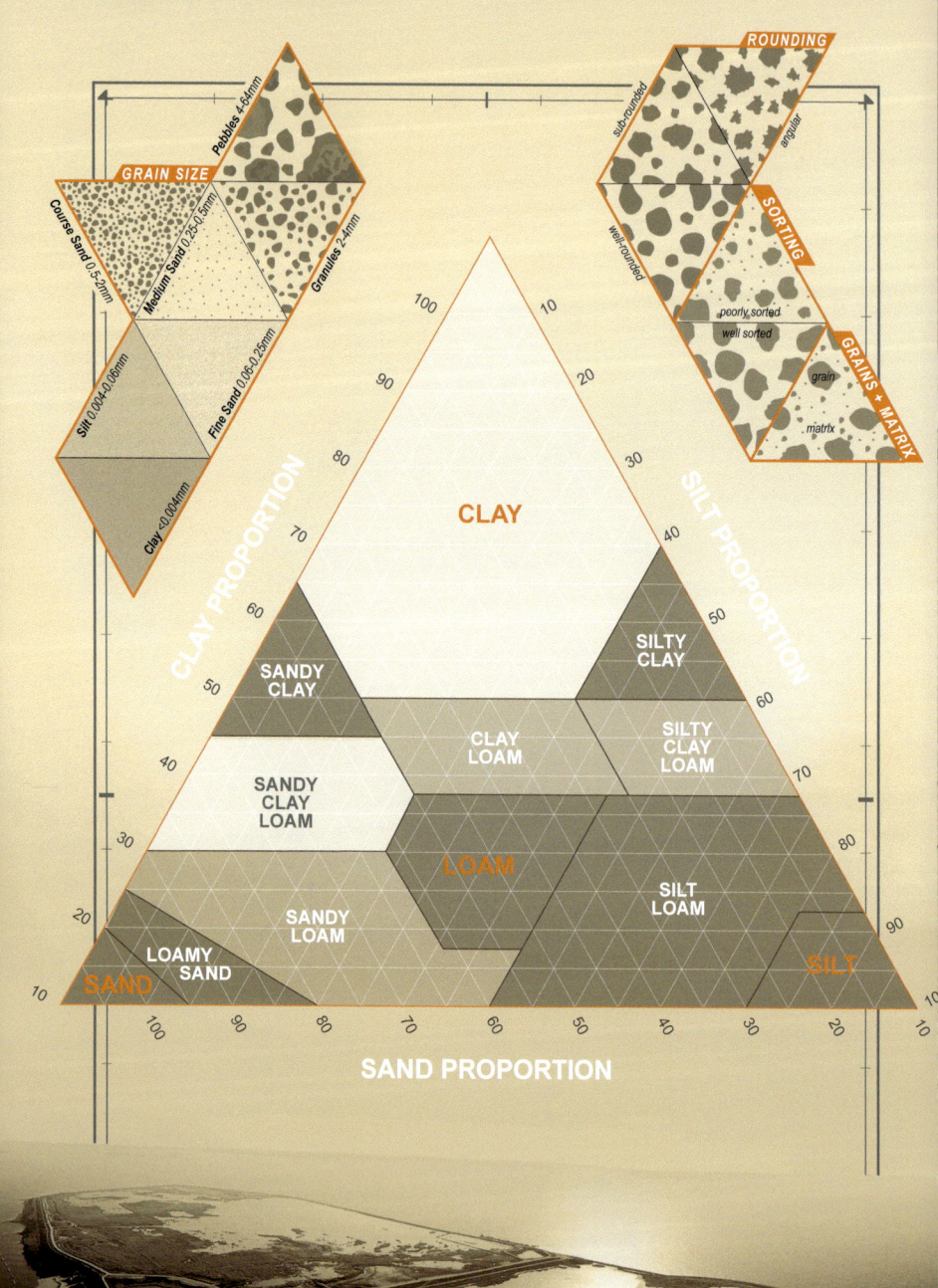

GRAIN SIZE

Pebbles 4-64mm
Course Sand 0.5-2mm
Medium Sand 0.25-0.5mm
Granules 2-4mm
Fine Sand 0.06-0.25mm
Silt 0.004-0.06mm
Clay <0.004mm

ROUNDING

sub-rounded
angular
well-rounded

SORTING

poorly sorted
well sorted

GRAINS + MATRIX

grain
matrix

CLAY PROPORTION

SILT PROPORTION

SAND PROPORTION

100 10
90 20
80 30
70 40
60 50
50 60
40 70
30 80
20 90
10 10

100 90 80 70 60 50 40 30 20 10

CLAY

SANDY CLAY

SILTY CLAY

SANDY CLAY LOAM

CLAY LOAM

SILTY CLAY LOAM

LOAM

SANDY LOAM

SILT LOAM

LOAMY SAND

SAND

SILT

Soil Composition + Quality

Public notice is hereby given, that the Mud Machine is now in operation, the Wardens of the Port of Baltimore, are ready to contract with any person who is desirous of purchasing the Mud raised from the Harbor at one dollar per Scow load (containing at least 500 cubic feet) when thrown over a wharf, and eighty cents when thrown over the Scow into the water.

By order of Saml. Young, Clerk,
Port Warden's Office
April 18, 1814[1]

With this notice the clerk of the Port of Baltimore Warden's Office asked a familiar question—will anyone take our mud? Because of the need to maintain channels for navigation all ports produce sediment. Some ports produce nice sediment; handsome sands and gravels useful for building roads, adding to concrete mixes or soil blends, or replenishing beaches. But most, like Baltimore, make mud.

Mud—the fine-grained silts and clays together with organic materials—stands alongside water itself as the most basic of materials in coastal and riparian zones across much of the world. In varying concentrations and contexts it forms the wetlands and mudflats, the beaches and dunes, the islands and riverbanks that cities like Manaus, Mumbai, Baltimore, and Buenos Aires rely on and fight against.

It is hard to recognize the mud if you find yourself in downtown Baltimore. Of course, all of the land that you see, the buildings and roads you inhabit, it's all built on mud. But we've covered it over with harder materials, finer things like granite and asphalt, pierced it with the conduits and tunnels that make a modern city. You catch glimpses of it in the old Middle Branch if you are up on the skyway of I-95 at low tide, all slow-moving sediments piling into little ridges and flats and sprouting phragmites. But to really see it you need to get out in the county, out where the old crabbing and dockworker neighborhoods are, where the salt water intrusion from sea level rise is killing off the trees and making farms back into wetlands.

HOW IS THE PORT OF BALTIMORE?

What I recommend is this: go in one of those institutional, white vans with the double swinging back doors and the vinyl seats. Once you get out of Baltimore city, you are surrounded by a kind of tableau of modern life, all horizontal big box stores and warehousing and on-ramps and grassy detention basins filled with cattails. It is hard to find food, so you bring your own. If you don't, your only hope is to find the Royal Farms gas station where you can get a good fried chicken sandwich made right there and just eat it on the curb.

The roads are choppy from land subsidence and lack of maintenance, and rail tracks cut across from time to time. These are signs of the work that is done around here, often obscured by fencing or shrubbery. Shipping, logistics, trucking, storing. Little towns, former company towns, are crossed by old rails and surrounded by brownfields and it all seems so generic and endless until you turn downhill and end up at some beautiful little cove, fringed with industrial relics and working life. It is J.B. Jackson's agrophilia laced over bumpy terrain and pushed right up next to Chesapeake Bay lands.

Baltimore is weird that way; hills and industrial plants sidled right up to mudflats and shallow bays and shipping channels. This is the serendipity of its geography and it explains a lot about the place. Like a lot of decent-sized cities of the Mid-Atlantic and Southeastern United States, Baltimore is built on the fall line; that transition between the rolling hills of the piedmont and the coastal plain where most of the rivers of any size have their last cataracts, or "falls" before becoming languid, silty beasts that stretch to the coast.

Atlantic Seaboard Fall Line

New York City NY

Trenton City NJ

Philadelphia City PA

Piedmont

Havre de Grace City MD   Wilmington City DE

Elkton MD

Baltimore MD

Washington City DC

Alexandria City VA

FALL LINE

Richmond VA

Petersburg City VA

Coastal Plain

Blue Ridge

Piedmont   Coastal Plain

FALL LINE

Crystalline basement

Slightly metamorphosed
Cambrian clastics

Clays and shales

Metamorphic gneisses and
schists

Coast line

Basemap sources: Esri, Maxar, GeoEye, Earthstar Geographics, CNES/Airbus DS, USDA, USGS, AeroGRID, IGN, and the GIS User Community

In places like Philadelphia, Richmond, Baltimore, and Raleigh this tumble downstream produces the hydraulic head that used to be necessary for milling. The natural falls and cataracts were also the upstream limit of navigation for most boats. These two facts are why industry—both manufacturing and shipping—clustered here, and left its imprint in the form of old warehouses, mills, and land use patterns. The difference is in Baltimore, there is no real coastal plain, just a hard transition right into the muddy tidal inlets of the upper Chesapeake. See, the valley of the Susquehanna River was drowned at the end of the last Ice Age, forming the Chesapeake Bay and bringing the littoral edge right up to the fall line. So Baltimore is the westernmost port on the East Coast. This little fact helps explain why the Baltimore and Ohio railroad came here, and still ends up mattering in the spreadsheets of supply chain dealers deciding how to truck sugar to Memphis and cars to Columbus.

THE PORT OF BALTIMORE

The port is known for the heaps of coal and sugar and grain it moves and stores in great, beautiful piles, for the shipping containers stacked in yards, and the multimodal centers and switchyards scattered around the harbor. But its real concern is mud. While the port does help build terminals, they don't operate them. Their main job is to work with the US Army Corps to make sure boats with 50′ deep drafts can make it to the wharfs in the harbor all the way from the Atlantic Ocean. It is a transportation organization after all, warehoused like most ports within the state-level Department of Transportation.

Being the westernmost port does come with a cost, and for Baltimore that cost is the maintenance of one hundred and seventy-five miles of navigation channel from the Virginia Capes to the Port of Baltimore. This concern requires the removal of about 5 million cubic yards of sediment each year that has fallen into the navigational channel. The removal happens through underwater excavation called dredging. It can be thought of as a kind of spatial formatting—a big kind—where we remake the variable and differentiating underworld to fit a generic dimension or tool. In this case, it's the 50′ deep draft of modern maritime shipping vessels.

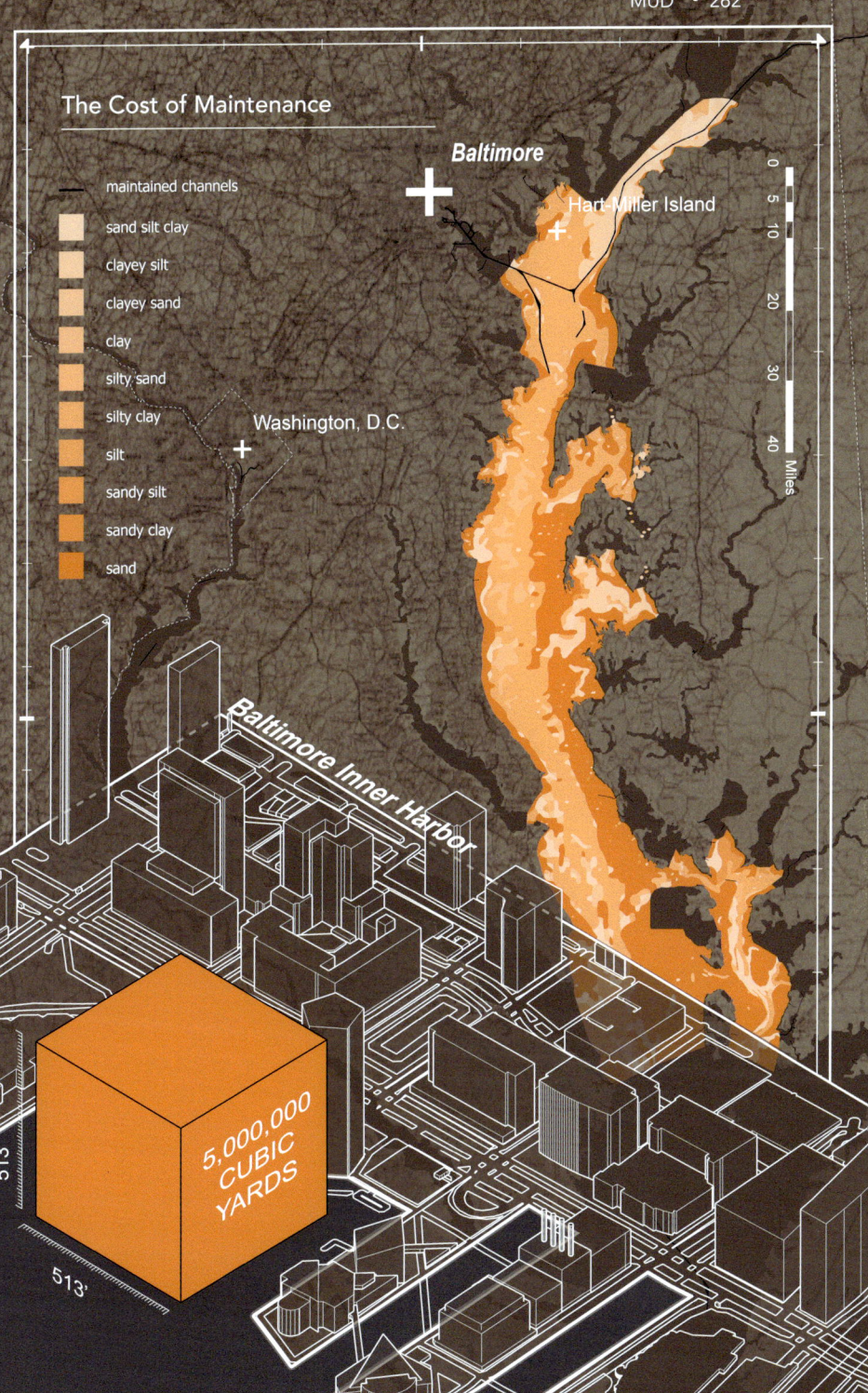

The Cost of Maintenance

— maintained channels

sand silt clay

clayey silt

clayey sand

clay

silty sand

silty clay

silt

sandy silt

sandy clay

sand

Baltimore

Hart-Miller Island

Washington, D.C.

0
5
10

20

30

40

Miles

Baltimore Inner Harbor

513'

513'

5,000,000 CUBIC YARDS

Most of the sediment dredged in Baltimore—about 85%—is mud. Made of fine-grained clasts instead of nice, hard sands and gravels. This matters, weirdly. You see, everyone wants sand and gravel. It is great for construction as fill and aggregate. It can be nice for beaches. A single grain of sand is about two hundred times larger than clay. It is easy to clean and looks nice. Because it is heavier it settles out of water quickly, leaving it nice and clear. Mud, on the other hand, is complicated. It makes murky water, its particles being so fine as to stay suspended for days even in a still tank. The flat-disk shape and positive charge of its primary component, clay, means the particles bond with all sorts of things, including industrial byproducts such as agricultural fertilizers which can cause algal blooms when concentrated in water bodies. These qualities also help clay retain moisture and nutrients, making it a critical component of healthy soils and wetland substrates.

Taken together in the context of a manufacturing center and port city, clay sediments act as a sink, a kind of residue or cultural byproduct of the industrial ingenuity and labor, lack of environmental knowledge, and the weathering and land use processes at work on the nearby terrain. And while clay-sediments are much cleaner in the Baltimore harbor nearly fifty years into the Clean Water Act, mud still connotes contamination and is treated as unsuitable for reuse; a problem to solve.

WHY SO MUCH?

The Virginia coast of the Chesapeake is often characterized by beautiful, sandy beaches, But there is a change somewhere just south of Annapolis. In the northern portion of the Bay the tendency toward long, broad beaches gives way to an edge of intricate inlets and brackish marshes. The bottom is soft, perfect for eelgrass and crabs. Sand becomes finer, and beaches are confined to small points or thin lines in a sea of mud. There are three big reasons that Baltimore is so muddy.

Because clay is small and flat and packs tightly, it bonds well with itself and so undergoes a process known as flocculation, or *clumping together*. This process of clays clumping together, becoming heavier and precipitating out of the water column is especially active in areas where salt water

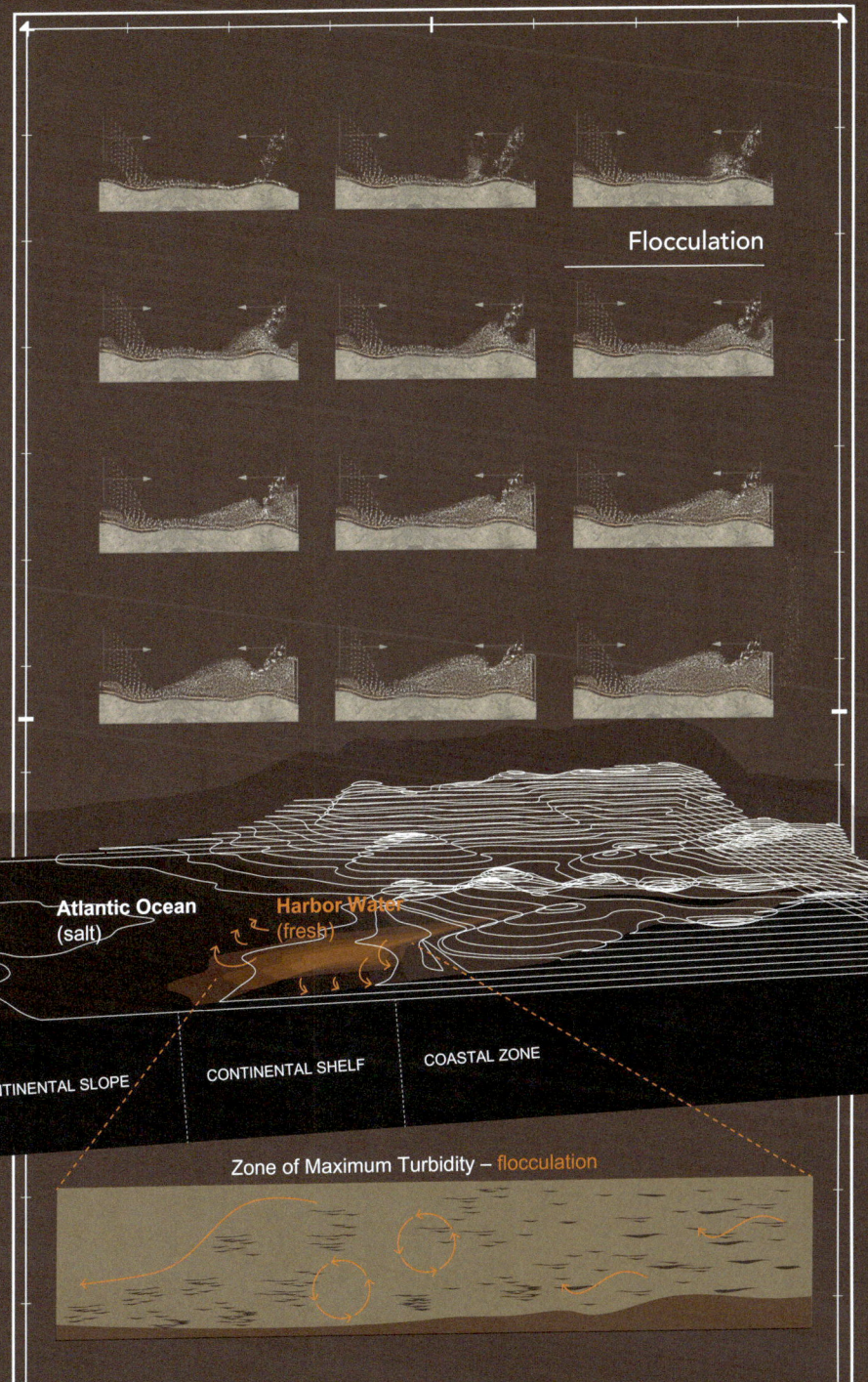

Flocculation

Atlantic Ocean
(salt)

Harbor Water
(fresh)

CONTINENTAL SLOPE   CONTINENTAL SHELF   COASTAL ZONE

Zone of Maximum Turbidity – flocculation

mixes with river inputs. This is in part because the salt water, being heavier, acts like a liquid bulldozer blade. It slides under the freshwater and stirs up the fine grains. In estuaries this area is known as the turbidity maximum. It shifts around with tides and storms but as the Susquehanna is the big tributary of the Bay, it is usually up around Baltimore and has been for a few thousand years.

A second geomorphological process present in almost all coastal landscapes is known as winnowing. Winnowing is originally an agricultural term meaning something like *to get rid of the undesirable part*. Along the coast it refers to the process by which daily tides and wind-driven waves pick up sediment in the near shore environment, which is typically a mixture of sand and gravel, silts, and fine clays. The heavier stuff tumbles along with the stronger, shore-bound waves and moves a bit closer to the shore, eventually becoming part of the beach-and-dune system. The finer stuff, however, is winnowed out. That is, it stays suspended in the water column, pulling back out to deeper water with undertow and low tide, where the velocity is slower and it can precipitate out. This process gives us beautiful beaches in places like North Carolina, Virginia, Florida, or Texas in the near term, and over eons it produces materials like bluestone in deeper water where the finer grains stack onto each other and condense under pressure.

The Virginia part of the Chesapeake has this character, with the winnowing producing elegant beaches punctuated by headlands and river mouths. However, in the northern reaches of the Chesapeake Bay, the winnowing is muted by the decreased fetch and smaller tidal prism and amplitude. The clay tends to settle out in the deepest water, which in the Chesapeake is along the former thalweg of the Susquehanna, now the shipping channel. Velocities below the surface are often slower in this deeper water, meaning that the clay and silt can settle out into a nice muddy layer at the bottom. This is the stuff that has to get excavated through dredging to the tune of 5 million cubic yards a year.[2]

WHERE DOES IT GO?

A thing exists in the midst of its wastes…

The aesthetic orientation man gives to the whole of his world represents a return enjoyment and to the elemental on a higher plane. The world of things calls for art…

All art is plastic.

-Emmanuel Levinas, *Totality and Infinity*[3]

There are modern monuments all around the Baltimore Harbor. For the last 50 years a system of landforms called *dredged material containment facilities* (DMCFs) have been created to house the mud dredged from the harbor. Greater in scale than the famous system of forts that protected the harbor in the early years of the Republic, these monoliths are scattered around the outer harbor, designed to hold tens of millions of cubic yards of mud in perpetuity. Created through a process of dredging, inflowing, dewatering, testing, trenching, and finally discharging the clear water after it is chemically verified as clean, the mud is compacted into high trapezoids behind protective dikes that promise future port terminals or land haphazardly reclaimed as upland habitat. These landscapes are monuments hiding in plain sight. Yes, they are a technical solution to the regulatory problems posed by the need to dredge and warehouse mud. But the DMCFs are also cultural landscapes whose simple geometric forms are evidence of ingenuity, the evolution of environmental protection, the necessity of maritime shipping, the physical fact of mud in a port town, and the cultural desire to make it go away.

The white institutional van will bring you to meet the mud, to a parking lot off of Riverview Road. It is on the Back River, a tidal creek out in the county whose lands are a patchwork of farms and coastal plain forest and houses on the coast. There's a little boat launch there, and this is where the boat leaves from early in the morning to go to Hart-Miller Island. It is operated by Maryland Environmental Services and brings the scientists, equipment operators, project managers, and any guests out to the island each day. After a 15-minute boat ride across the Chesapeake you will be convinced this is the best way to start a day. About then it is time to get off at the dock of Hart-Miller Island.

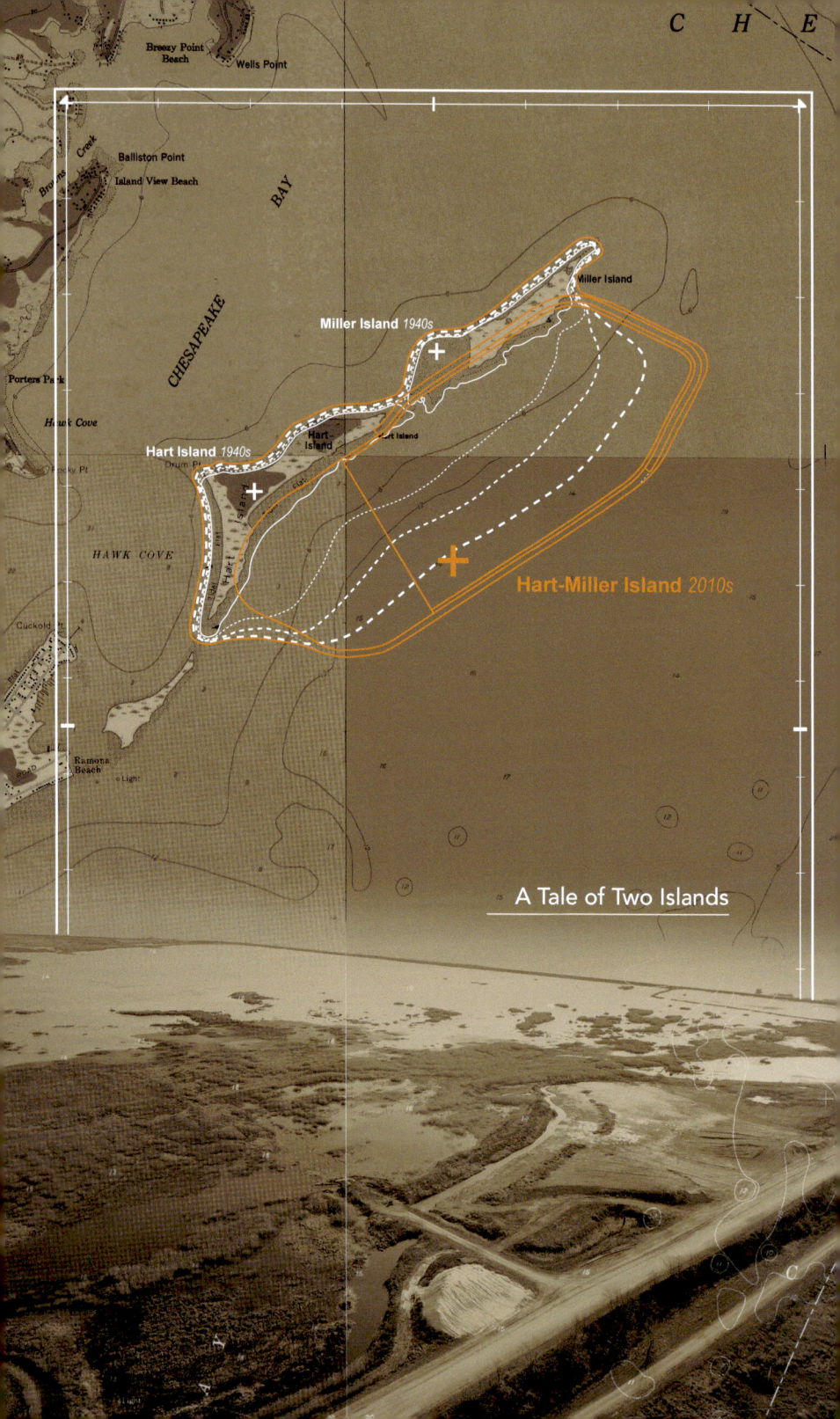

C H E

Breezy Point
Beach
Wells Point

Balliston Point
Island View Beach

CHESAPEAKE
BAY

Browns Creek

Porters Park

Hawk Cove

Rocky Pt

Miller Island *1940s*

Miller Island

Hart-
Island

Hart Island

Hart Island *1940s*

Drum Pt

HAWK COVE

Hart-Miller Island *2010s*

Cuckold Pt

Ramona
Beach
Light

A Tale of Two Islands

In the past Hart and Miller Islands were separate, forming a chain with the still existing Pleasure Island just to the south and offering boating, birding, camping, and crabbing for at least a few thousand years. In the office trailer I see a map attesting to this duration, one that indicates some probable archaeological sites on Pleasure Island that were never excavated. In the 1960s Hart and Miller Island began to recede below the tides as changing land use practices and intensifying storms battered the topography and the sand supply dwindled, trapped behind the Conowingo Dam on the Susquehanna. The remnants were repurposed as foundations for a monolith of mud through the construction of a tremendous perimeter dike.

Opened to receive sediment in 1984 after 14 years of construction and lawsuits, Hart-Miller Island is now a low trapezoid of mud rising out of the Bay.[4] Forty-four feet high and 1,100 acres in area, locust and poplars growing out of the armored sides covered with three-foot diameter stone and rip rap. What a wild idea! It wasn't a bad one. The Clean Water Act and state laws in the 1960s prohibited the discharge of pollution in state and federal waters. Because ports tended to cluster dirty manufacturing industries, and the clays would bind to or mix with many of their byproducts such as petrochemical compounds and heavy metals, the mud was real dirty. Additionally, the agricultural runoff that brought phosphorous and clay to the Bay was producing algal blooms that were impairing the bountiful fish and mollusk populations. Thus, Hart-Miller Island. These "dredge material containment facilities" were honest, if not smart. They would contain the stuff, looking and acting almost like a modern-day landfill. About one hundred million cubic yards now sit here. If you stacked it vertically on a football field it would rise almost eleven miles. For 30 years the port put its mud here until the facility closed in 2005.

On top of Hart-Miller the island is as flat as a pancake with no perceptible variation in the topography, though you wouldn't know it at first glance because the phragmites is eight feet tall and thicker than a corn field. The western three hundred acres is a bit different. Lower, with a pool and some wetland areas. It is the "South Cell" and was filled and remediated first. At the far eastern end of this inelegant lozenge-shaped island there is a pool, evidence of a bit of grade difference after all. This pool grows and shrinks with the seasons and is the bane of the operators

CASPIAN TERN
Hydroprogne caspia

SPOTTED SANDPIPER
Actitis macularius

SWAMP S
Melospiza

Ecologies of "The Pudding"

WILLOW FLYCATCHER
Empidonax traillii

migration

breeding

molting

JAN

FEB

MAR

APR

MAY

JUN

JUL

AUG

SEP

OCT

NOV

DEC

10' ORIGINAL DIKE HEIGHT (1984)

here. Originally they had intended to simply discharge the water and then call it a state park, allowing the port to wash their hands of it and move on. But for that mud. You see, the marine sediments, once they sit up in the air like this, the salts oxidize and produce a highly acidic residue. The soil pH ranges from 3.0 to 3.8 in most places, and any water that runs over it is far too acidic to discharge into the Chesapeake Bay under state law. It is almost as if the mud is protesting: "you shouldn't have made me into a mountain. I shouldn't be like this."

Take a walk on it. A fine, powdery white. Wind will pick it up and blow it in your eyes, your nose, and fill your pores. This is the stuff that has oxidized. Use a tubular soil sampler and you'll see various stages of oxidation creating little pyrites and sulfurous-smelling bits. Jump up and down, or better yet, drive some heavy equipment over it, and you will feel the earth shake. This earthquaking is you floating on a forty foot high pillow of what the foremen and equipment operators affectionately refer to as "the pudding." The dredged sediment was put in wet, basically, and except for the top few feet, has never been able to really dry because the sides were built for containment, and the surface water cannot reliably be discharged.

Take advantage of the wetness. Cut down into it. It is a complex black. Absent oxygen after the top six or eight inches, it is a smooth, stifling gel. You can walk on it, pile it. It will hold its shape. Steep slopes and crisp forms. It is almost modeling clay. It is almost archaeology. Full of artifacts, an artifact itself. All the labor of the port, the factories, the floods, the crabbing, the fishing. Some bit of those processes made it here, leaving signatures in the sediment.

LANDSCAPES OF LOCAL POLITICS

Hart-Miller Island has a specific history, but it isn't really unique. These islands are everywhere. Poplar Island in the middle of the Chesapeake Bay is a $1.4 billion infrastructural monument to mud storage and habitat creation. Ferry Slip Island and South Pelican Island on the Cape Fear, south of Wilmington, are old dredge islands that have morphed into great bird stopover habitat. Hog, Barnwell, Jones, Cockspur Islands in the Savannah River, on the South Carolina side.

You can trace this down most of the coast, all throughout the Gulf, and keep on following it all the way down to Argentina, where a particularly extreme and hopeful case can be found at the Reserva Ecologica Costanera Sur in Buenos Aires.

That city sits on the edge of the Rio de la Plata, a broad estuary. The edge between water and city was formerly shallow mudflats and it was here in the late 1970s that the military dictatorship then in charge of the country decided to make an island with a new administrative headquarters. It made a lot of sense, really, if you're a cruel, murderous, and therefore unpopular dictatorship—the access to the nerve centers of the historical downtown, and the access control afforded by an island, all in one act of landscape-making.[5]

Well, now the island is the *Reserva Ecologica*. This agglomeration of demolition debris from urban highway projects, dredged material from the port, and alluvium from the Rio de la Plata aggregated through processes of social and geological sedimentation, eventually became a legally protected, public terrain dedicated to recreation and environmental conservation. Not unlike Baltimore's Hart-Miller, but right downtown. It is the result of government directives redirecting the material excesses of a massive industrial capital, intersecting the autonomous tendencies of regional plant biomes and the great Paraná River during the rainy season, all being shaped by the desires of local emergent publics and mediating practices of professional landscape designers, gardeners, and engineers over time.

Most of these proposals called for versions of historical types—park spaces, towers, promenades, super blocks--developed for solid terrain according to prevailing aesthetic and political-economic ideas of eighteenth and nineteenth century Europe and North America.[6] This tendency to overlook the *ribera* and its soft strand plains, unique vegetation, productive benthic habitats, as well as its attendant cultural ingenuity masked as inefficiency and excess is a through line in the history of the *ribera*, from the dictatorship to Le Corbusier's 1938 vision for Buenos Aires to the Malaspina Expedition of 1789.[7] For Malaspina, the city of Buenos Aires was a disappointment, a sentiment later echoed by Darwin when he remarked that "the Plata looks like a noble estuary on a map, but in truth it is a poor affair. A wide expanse of muddy water has neither grandeur nor beauty."

In truth it wasn't the landscape that was lacking, but their aesthetic values.[8] Of course, naturalists at the time were male, European dilettantes with a strong commitment to property and picturesque aesthetics, and this didn't really square with mudflats and marshes. These landscapes offered no distinctive vantage or scenic juxtaposition from which to dramatize a panorama or elevation, only a relentless horizontality of land and water and the subtle shifts of mud, plants, and human uses.

But today, by a coincidence of history and place and some well-timed maneuvers by locals and experts, these things are valued. In 1983 the fall of the dictatorship coincided with a major flood, and with the work of activists, scientists, neighbors, and professionals in the last 35 years the *Reserva Ecologica* has resulted in a kind of hopeful, shabby example of what is possible in these muddy places.[9]

Maybe this potential exists because they are outside of our recent cultural boundaries. Like Malaspina, we've discounted them, or tried to get away from them, or at least found them confusing. When we say we want a park, we know what we mean, and why. It will be picturesque, somehow. It will have some programming, probably a bit of active recreation. Just what and where, how much; that is probably worth a few community meetings, a bit of consultant fee to work out. But we're not reinventing anything here. The same holds for the plaza, the playground, the courtyard. We know what the project is about, or at least those doing the deciding do. But it isn't so clear in these muddy places. Sometimes the 'what' isn't clear at all, much less the why or the how.

The processes of landscape formation that have formed the *Reserva* and its publics since that fateful year of 1983 are instructive, though.[10] The first was started in 1992 when landscape designers Robin Moore and Nilda Cosco were brought in by local NGOs to hold a public meeting, study the *Reserva*, and make recommendations. The resulting report provided a basis for a set of incisions—small design interventions such as railings and paths and overlooks that would ameliorate the effects of the quarter-million visitors per year on the unique and evolving ecosystems.

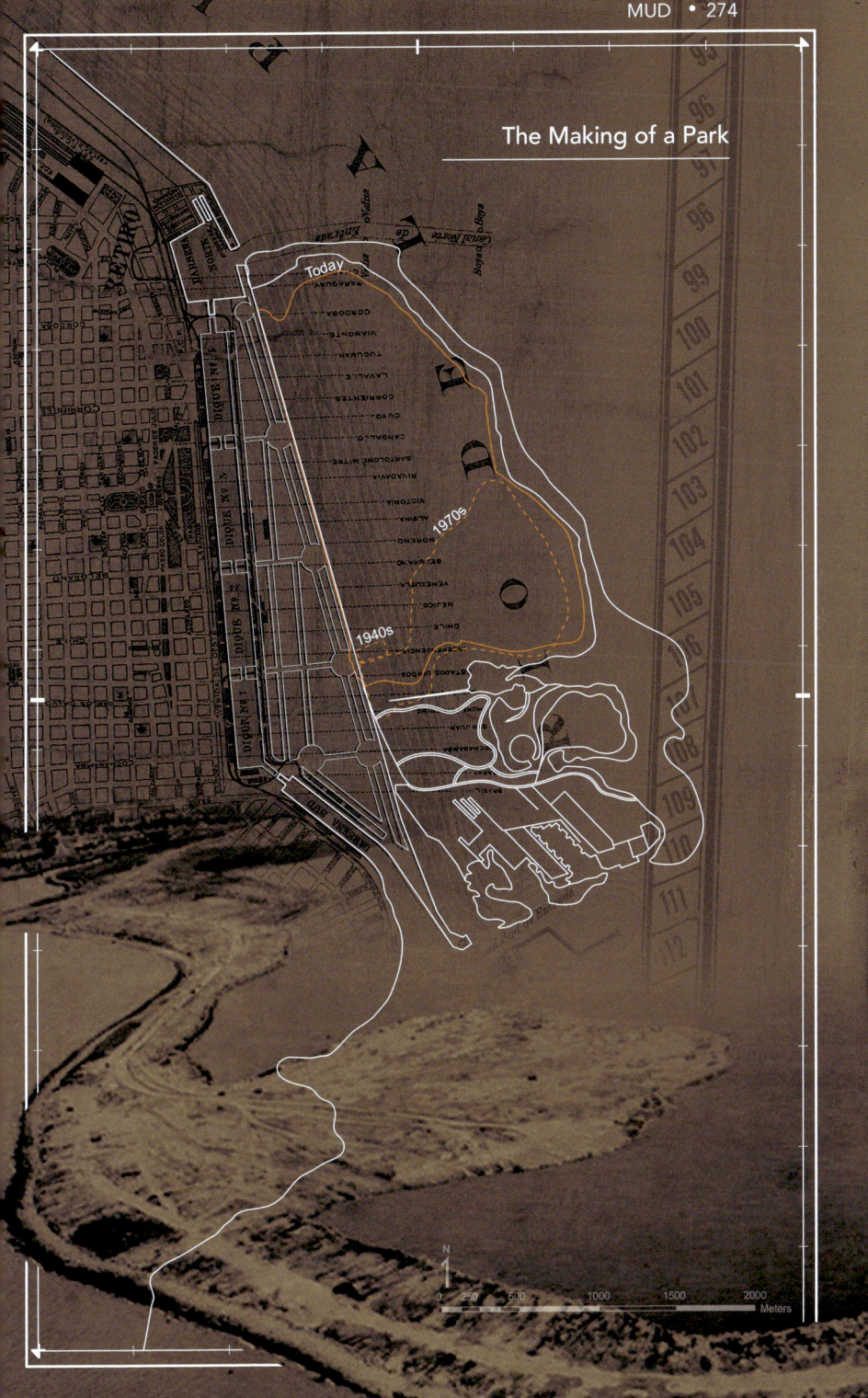

The Making of a Park

Today

1970s

1940s

This report was prepared for the Argentina Wildlife Federation (Fundacion Vida Silvestre Argentina). Though seemingly simple, it was a radical stance given the dominant conversation at the time and its focus on architectural tropes and picturesque aesthetics. The Moore and Cosco plan eschewed these moves in favor of the plant communities emerging from the mud and existing topographic high zones created by the rubble. It preserved the experiential history and gave shape to the desires of the new public forming in support of the *Reserva*, allowing it to effectively resist the municipal government's efforts to develop and sell the new lands for commercial and residential real estate. By leaning on ecological arguments related to species diversity and ecotone uniqueness, the *Reserva*'s constituency was able to maintain the characteristic integrity of the place while confining development pressure to the old adjacent former port.

The mud had helped conjure a new public of users, constituents, and defenders that was given a voice through the proposal, binding together environmental concerns about biodiversity together with an aesthetics of mud. This aesthetics doesn't valorize picturesque tropes and clean lines of architectural urbanism. Rather, it revels in the subtle, disquieting adjacencies of rubble and pulpy riparian vegetation, in sediment that flows across administrative sectors and jurisdictional boundaries, forming new alliances between amphibians, minerals, plants, and human communities.

Shortly after the Moore and Cosco proposal a new NGO organized—Fundacion Ciudad. One of their first acts was to launch a seven-year long "City and River" event series meant to incubate and develop ideas related to the waterfront in Buenos Aires. These events were based around the simple but then-radical idea that the waterfront should be public and accessible. Referred to by the rather anodyne euphemism "adaptive co-management," this process was a riotous and uneven one, predictably marked by loud-talkers, political posturing, and ineffectual bureaucratic grinding as much as ingenuity and collaboration. Nonetheless, by 2004 this process resulted in shared goals and a plan to remediate eighteen hectares of brownfield within the *Reserva*. The area had been the site of the Covimet aluminum factory whose construction and operation had left the land compacted and rigid.

The new proposed project would demolish the old building site and redirect forty thousand cubic meters of excavated material from new subway works in the city to cap the brownfield and be shaped to create landforms. These earthworks were intended as a kind of exaggerated, localized reproduction of the original bluffs where Buenos Aires was founded and in a similar way they would offer prospect above an expansive, muddy landscape. The landforms were subdivided many times, shaped with 40% slopes and leveled tops 14 meters above the pre-existing grade, with their main orientation echoing the throughput logic of the industrial site. Maintenance practices thus far have provided for the tops of the landforms and the spaces between to be mowed, and 1,500 trees were used to vegetate the slopes. This move created a system of overlooks perched above the original dikes with a dry-mesic forest environment that offers a sense of enclosure and promises good mountain biking.

The strategy here is in keeping with the nature of the *Reserva* and is indicative of the aggradation process that hatched it. It is crude and large, repetitive and subtly differentiated. The collaborative project to repurpose the waste material linked the desires of the public works department and construction contractors producing an excess of excavated material together with that of a local public and risk factors caused by remnants of an old aluminum brownfield. This action of industrial aggradation through dumping and shaping combined with planting and maintenance is leading to an increase in habitability while recontextualizing traces of the old factory site.

ART IN THE AGE OF ISLANDS

Places like the Reserva Ecologica or Hart-Miller or Poplar Island are uncanny, strange yet familiar, conjuring trepidation and revulsion through allure. They have island-like characteristics, what with being in the middle of the water, attracting birds and plants and fishes and amphibians. I am always tempted to see them through a critical lens. "You are not an island so much as a monument to our hubris, enabled by techno-scientific knowledge and fossil-fueled power. You were made for a purpose, to solve the problem of where to put all the mud." And that is true. It would be nice to leave it there, self-satisfied, but for that mud. Like most solutions outside of the abstracted confines of math books, the mud islands have spawned new problems.[11]

The mud persists as this kind of obstinate material, one that seems to insist on the fact that magic does indeed exist in the world. How do you predict the migration of marshes, the protection needed from the next hurricane? How do you describe and model the movement of mud in a nearshore system? Why do these huge monument-islands even exist? No one knows the answers to these questions, really, though it would be valuable if we did. And we've tried. But the mud reminds us that the world is far weirder than we usually give it credit for, at least in our professional roles. The mud remains autonomous, irreducible, existing beyond our policies and tools and best intentions. Heraclitus was right—nature loves to hide.

The age of islands is probably ending. We won't build many more of the huge containment facilities in the water in the future. Hart-Miller and her cousins are going out of style. They will stand as monuments to a time when we valued our coastal landscapes, but didn't realize how important and valuable mud was, and did wild things to them both. Now the islands are expensive, and the sediment is cleaner.

And yet there are things we might still do with the problem, even if we can't solve it. As the islands fill up, we can figure out smarter ways to shape their boundaries. We need good cheap ways to create undulating topography on the top and healthy soil with the sediment on site, during the placement process. So we don't have to truck it in from somewhere else, strip-mining another healthy landscape to restore a scar. And the edges where these island-things meet the water could be better. Yes, they could provide habitat and such, but more intriguing is how to make them *habitable*. We don't absolve our sins by making everything for the terns, so we might as well make a few of them the monuments they are. Have you ever walked on the armor stoned edge of a containment facility in the summer, all jagged boulders and goldenrod and thistle? What a riot. Moving, even. Also, quite dangerous. Designers could make it a bit less so, in just the right way that honors the strangeness of these ziggurats.

Perhaps more important, we need that sediment for other things, and are figuring out how to use it. Bay-bottom crab habitat and oyster restoration, wetland creation, carbon sinks, wave attenuation, shoreline risk reduction, storm surge protection. These practices are critical now, and have to be scaled up. The mud will have a role in it. In our modern way we accidentally created monuments in one place as a neatly partitioned and knowable solution to a problem in another. In the future, the Chesapeake Bay and estuaries and river mouths up and down the coasts will be monuments themselves. Far subtler and more vast than any of these islands, a place where habitability multiplies even in the face of risk and loss and overwhelming, uncanny beauty. If we have taken mud for granted until now, it is only because it is so fundamental, so basic, that we cannot imagine our favorite places without it.

ENDNOTES

1   *American Commercial and Daily Advertiser*, April 19, 1814, Maryland State Archives SC3392.
2   2019 Dredged Material Management Program, "Annual Report to the DMMP Executive Committee." November 8, 2019.
3   Emmanuel Levinas, *Totality and Infinity: an Essay on Exteriority*, trans. Alphonso Lingis (Dordrecht: Kluwer Academic Publishers, 1991), 139-140.
4   Shin, Raymond K., "A Costly Standoff: The Hart and Miller Islands Controversy" (2007). *Legal History* Publications, 38. Accessed 17 October 2020. https://digitalcommons.law.umaryland.edu/mlh_pubs/10.
5   Brian Davis and Erin Putalik, "The Reserva Ecologica: Three Streams of Material Excess in Buenos Aires," in *New Constellations, New Ecologies, Annual Meeting Proceedings*, eds. Edward Mitchell and Ila Berman, (101st ACSA Annual Meeting Proceedings, 2013), 29-36.
6   Brian Davis, "The Asymmetry of Landscape: Aesthetics, Agency, and Material Reuse in the Buenos Aires Reserva Ecologica," *Journal of Landscape Architecture,* Vol. 13, Issue 3 (2018): 83.
7   Brian Bockelman, "Along the Waterfront: Alejandro Malaspina, Fernando Brambila, and the Invention of the Buenos Aires Cityscape, 1789-1809," *Journal of Latin American Geography*, Volume 11, Special Issue, (2012): 61-88.
8   Brian Davis, "Muddy Materialisms" in *Viscosity: Mobilizing Materialities*, eds. Emily Eliza Scott, Karen Lutsky, Ozayr Saloojee (Minneapolis: University of Minnesota School of Architecture, 2019): 35-46.
9   Davis and Putalik, "The Reserva Ecologica," 29-36.
10   Brian Davis, "The Asymmetry of Landscape," 84-85.
11   Rob Holmes, "The Problem with Solutions," *Places Journal,* July 2020. Accessed 17 October 2020. https://doi.org/10.22269/200714.

BIBLIOGRAPHY

2019 Dredged Material Management Program, "Annual Report to the DMMP Executive Committee." November 8, 2019.

American Commercial and Daily Advertiser, April 19, 1814, Maryland State Archives SC3392.

Bockelman, Brian. "Along the Waterfront: Alejandro Malaspina, Fernando Brambila, and the Invention of the Buenos Aires Cityscape, 1789–1809," *Journal of Latin American Geography*, Volume 11, Special Issue (2012): 61–88.

Davis, Brian. "Muddy Materialisms," in *Viscosity: Mobilizing Materialities*, eds. Emily Eliza Scott, Karen Lutsky, Ozayr Saloojee, Minneapolis: University of Minnesota School of Architecture, 2019.

Davis, Brian. "The Asymmetry of Landscape: Aesthetics, Agency, and Material Reuse in the Buenos Aires Reserva Ecologica," *Journal of Landscape Architecture*, Vol. 13, Issue 3 (2018): 78–89.

Davis, Brian and Putalik, Erin. "The Reserva Ecologica: Three Streams of Material Excess in Buenos Aires," in *New Constellations, New Ecologies, Annual Meeting Proceedings*, eds. Edward Mitchell and Ila Berman, 101st ACSA Annual Meeting Proceedings, 2013.

Holmes, Rob. "The Problem with Solutions," *Places Journal*, July 2020. Accessed 17 October 2020. https://doi.org/10.22269/200714.

Levinas, Emmanuel. *Totality and Infinity: An Essay on Exteriority*, trans. Alphonso Lingis. Dordrecht: Kluwer Academic Publishers, 1991.

Shin, Raymond K. "A Costly Standoff: The Hart and Miller Islands Controversy" (2007). *Legal History* Publications. Accessed 17 October 2020.

IMAGE CITATIONS + CREDITS

Listed by Page Number

253   Photo: Matthew Seibert

255   US Department of Commerce, National Oceanic and Atmospheric Administration, National Ocean Service, Coast Survey. " Chesapeake Bay Sandy Point to Susquehanna River." Scale 1:80,000 [map]. 2020. "Navionics ChartViewer." Accessed October 27, 2020. https://webapp.navionics.com/?lang=en#boating@10&key=lggrE%7CbdcJ.

       Research Assistant: Tian Wang

257   Photo: Matthew Seibert. "Guide to Texture by Feel | NRCS Soils." Accessed October 31, 2020. https://www.nrcs.usda.gov/wps/portal/nrcs/detail/soils/edu/?cid=nrcs142p2_054311.

       Research Assistants: Leah Kahler, Tian Wang

260   Basemap sources: Esri, Maxar, GeoEye, Earthstar Geographics, CNES/Airbus DS, USDA, USGS, AeroGRID, IGN, and the GIS User Community

       Research Assistant: Tian Wang

262   Axon: © OpenStreetMap contributors, https://www.openstreetmap.org/copyright; "Data Sets of the Surficial Sediment Distribution of Chesapeake Bay, Maryland." Accessed October 27, 2020. http://www.mgs.md.gov/coastal_geology/baysedata.html

       Research Assistants: Leah Kahler, Tian Wang

264   Erin Davis. "Flocculation & Sedimentation." Education, August 30, 2015. https://www.slideshare.net/ERINDAVIS4/flocculation-sedimentation.

       Research Assistants: Leah Kahler, Tian Wang

267   Photo: Matthew Seibert. US Geological Survey. "Middle River, MD." Scale 1:24,000 [map]. 1951. US Geological Survey. "Gunpowder Neck, MD." Scale 1:24,000 [map]. 1949. US Geological Survey. "Sparrows Point, MD." Scale 1:24,000 [map]. 1969. US Geological Survey. "Swan Point, MD." Scale 1:24,000 [map]. 1953.

       Research Assistant: Tian Wang

269   Photo: Matthew Seibert. "Birds of the World - Comprehensive Life Histories for All Bird Species and Families." Accessed October 31, 2020. https://birdsoftheworld.org/bow/home.

Research Assistants: Leah Kahler, Tian Wang

274   Ernesto Escalante. "Nuevo plano de la ciudad de Buenos Aires." Scale 1:32,000 [map]. 1912.

Research Assistant: Tian Wang

META

Material as Physical History
of a Relationship

by Elizabeth Hénaff

BOLITE

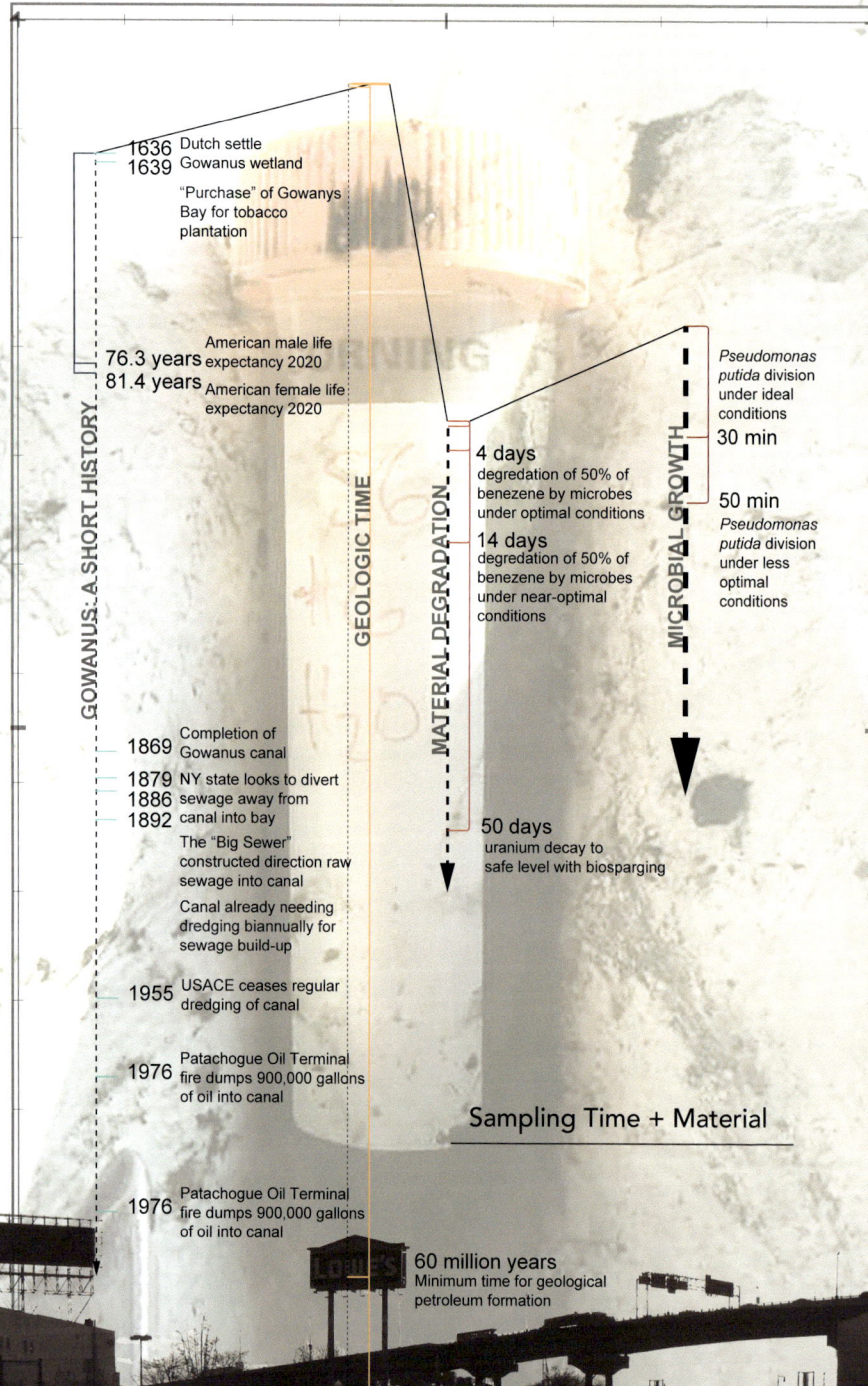

GOWANUS: A SHORT HISTORY

1636 Dutch settle
1639 Gowanus wetland

"Purchase" of Gowanys
Bay for tobacco
plantation

76.3 years American male life
expectancy 2020
81.4 years American female life
expectancy 2020

GEOLOGIC TIME

MATERIAL DEGRADATION

4 days
degredation of 50% of
benezene by microbes
under optimal conditions

14 days
degredation of 50% of
benezene by microbes
under near-optimal
conditions

MICROBIAL GROWTH

Pseudomonas
putida division
under ideal
conditions
30 min

50 min
Pseudomonas
putida division
under less
optimal
conditions

1869 Completion of
Gowanus canal
1879 NY state looks to divert
1886 sewage away from
1892 canal into bay

The "Big Sewer"
constructed direction raw
sewage into canal

Canal already needing
dredging biannually for
sewage build-up

1955 USACE ceases regular
dredging of canal

1976 Patachogue Oil Terminal
fire dumps 900,000 gallons
of oil into canal

50 days
uranium decay to
safe level with biosparging

Sampling Time + Material

1976 Patachogue Oil Terminal
fire dumps 900,000 gallons
of oil into canal

60 million years
Minimum time for geological
petroleum formation

4.74 billion years
half-life of Uranium-238

nce a verdant estuary home to the Lenni Lenape people, the Gowanus Canal's soft marshy edges are now ten- to forty-foot sheer walls of corrugated steel, creating the hard edges necessary for more than a century of industrial exchange between barge and Brooklyn's factories. Emblematic of many such sites across the country, the industries that have now moved elsewhere have left a legacy in the form of toxic waste accumulated in the sediment of the canal. Gowanus has kept those walls of steel and concrete, but now high-end condos, cocktail bars, and at least one shuffleboard club have appropriated the industrial shells of former industry like an army of opportunistic hermit crabs. This new look is as new as its Superfund designation, but the communities and families that have lived in the area for decades have had no choice but to learn to live with this toxic legacy. The slow and dedicated work continues in the form of ongoing advocacy as the neighborhood rapidly changes. A Whole Foods, boasting the city's largest rooftop farm, and crowded art galleries are gaining ground on this terrain of complex histories these days.

It's far too easy to avoid the history of toxicity and extraction when gossiping over a boulevardier, planning a new startup

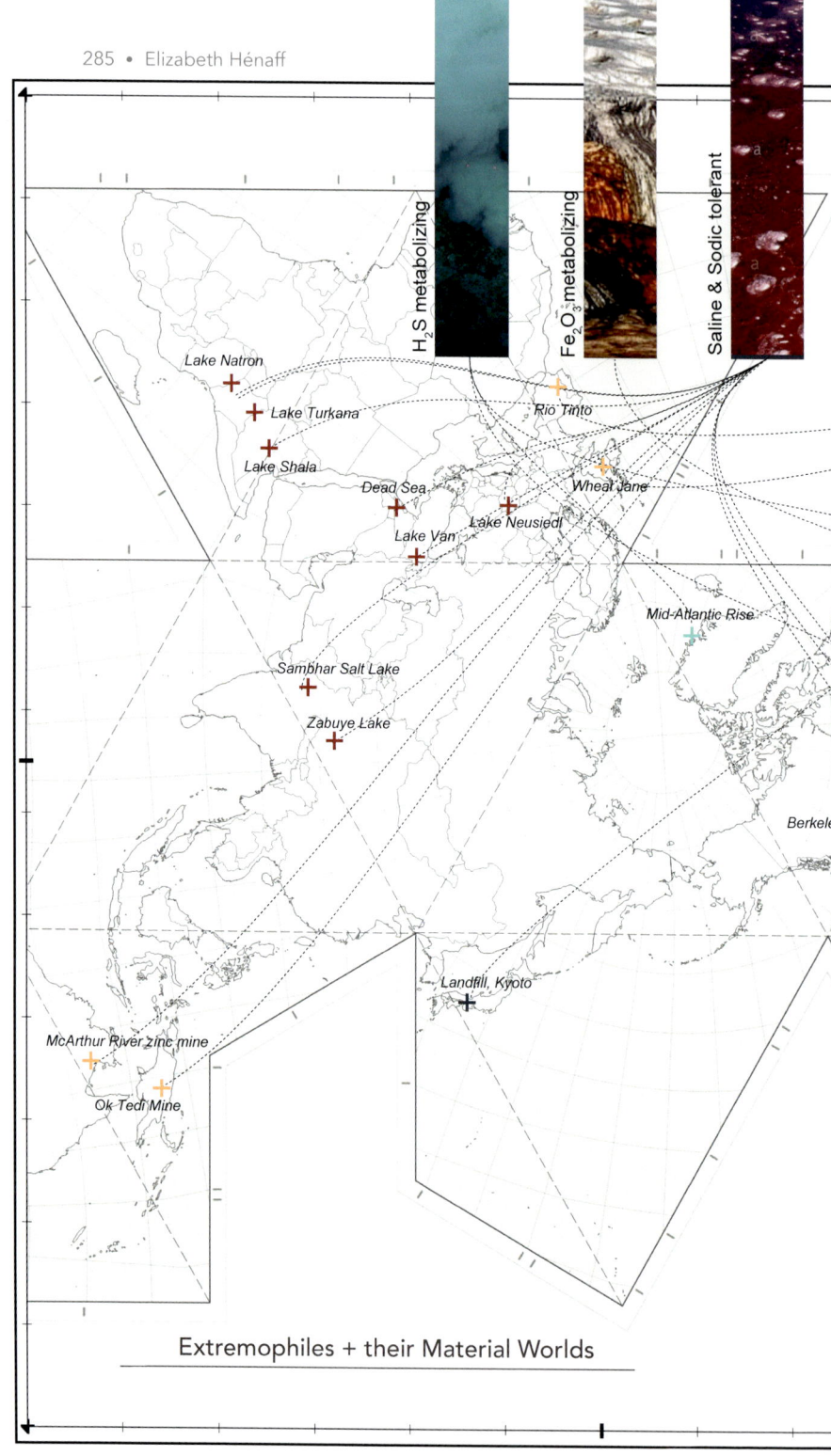

H₂S metabolizing

Fe₂O₃ metabolizing

Saline & Sodic tolerant

Lake Natron

Lake Turkana

Lake Shala

Rio Tinto

Dead Sea

Wheat Jane

Lake Van

Lake Neusiedl

Mid-Atlantic Rise

Sambhar Salt Lake

Zabuye Lake

Berkele

Landfill, Kyoto

McArthur River zinc mine

Ok Tedi Mine

Extremophiles + their Material Worlds

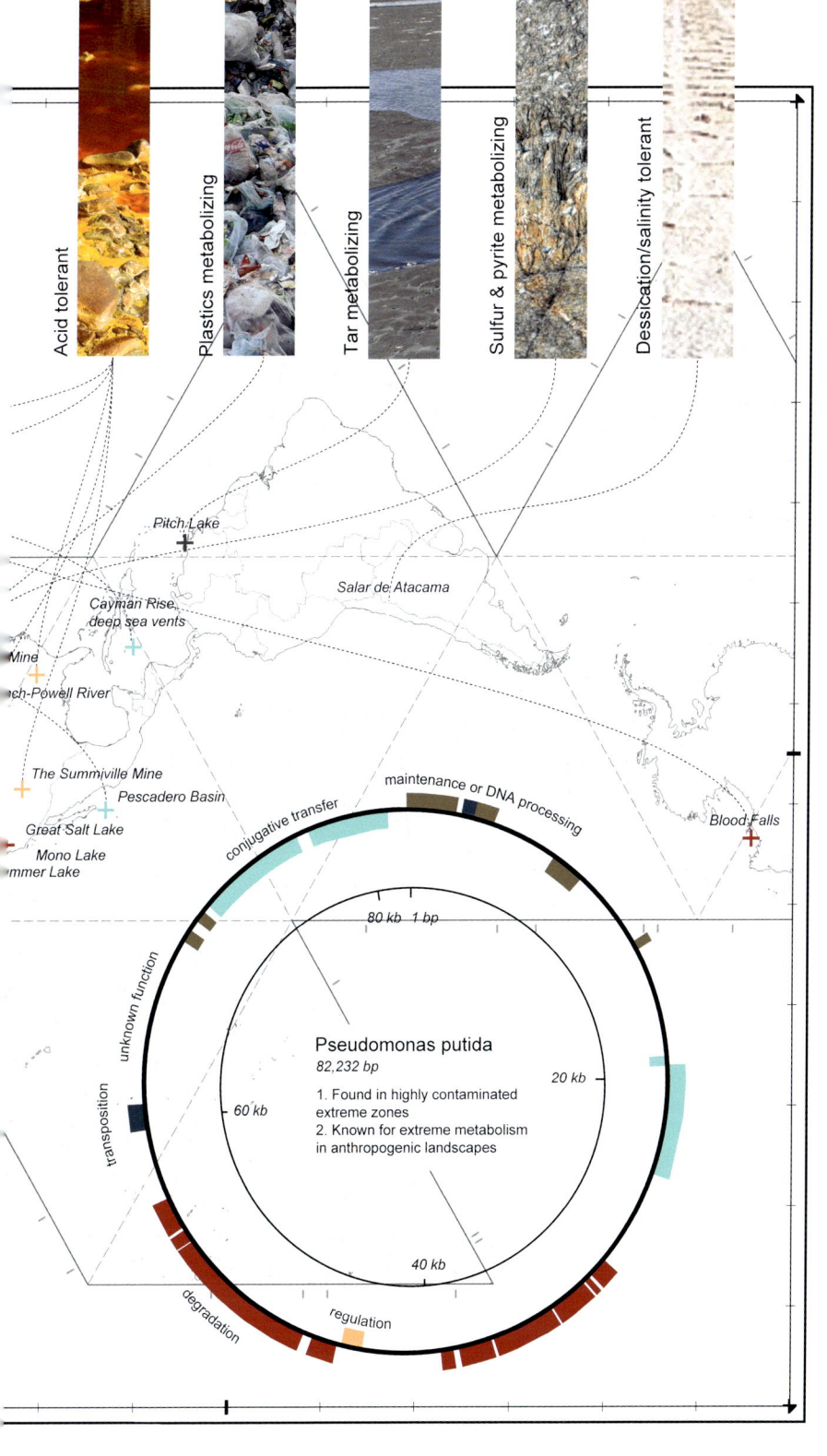

Acid tolerant

Plastics metabolizing

Tar metabolizing

Sulfur & pyrite metabolizing

Dessication/salinity tolerant

Pitch Lake

Salar de Atacama

Cayman Rise
deep sea vents

Mine

ch-Powell River

The Summiville Mine

Pescadero Basin

Blood Falls

Great Salt Lake

Mono Lake

mmer Lake

conjugative transfer

maintenance or DNA processing

unknown function

80 kb 1 bp

Pseudomonas putida
82,232 bp

1. Found in highly contaminated
extreme zones
2. Known for extreme metabolism
in anthropogenic landscapes

20 kb

transposition

60 kb

40 kb

degradation

regulation

over cappuccinos, or bickering over band names with a flight of craft kombucha. Until you have to walk across the Gowanus Canal, that is, or until the winds and tides align to bring a sulfurous waft of organic waste and oil. Crossing the last remaining wooden retractable bridge in Brooklyn (one of the only four remaining in the US), you peer into an impressionistic canvas of oil sheers, drifting slowly along with particulate matter of dubious origin.

If you dare to venture into the waters in search of microscopic life, it's advisable to don the biohazard suit of armor: Tyvek coveralls, rubber boots, gloves, goggles. Often referred to as "black mayonnaise," the Gowanus sediment reeks of a damaged planet: petrochemical remnants of the industrial era and freshly decaying organic matter. This Gowanus special is a toxic cocktail of solvents and sewer overflow.

As a biologist, I'm interested in the interaction of living beings and their environments. I have spent a lot of time thinking about organisms, environments, and their interactions: anything from the dance between plants and gravity, to how genomes respond to stress, to human cancer. But the first time I really started unpacking the continuum of organism and material environment was one crisp Winter weekend in 2014 in Brooklyn, shortly after I had arrived in New York City with a newly minted computational biology PhD.

SAMPLING EXPEDITION

The Tyvek hazmat suit crinkles and I'm awkwardly stomping in rubber boots five sizes too big. The rest of the team and I are on a different mission than the perambulating moms headed to Whole Foods or the lithe climbers coming from Brooklyn Boulders. We're here on a hunt for microorganisms in the sludge and the stories they hold within.

Our team, comprised of two landscape designers, a biotechnology entrepreneur,[1] and me, the bioinformatician, make our way down to the Gowanus Canal shore. You can only access the canal from a few points. Most of its perimeter is guarded by the chain link and barbed wire of the manufacturing sites abutting the bulwarks that frame the body of water. These embankments have drawn the once flowing path of the creek into a tightly engineered canal.

To launch, we have been granted access to a little stretch of rocky shoreline accessible by a set of crumbling concrete stairs. The canal is just under 1.5 miles long, and public access is strictly prohibited due to the public health risks.[2] Any entry points to the canal are owned by the manufacturers that run along it, the one exception being the new condo, a tall developer-modern affair with tiered patios and a parking lot (both hot commodities in Brooklyn). The complex also boasts a floating dock for residents' use only, though residents sunning in their swimwear is unimaginable from inside my plastic suit. SCAPE, a landscape architecture firm, has proposed a project that would restore the historic estuarine conditions of the canal,[3] so it's possible that someday, multicellular creatures might swim in Gowanus again. That vision seems far off from where our team stands today, but community groups like the Gowanus Canal Conservancy[4] are steadily organizing and advocating for the legislative framework necessary to make it happen.

We splash our borrowed canoes[5] into the water, settling into them and the rhythm of rowing. We have a hopeful map, locating the fourteen different locations we plan to stop and take sediment samples from the bottom of the canal. The sites were chosen to represent a diversity of conditions of depth, light, and salinity. As a tidal estuary, the canal sees two high tides a day, with more brackish water near the connection to the East River than upstream. Some sites were under bridges, which never saw the light of day, others were located underneath anywhere from five to twelve feet of canal water. A canoe holds three people, each with a predetermined role. The person at the stern wields the fifteen-foot-long PVC pipe, digging it into the muck, capping it, and drawing it up. The person at the bow steers the mouth of the pipe, catching some of the slurry of sludge in a test tube as it is released, while the middle seat stabilizes the vessel during the process and logs the sampling data in our custom-built mobile app. We are four teams rotating two canoes and after a few hours, we recover all the samples we need. The test tubes are packed into biohazard bags and taken back to the Mason Lab[6] at Weill Cornell Medicine to be stored at -80°C.

Later in the comfort of the lab, we use metagenomic sequencing to extract and read DNA from the samples we collected. The analysis showed that some species of bacteria in the canal are similar to what are found in a marine environment. Although difficult to recognize, this makes sense from its geologic designation as a tidal estuary.

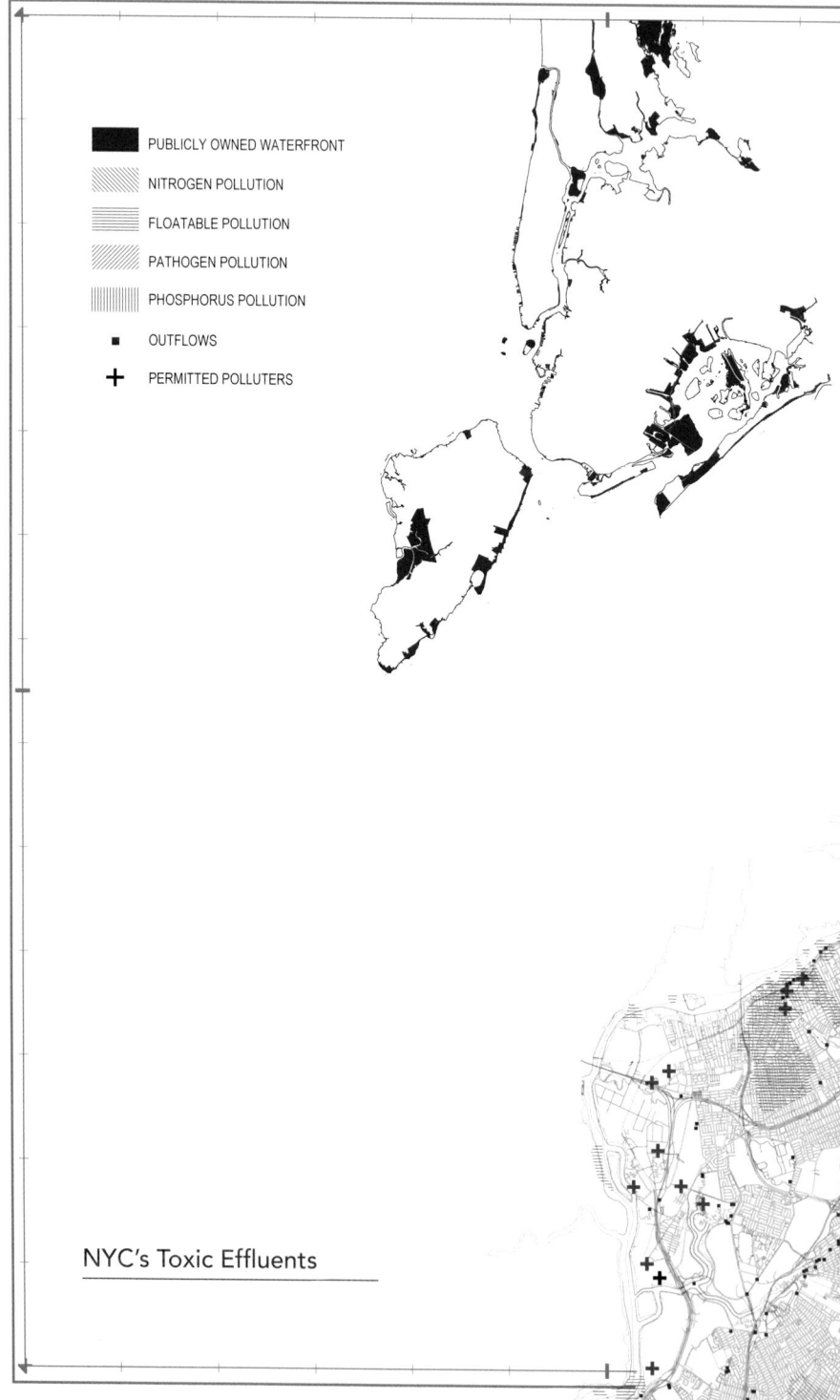

PUBLICLY OWNED WATERFRONT

NITROGEN POLLUTION

FLOATABLE POLLUTION

PATHOGEN POLLUTION

PHOSPHORUS POLLUTION

■ OUTFLOWS

✛ PERMITTED POLLUTERS

NYC's Toxic Effluents

0   1.25   2.5   5   7.5   10
                                                    Miles

N

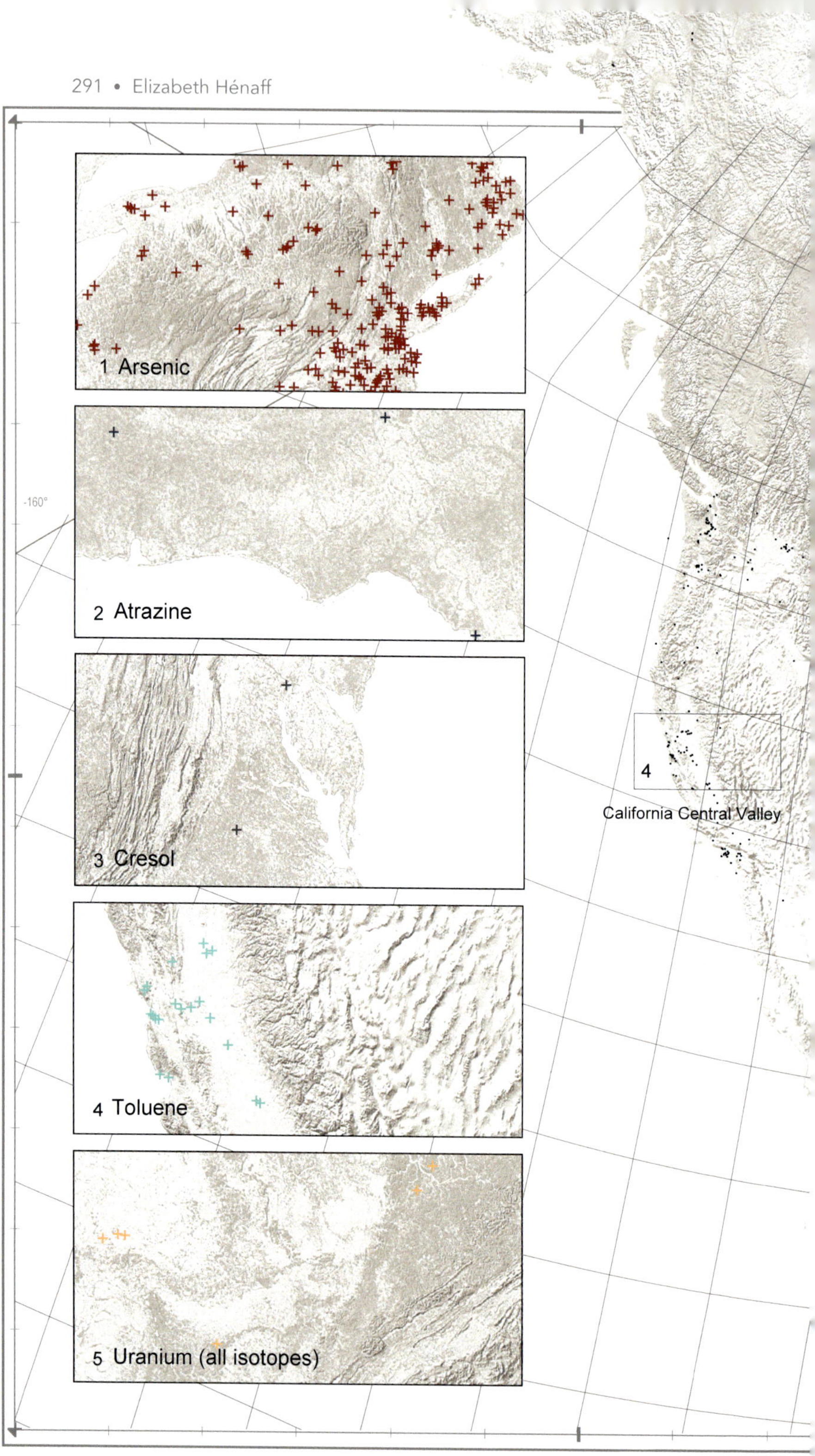
1 Arsenic
2 Atrazine
3 Cresol
4 Toluene
5 Uranium (all isotopes)
-160°
4
California Central Valley

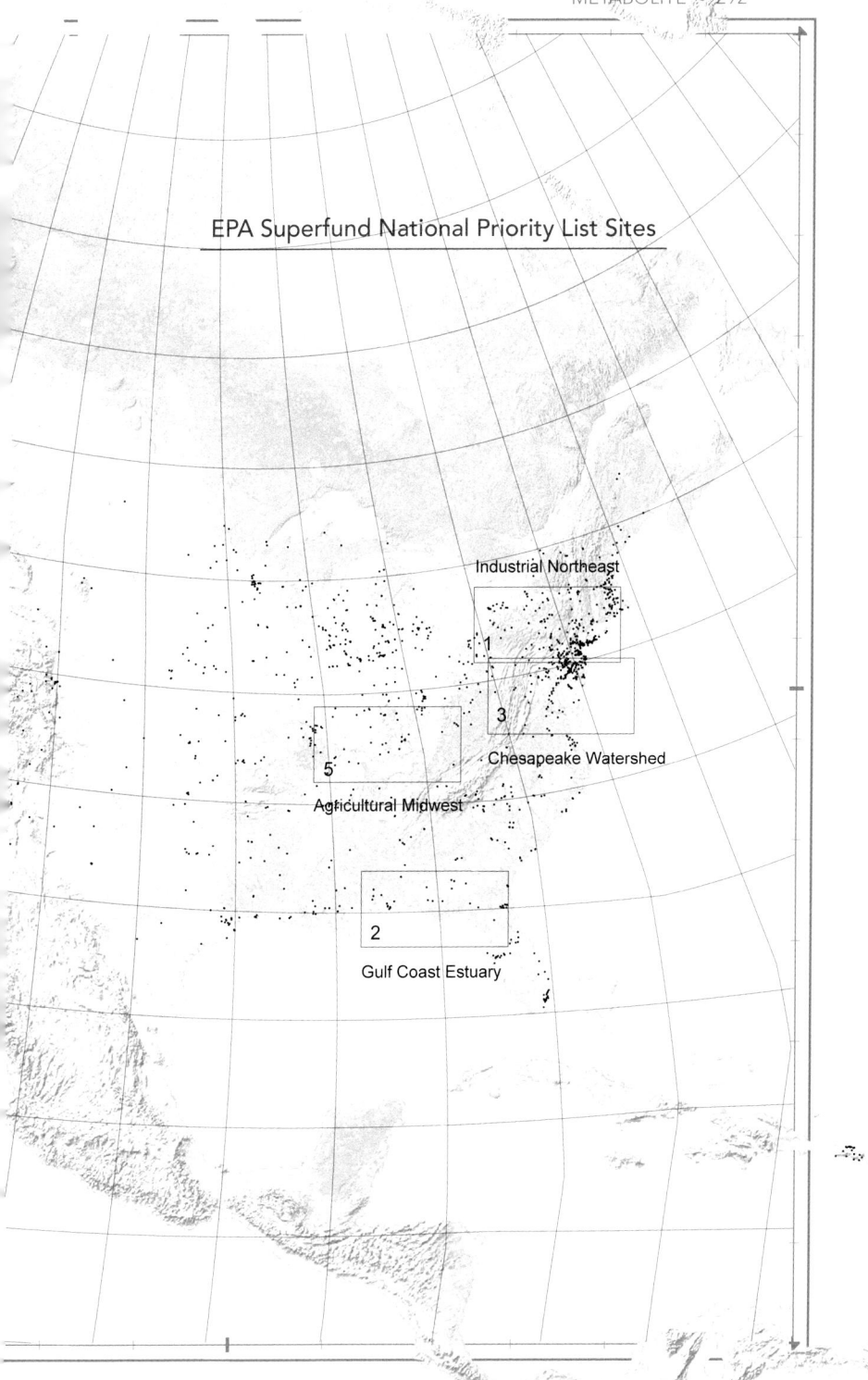

EPA Superfund National Priority List Sites

Industrial Northeast

1

3

5

Chesapeake Watershed

Agricultural Midwest

2

Gulf Coast Estuary

Most surprisingly, the DNA analysis shows that the microbial community as a whole is able to break down some of the toxic compounds with which it's challenged. Over almost two centuries, the local microbiome has adapted to the impacts of human intervention and is slowly remediating our toxic legacy.

LAYERS OF GOWANUS HISTORY INSCRIBED IN SEDIMENT

Our team's act of extraction was just the most recent in the history of the Gowanus area. With human settlement documented as early as 12,500 years ago,[7] at the time of European contact the area was inhabited by the Lenni Lenape group of Algonquin people. Also known as the Delaware, the local members of this once-widespread population speak a dialect known as Munsee.[8] The body of water named Gowanes Creek was turned into a canal in 1853 by dredging and widening its course and shoring its banks. Under the weight of the city's thrusting angles, it became the Gowanus Canal. Serving as transport for goods and materials through Gowanus Bay to the Hudson and the Atlantic, it also functioned as a dump site for waste generated by the industries alongside it. The environmental history of the canal today is emblematic of the many post-industrial Superfund[9] sites across the country. These sites were once important spaces for production and manufacturing that have since changed locations, leaving behind them a material, economic, and social legacy of toxicity.

The pre-industrial makeup of the Gowanus area would have been identified as a tidal estuary—birds, fish, footlong oysters. Today, it is far harder to read this ecology beneath its composite of anthropocentric histories. The robust ecosystems that enticed the capital expansion of the Dutch West India Company in the region, ultimately leading to the systematic eradication of Indigenous lifeways in this geography, are now largely erased and paved over, seemingly devoid of any of the diverse lives it hosted. But the Anthropocene has not entirely wiped away the pathways and communities of microbial life.

Since the mid-nineteenth century, these Gowanus industries have included various forms of petrochemical processing: coal pockets, asphalt, coal and coke tar, gas and oil companies, as well as machine shops and lumber yards, chemical and paint plants, tin goods and iron foundries and power plants.[10] These products and also the material waste—coal tar, solvents, spilled oil accruing in the canal's sediment—were the outputs of the factories. If these industries were organisms, these products would be the metabolites of the neighborhood.

The PVC pipe extraction system we used is similar to how one of the bartenders ashore might sample a fancy cocktail before serving it: a straw in the drink, capped with their finger and drawn out to sip the liquid. The sludge is dark, viscous, and reeks of gasoline. One of the main contaminants is a group of byproducts from petrochemical industries: coal tar extraction, coal gas production, and oil refining.[11] Before container shipping moved the industry to New Jersey, this area was a center for petro-energy production. Indeed, Consolidated Edison, one of the two contemporary power companies that serve New York City, inherited both the land and the liability from the previous coal-based energy facilities. As part of the Superfund legal process, they are held liable for the contamination and will be partially responsible for funding its cleanup.

That cleanup work has already begun. In 2019, only a few years after my first encounter with Gowanus sludge, I led a walking tour of Gowanus with a group of students[12] that was curtailed by fallout from the Superfund remediation pilot work. When we reached the usual Whole Foods entrance, we were forced to make a detour because the waterfront esplanade was closed. Its brick paving had buckled and slumped down, the infrastructure likely impacted by the dredging for the pilot EPA study happening in the nearby turning basin.[13] Instruments like crack meters and laser survey prisms dot the landscape, demonstrating both an awareness of these hazards and the need to quantify them. It is no surprise that extracting tons of mud[14] from the bottom of the canal would affect the surrounding infrastructure and the shape of the terrain.

The Whole Foods opened in 2013, only three years after the EPA designated the site as Superfund. As the harbinger of gentrification, Whole Foods' arrival was contested by residents and advocacy groups alike. Now one of the largest grocery corporations, Whole Foods claims to include local producers and artisans in their supply chain. The shopping carts and compostable produce bags will one day be like the amphorae of ancient Rome,[15] but the need and economic motivation of trading food and goods from local and faraway sources is the same. The store does, however strangely, offer one of the only accessible waterfront spaces and its esplanade has become a waterfront commons,[16] rare in this part of Brooklyn.

The neighborhood's demographic is under rapid transformation, and the land itself is undergoing a comprehensive rezoning process. As far as heavy industry, only one cement plant remains. A main concern for the community groups watchdogging this upswing in new residents is to ensure that the sewage infrastructure meets the demands of residential development. The combined waste system in most of Brooklyn collects raw sewage and rain effluent through the same set of pipes, which overflow during heavy precipitation events into the bodies of water surrounding the city. Gowanus by Design,[17] a non-profit design studio founded by a local community board member, has been leading data-driven discussions in advocacy for improved infrastructure, including sewage management, to be included in development plans.

Even before being recognized as a Superfund site in 2010, the canal has existed as an extreme environment since the first industrial waste was discarded there. For scholars of the microbiome, extreme environments like the Gowanus teach us a lot about the rules of life, as anything that lives there has been forced to adapt to inhospitable material conditions. Extreme environments sit at the tail ends of the bell curve: defined by temperatures, altitudes, and atmospheres unfathomable to human life. They can be natural (deep sea vents, the Poles) and also anthropogenic (the inside of nuclear reactors).[18] These abnormal conditions have given rise to novel lifeforms and genius adaptations. The extremity of the Gowanus lies in its toxic sludge: poisonous to most life forms, hospitable substrate to rare others. As a result of almost two centuries of use as an industrial thoroughfare in a rapidly expanding Brooklyn, the Gowanus sediment is now home to the chemical afterlives of the tanneries, paper mills, manufactured gas plants, and chemical plants it once serviced so dutifully.

Our DNA analysis of the canal's sediment samples shows that the metabolism of the microbial community includes some interesting bioremediation activity. For instance, the sediment microbiome degrades cresol (a petrochemical byproduct), toluene (an industrial solvent), and fixes heavy metals including arsenic.[19] Each of these metabolites can be traced to historical records of factories in operation at the site. This means that the present-day, living microbiome of the canal maintains a history of human intervention at the site, a molecular echo.[20] Arguably, a more precise record than written histories,[21] that is, if you know how to read it. It is precise from a more-than-human perspective, meaning that it highlights the components of the environment that are perceived by, and meaningful to, nonhuman organisms.

As such, the present-day microbial map of the canal defines a material record of human and nonhuman intervention at this site. What is made legible by this map is how the living component of the environment is being sculpted by the material properties of that environment, as it leaves distinct traces of accumulated metabolites—molecules that have been acted upon by living organisms. The metabolisms we see in the canal microbiome are proxy indicators for its abiotic material properties, and mapping these living populations is a form of material cartography.

These maps tether the complex socio-economic continuum of human health in the neighborhood with the geo-material history of Gowanus. These maps not only function across geography and time, they provide insight from the smallest DNA molecules to far-reaching extraterrestrial human inhabitation. They map impacts of multinational corporations, deep ocean topographies, industries, and economies. They show us that materials are not fixed, rather in a constant state of transformation, of becoming, inextricable from the living metabolisms they enact.

METABOLITES AS MATERIAL HISTORIES
My inclination (as honed by my training in the study of living organisms) is to think of the role of the microbe as that of the actors mediating material agency. For this essay, I experiment with decentralizing the microbes as the focus of agency, in order to make way for the metabolite, material that has been transformed by a biological process. I outline the materials related to industrial development and their agency in human ecologies, economies, and histories from the point of view of the resident microbiome.

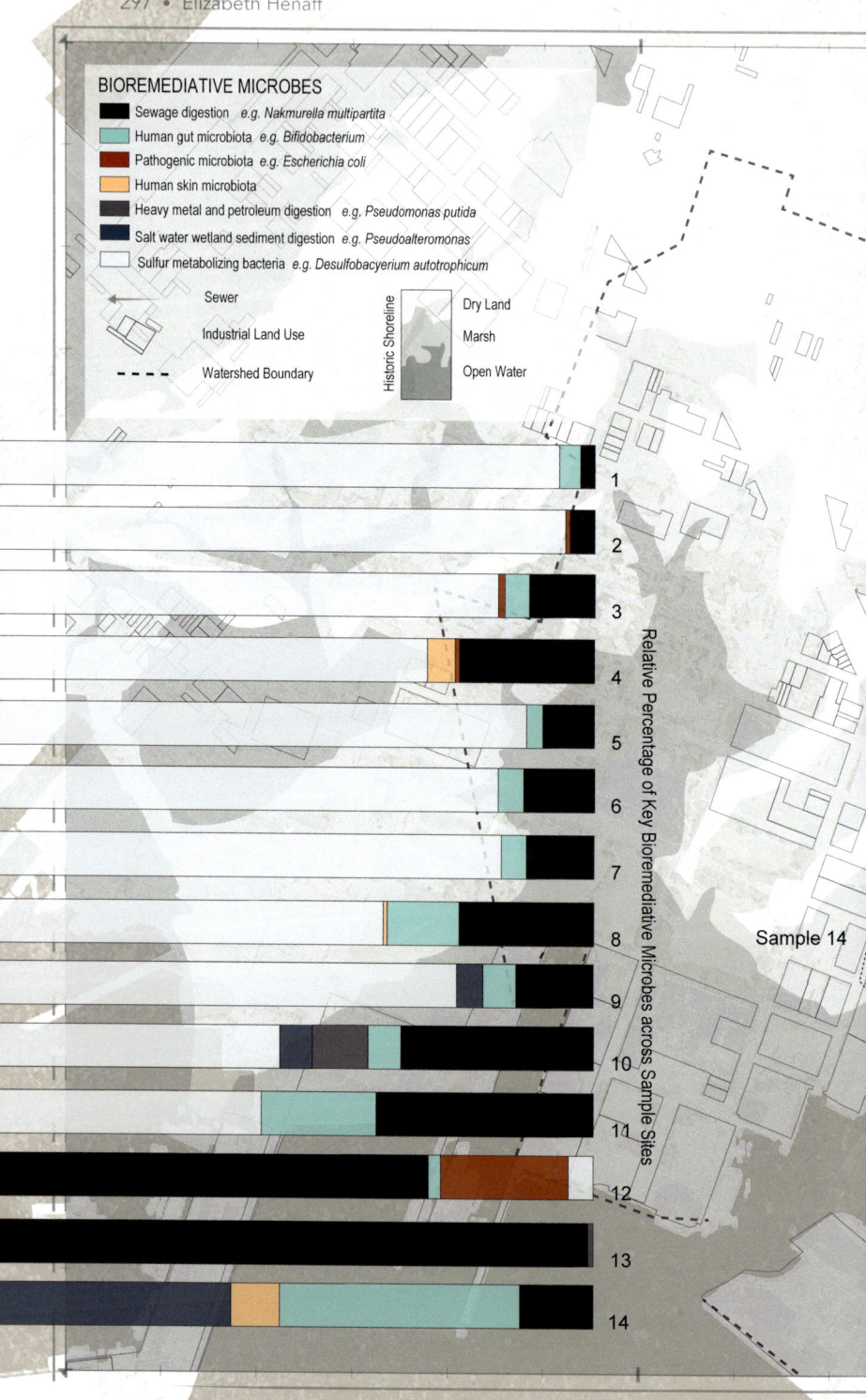

BIOREMEDIATIVE MICROBES

- Sewage digestion *e.g. Nakmurella multipartita*
- Human gut microbiota *e.g. Bifidobacterium*
- Pathogenic microbiota *e.g. Escherichia coli*
- Human skin microbiota
- Heavy metal and petroleum digestion *e.g. Pseudomonas putida*
- Salt water wetland sediment digestion *e.g. Pseudoalteromonas*
- Sulfur metabolizing bacteria *e.g. Desulfobacyerium autotrophicum*

Sewer

Industrial Land Use

Watershed Boundary

Historic Shoreline

Dry Land

Marsh

Open Water

Relative Percentage of Key Bioremediative Microbes across Sample Sites

Sample 14

1
2
3
4
5
6
7
8
9
10
11
12
13
14

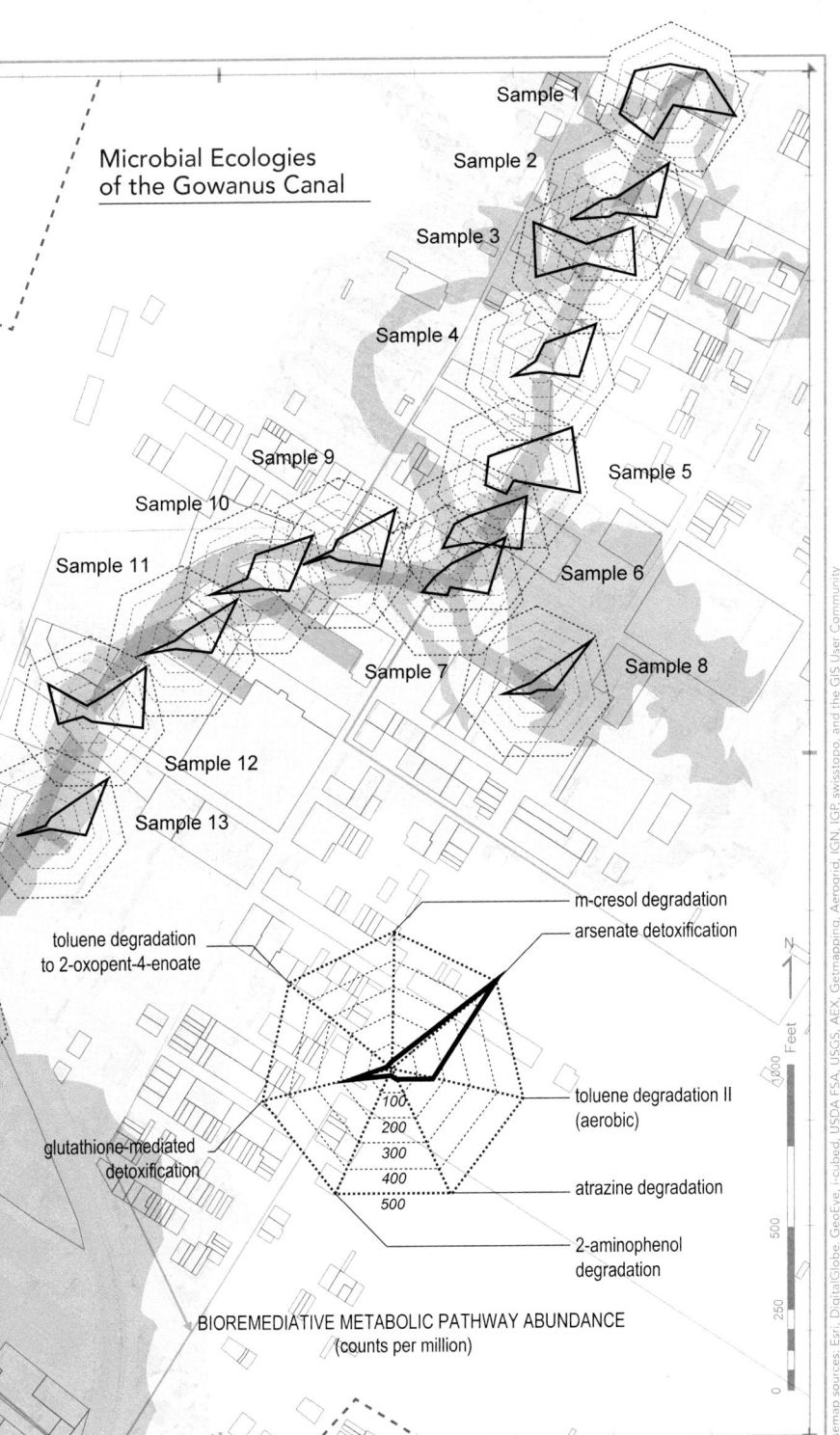

Microbial Ecologies
of the Gowanus Canal

Sample 1

Sample 2

Sample 3

Sample 4

Sample 9

Sample 10

Sample 11

Sample 5

Sample 6

Sample 7

Sample 8

Sample 12

Sample 13

m-cresol degradation

arsenate detoxification

toluene degradation
to 2-oxopent-4-enoate

toluene degradation II
(aerobic)

glutathione-mediated
detoxification

atrazine degradation

2-aminophenol
degradation

100
200
300
400
500

BIOREMEDIATIVE METABOLIC PATHWAY ABUNDANCE
(counts per million)

Basemap sources: Esri, DigitalGlobe, GeoEye, i-cubed, USDA FSA, USGS, AEX, Getmapping, Aerogrid, IGN, IGP, swisstopo, and the GIS User Community

Feet

1,000

500

250

0

The material stories laid out in the preceding chapters are all undergirded by teeming microbial assemblages. My training in metagenomics allows me to offer a coda of sorts, one that recalls the preceding chapters and the metabolites found in the assemblages described by other onto-cartographers.

In these pandemic years, we are all too keenly aware of the invisible biological component of our environment. Though viruses are fundamentally different from microbes, the objects and rituals that have become part of our daily lives—the masks, the excessive hand-washing—manifest a new relationship to the invisible biology around us. We all know that there are microbes everywhere, in any material condition you can imagine. You can find them in the most quotidian of spaces—your office,[22] the subway[23]—but also in the most severe natural environments. They are also found in sea vents exposed to extreme temperatures and lack of carbon and oxygen; high altitude salt flats where plant life cannot sustain the radiation and drought; even in the enormous burning methane pits of Turkmenistan.[24] You can also find them in extreme anthropogenic environments: inside nuclear reactors, in the cleanest of clean rooms where spaceships are assembled, including on the outside of these when they return from their mission.

But what exactly is a *metabolite*? Any material that has been acted upon by an organic process is, by definition, a metabolite. The name refers to a material that is metabolized—either ingested or secreted—by an organism. Metabolites operate in a series of intertwined loops. At some point in its cycle, it is likely to have been processed by microorganisms:[25] ingested, only to be transformed and then secreted, followed by ingestion by yet another organism. The process of life as continual modification: ingestion, transformation, secretion, repeat is referred to in scientific terms as a metabolic pathway; it is a diligent and precise series of intracellular reactions that transform one molecule (material) into another. For example, human metabolism transforms last night's glass of wine (ethanol) into water and carbon dioxide by the morning. At a cellular level, each reaction—the transformation of one molecule into another—is performed by a particular enzyme or protein, and this set of reactions comprises a metabolic pathway. So, the material of ethanol becomes relationally enmeshed with the material of carbon dioxide by a series of transformations: the ethanol degradation pathway.[26]

In general, if there is a material, you will find a microbe that is doing something with it. A lot can be learned about the functional properties of a substrate (such as garden soil, or the water of a favorite swimming hole, or a nearby Superfund site)[27] by interrogating the metabolic potential of the microbes that live in it.

I see the metabolite as a material history of a series of relationships because microbes respond to their environments through changes to their metabolism. Indeed, if the materials that comprise an environment include heavy metals, the microbe is pushed to evolve to withstand the metals. If the pH of the substrate is low, it will adjust to acidic environments. Measuring these properties provides a type of sensor for abiotic conditions.[28] Documenting these metabolisms gives you a material metric (the metabolite), an artifact of metabolic transformation.

In the context of microbial communities, it is possible that different actors (microbes) within the community are performing different parts of that pathway. The material transformations of metabolism are therefore born out of a relationship between a community of microorganisms and their environment, rather than that of a singular actor to a particular place in time. The agency of these metabolites exists by virtue of their entanglement in all the living things that process them; they are used as food or fuel, or secreted as waste or commodity.

Microbes evolved in a material world.[29] In that respect, nothing has changed. They continue to respond *to and within* their material assemblages. Materials impact microbes, and microbial communities impact materials. This happens on the scale of the individual microbe, but more meaningfully so at the scale of microbial communities. These communities are assemblages of individuals of various microbial species that define a relationality among themselves, and, as a whole, with their environment. Still, microbes defy categorization by species as these relations are fluid and mutable.

The agency of the material world is inextricably microbial. Indeed, this microbial world, including its metabolism of materials, is the context in which humans evolved. There were

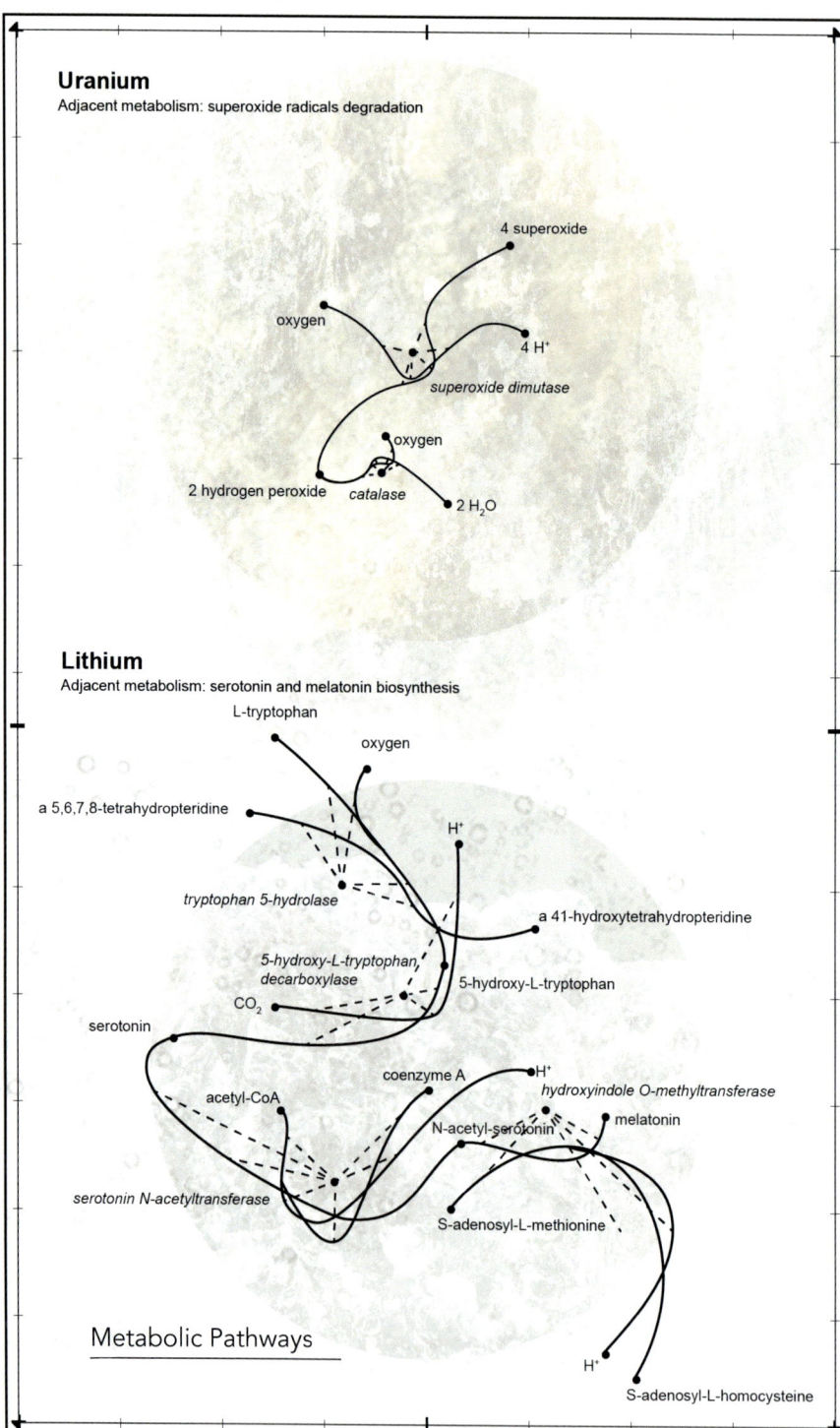

Uranium
Adjacent metabolism: superoxide radicals degradation

4 superoxide

oxygen

4 H⁺

superoxide dimutase

oxygen

2 hydrogen peroxide   *catalase*   2 H₂O

Lithium
Adjacent metabolism: serotonin and melatonin biosynthesis

L-tryptophan

oxygen

a 5,6,7,8-tetrahydropteridine

H⁺

tryptophan 5-hydrolase

a 41-hydroxytetrahydropteridine

5-hydroxy-L-tryptophan
decarboxylase

5-hydroxy-L-tryptophan

CO₂

serotonin

coenzyme A   H⁺

acetyl-CoA

hydroxyindole O-methyltransferase

melatonin

N-acetyl-serotonin

serotonin N-acetyltransferase

S-adenosyl-L-methionine

Metabolic Pathways

H⁺

S-adenosyl-L-homocysteine

at least a billion years of microbial life before multicellular organisms, let alone humans, arrived.[30] Multicellularity is the assemblage of specialized cells acting in concert, forming a single organism. It first originated in the ocean, from single bacterial cell associations—similar to biofilms we observe today (like lichen, or a kombucha scoby, or a cheese rind). These became gradually more organized, until the proverbial fish walked out of water. These fish were not departing from the microbial womb, rather shifting through and stretching the microbial continuum.

All multicellular organisms rely on symbiosis with microbes for essential functions of metabolism and defense. Microbial communities are the interlockers between multicellular organisms and their material environments, and any relation of humans to the material world is a relation to a metabolite and the microbes that processed it. This interdependence, or relationality, is the case from the scale of an individual microbe to a fish or a human. And, very importantly for the work of this atlas, is also at the telescoping scale extending from planetary assemblages to the economies and ecologies we inhabit.

CODA

These onto-cartographic observations of the metabolic pathways of the microbiome allow us to see a model for material agency,[31] one that works well beyond the existing anthropocentric paradigm. In looking closely at a specific series of metabolites—the metabolic relationships surrounding the other materials described in the previous chapters then coming back to Gowanus—my goal is to show the agency they maintain in the given assemblage they are part of, and how microbes tend to play along in that assemblage. Here we traverse scales and orders of magnitude from the molecular to the cosmic through the monumental, communal and highly personal.

URANIUM

Death by radiation is not inevitable, and organisms have a diversity of levels of radiation tolerance. Microbes have been found to inhabit naturally high-radiation areas such as deep-sea hydrothermal vents. Bacteria and fungi have also been found inside nuclear reactors and inhabiting the cooling waters of reactors.[32] Some microbes (such as the aptly named *Deinococcus radiodurans*), are able to withstand levels of radiation that, for humans, would be catastrophic. This shows that histories of genetic damage are recoverable in some organisms, and these mechanisms alter the agency of radioactive materials. The metabolite in this assemblage is the DNA molecule itself—the Deoxyribose Nucleic Acid, long polymeric molecule, storage of genetic information—being damaged. Radiation affects material DNA structure, synthesis, and repair. The mechanisms of resistance fall in two general categories of metabolisms: DNA repair (fixing damage that occurs from radiation) and free radical absorption (reducing the extent of DNA damage). A microbial protein able to absorb ionizing radiation has been isolated and used successfully to impart radioresistance in mice.[33] This is one of the several approaches considered for radiation therapy in acute exposures.[34]

Scientists are studying radioresistant microorganisms with the goal of mining their unique repair mechanisms for human genetic modification technologies. This is under consideration in the context of long-term space flight, as astronauts are exposed to high levels of solar cosmic radiation, a significant hurdle for multigenerational space travel. This earthly decamping is sometimes framed under the narrative of ecosystem collapse after nuclear warfare, ozone depletion, or other cataclysmic anthropogenic events. Escaping the radiation of nuclear winter means braving the radiation of space. This scenario is one of the only ones that bioethicists consider in justifications for heritable human genetic modifications.[35]

Here microbial metabolisms are taking us on a voyage across orders of magnitude of scale, mediating our relationship to the cosmos from their microscopic vantage point.

LITHIUM

Some psychotropic drugs, such as lithium, are hypothesized to modulate gut microbiome composition.[36] Gut microbes also dictate the absorption of drugs, including the elemental material of lithium. Disruptions in gut microbiome composition are correlated with many psychological disorders, such as Autism Spectrum Disorder, schizophrenia, and Major Depression Disorder or clinical depression. Microbes regulate the synthesis of neurotransmitters such as serotonin,[37] affecting all sorts of physical phenotypes,

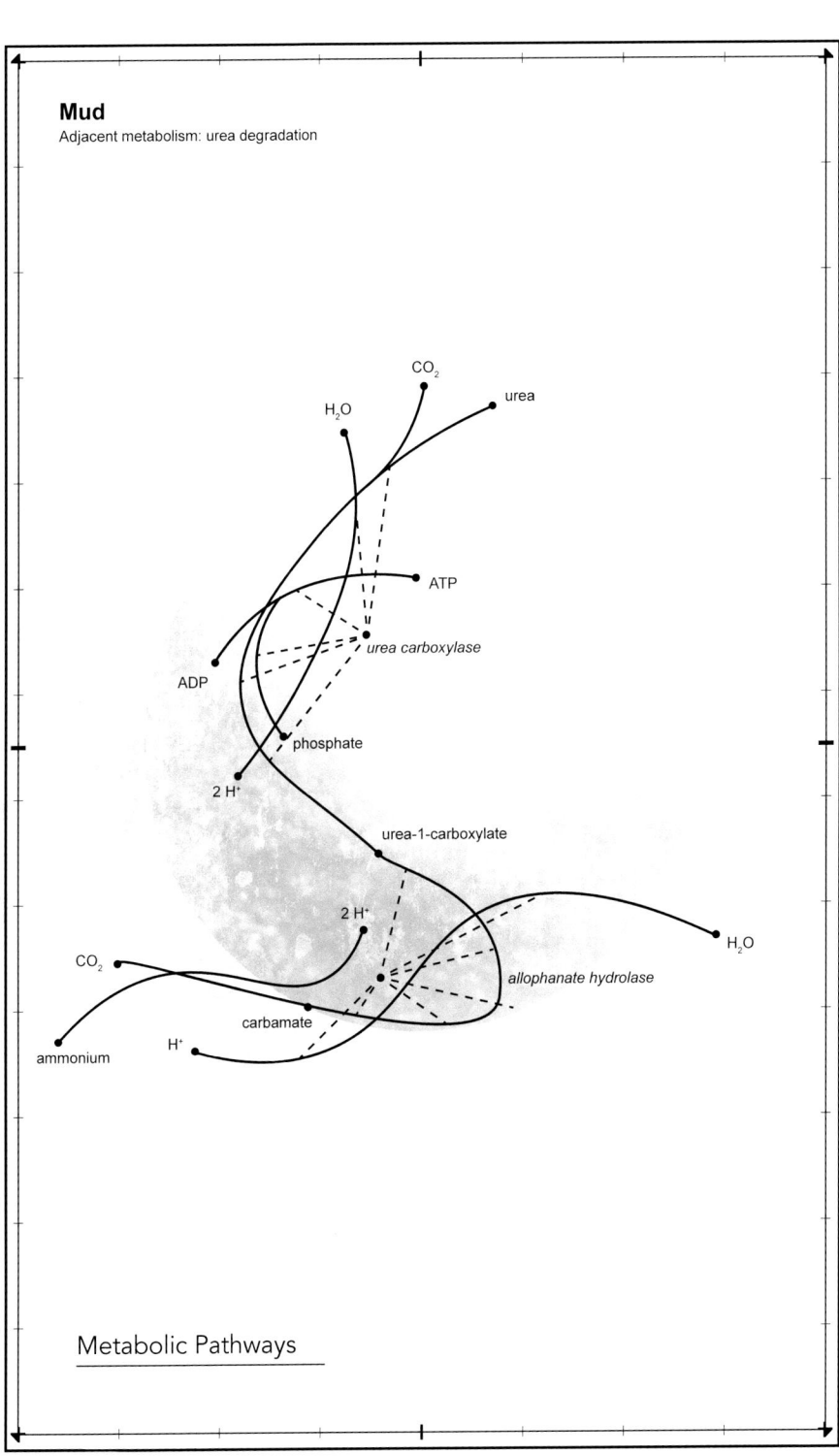

Mud
Adjacent metabolism: urea degradation

CO_2

H_2O

urea

ATP

urea carboxylase

ADP

phosphate

2 H$^+$

urea-1-carboxylate

2 H$^+$

H_2O

CO_2

allophanate hydrolase

carbamate

ammonium

H$^+$

Metabolic Pathways

including mental health. In mouse models for anxiety and schizophrenia, these phenotypes are recovered with gut microbiome changes. People with bipolar disorder often have gut dysbiosis, as well as individuals with autism. As an actor in this assemblage, lithium has a proven effect on the composition of gut microbiomes. This microbial mediation between the inner and outside worlds happens at the most vulnerable nexus, that of mental health, and reminds us that despite our mechanistic models, bodies are not machines.

MUD

Mud has been used in many ways as a building material—for industrial "monuments" and also modest habitations—and often the solidification process is aided by microbes. The process of microbially indurated rammed earth construction involves solidification of the substrate used (earth, sand, mud) by Microbially Induced Calcite Precipitation.[38] These microbes precipitate calcium carbonate—the material of seashells—which fills the cavities in the substrate and solidifies it. The metabolite in this case is the calcium carbonate, which solidifies the rammed earth structures when precipitated by the microbes. Rammed earth has been used to create massive monuments such as the Great Wall in China, but also homes such as those found in cities in Portugal or Vietnam. Microbially assisted construction is also happening in the same Chesapeake Bay as the mud monuments. Researchers and artists are fostering microbially-made oyster reefs to be transient structures to reseed oyster populations.[39] These structures rely on bacterial cementation that mimics shellfish mineralization (or coral reef formation) and are transient monuments to reconstruction, as they will be destroyed upon successful recolonization by oyster populations. The metabolites of calcium carbonate hold together soil and mud, defining habitable environments for human and more-than-human assemblages.

SAND

The material of sand is a metabolite. Synthesized in a marine assemblage, it is fodder from the coral holobiont,[40] eaten and excreted by the fish, and released as particles into the water. As sand, it accumulates on beaches and in deep sea reservoirs. This metabolite and the microbial symbionts of the corals are the key actors in replenishing the reservoirs

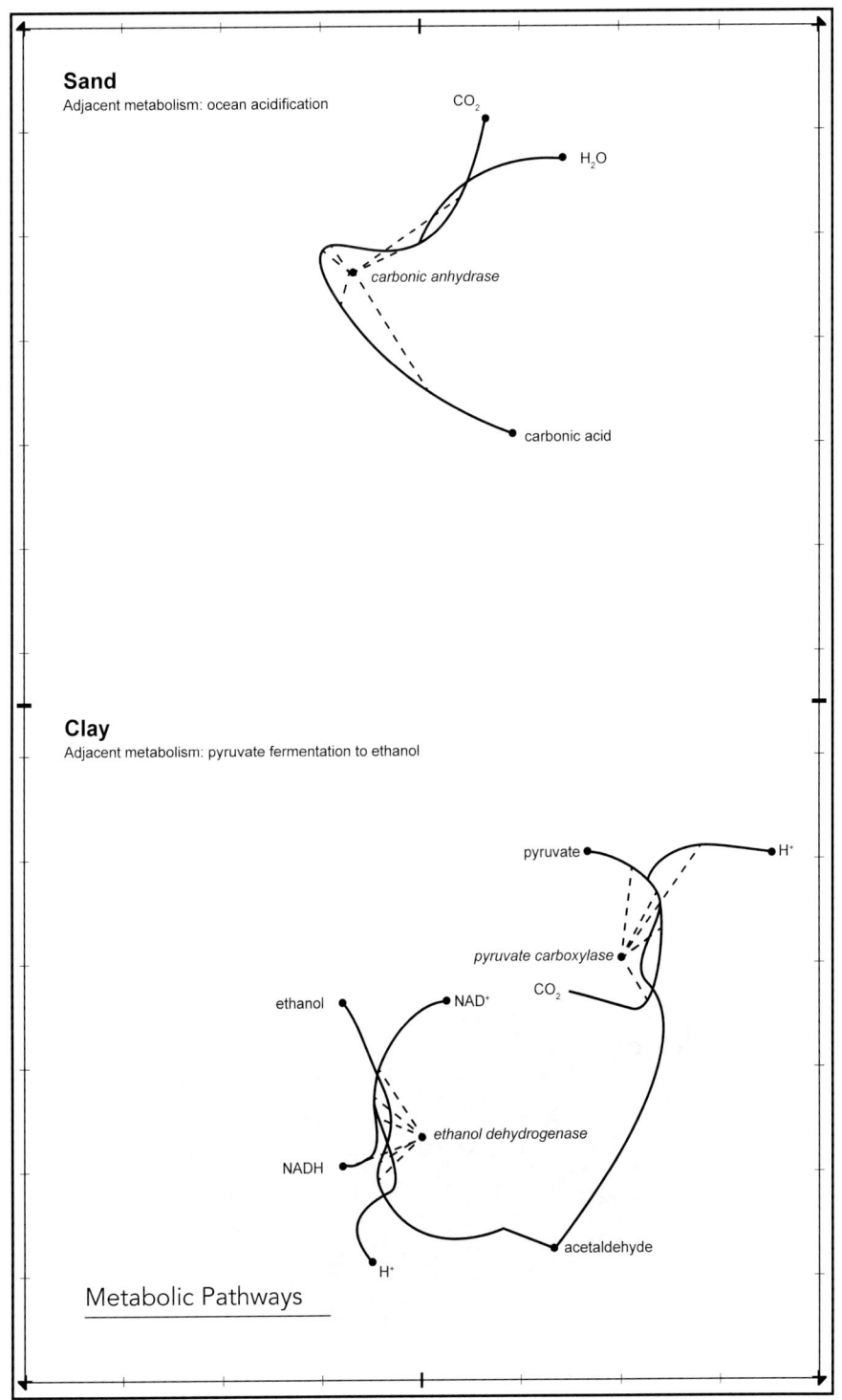

Sand
Adjacent metabolism: ocean acidification

CO$_2$

H$_2$O

carbonic anhydrase

carbonic acid

Clay
Adjacent metabolism: pyruvate fermentation to ethanol

pyruvate

H$^+$

pyruvate carboxylase

CO$_2$

ethanol

NAD$^+$

ethanol dehydrogenase

NADH

H$^+$

acetaldehyde

Metabolic Pathways

of sand that are mined to nourish the depleted Florida beaches.

The white sand found on Florida beaches and throughout the Caribbean originates from the surrounding coral barrier reefs. The coral skeleton, composed of calcium carbonate—like seashells and like the rammed earth described above—is broken down and eaten by fish, excreted and eventually makes its way onto beaches and into underwater reservoirs.[41] Coral is a holobiont, and it depends on its microbial symbionts for growth and photosynthesis.[42] Sand comes from corals, and corals need microbes.

Part of the sand depletion story is ocean acidification which kills the coral symbionts and, in turn, kills the coral. This idea of the offshore sand reservoirs as depletable reserves of material is underpinned by an idea of reserve as locus rather than an assemblage. Seeing the sand reservoir as a finite resource dissociates it from the coral ecosystem assemblage it is part of and introduces it into the engineering assemblage of the jetties and shoreline constructions. Here the coral holobiont connects us to deep sea and deep time through its metabolites that feed the beaches of our leisure.

CLAY

Food trade is enabled by the ability for food storage, itself dependent on food preservation. Fermentation is one of the most ancient forms of food preservation, and food tells the story of a region. Clay pots were and still are used for food fermentation, including that of vegetables and wine. The clay as material has an important role in creating a container for this method. The vessel, with its ability to create an adequate micro-environment, is the power behind this old technology. In fact, before the discovery of microbes and microscopy, the vessels themselves were seen to be the living force behind fermentation, and the process was seen to be specific to a vessel.[43]

The oldest known vessel designed for fermentation, dated to ~9000 cal BP,[44] was a trench found in Sweden,[45] with remains of fermented fish bones. Old Egyptian amphoraes still hold DNA of ancient yeast.[46] Clay containers are able to preserve the genetic information of the domesticated microbes they nurtured for thousands of years. Just as the clay vessels are specific to a place, or terroir, so is the food that they contain. For fermented foods, this terroir is linked to the microbes that fermented it and the metabolites of our daily meals give us a visceral sense of place.

CRUDE

The agency of crude lies in its energy potential, notably, the high-energy hydrocarbon molecules that we burn for fuel. Its agency also lies in its toxicity—it is these same energy-dense molecules that are toxic to people and ecosystems. This agency is wrought in the catastrophic spills that have paced the history of fossil fuel extraction and dependence: Exxon-Valdez in Alaska in 1989, Deepwater Horizon in 2010, and now the recent Dakota Access Pipeline (DAPL) spills. Fungi and bacteria are the only actors able to degrade these hydrocarbons to innocuous compounds for the multicellular members of the affected ecologies.

The spill at the DAPL and the contamination at the Gowanus Canal are just two of many instances of extracted petro-resources out of place. When oil spills occur, local microbiomes can adapt to degrade the hydrocarbons and use them as energy sources. This has been described extensively in the context of the Deep Water Horizon spill.[47] This adaptation can happen spontaneously, or be seeded by introducing microbes extracted from similarly contaminated sites that have already had the time to adapt.[48] Chemical mitigation strategies—such as the use of dispersants—not only hide the extent of the spill and are more toxic than oil itself but also inhibit the microbial growth that can remediate the oil concentrations.

Microbes are of great concern to the oil industry, due to their ability to consume oil in pretty much any of its forms. They degrade fuel stores and their biofilms clog lines, but this ability can also be harnessed. In particular, microbial metabolisms are exploited to aid in hydraulic fracturing extractions,[49] either by pumping in nutrients to stimulate the growth of certain microbes, or by inoculating with microbes directly, both with the goal of altering crude viscosity and availability.

Human extractive practices determine whether microbes that consume hydrocarbons are good or bad. When considering the assemblages affected by the toxic nature of oil, we see Native American populations disproportionately affected by the destruction of ecosystems that their lifeways rely on, and they are consistently inadequately

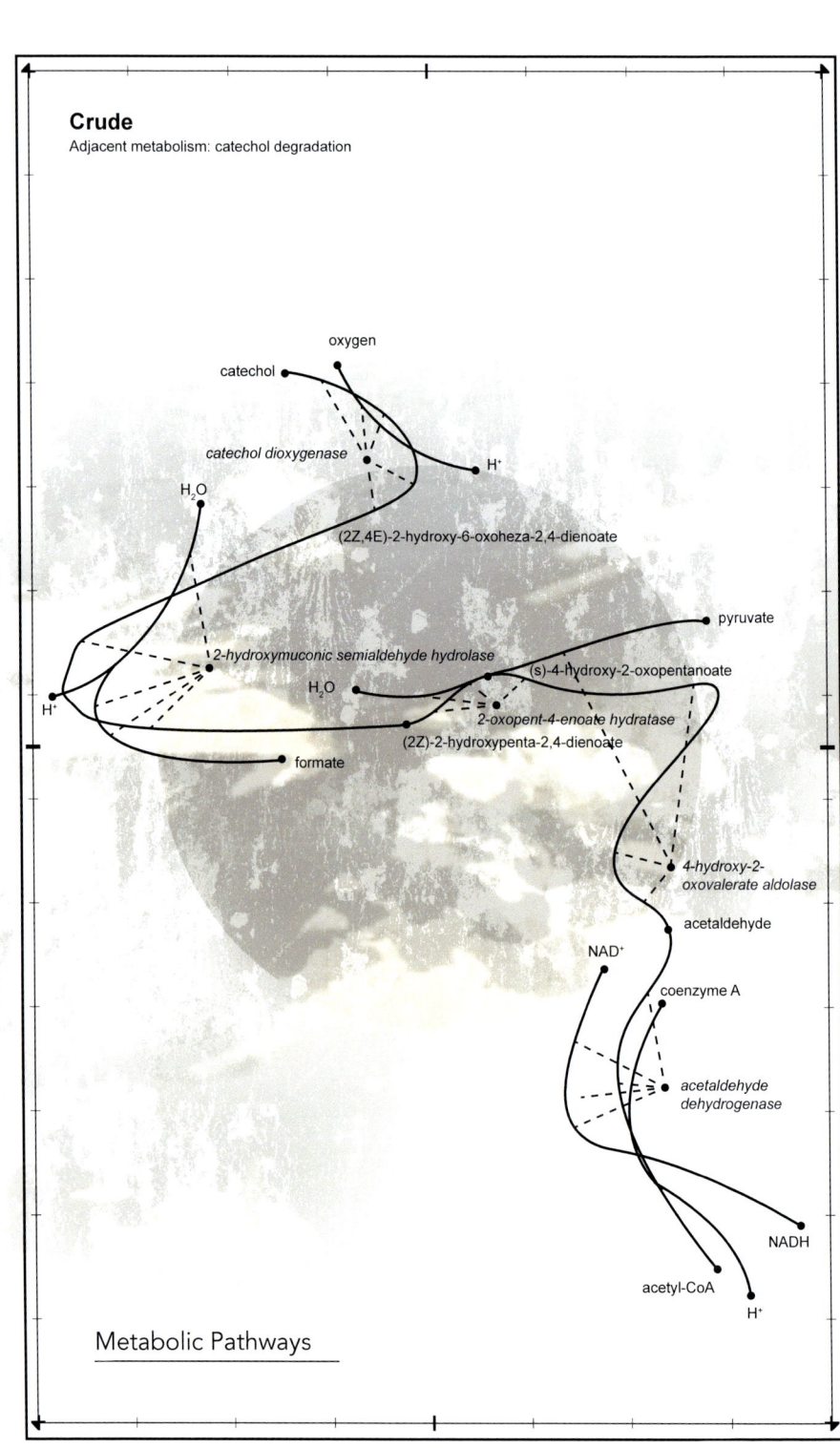

compensated for these losses. In effect, the mere notion of economic compensation for cultural and community loss is a cultural construct of the Western capitalist judicial system that acknowledges only harm to the individual.[50] Perhaps the framework of metabolites helps provide an argument for collective futures.

TOLUENE

Back to the Gowanus Canal and its resident microbiome telling the history of human intervention at this site. Our DNA analysis shows that the metabolism of the microbial community as a whole includes some interesting bioremediation pathways. For example, the sediment microbiome degrades cresol (a petrochemical byproduct), toluene (an industrial solvent), and fixes[51] heavy metals including arsenic. Each of these metabolites can be traced to historical records of factories in operation at the site. Let's look into toluene, and trace its agency then and now.

Toluene is widely used as an industrial additive and solvent. This liquid is usually obtained from gasoline refinement and coke production from coal. It can be considered a metabolite of the petrochemical industry—the result of some reactions, and input for others. Quite literally, it can be used for fuel and has higher octane ratings than gasoline. Toluene and its related compounds can be degraded by bacteria, through various metabolic pathways. Most of these ultimately lead to pyruvate, a small molecule that is at the nexus of many metabolic pathways, including sugar metabolism.

In short, toluene can be used by these microbes as an energy source just like sugar, but with many intermediate processes. It is a petrochemical product, a resource out of place resulting from the extractive endeavors of the fossil fuel industry and is now feeding these microorganisms. Despite the competition in the community for these resources, their consumption is adaptive rather than extractive.

The microbes that degrade toluene are in the taxonomic range of Proteobacteria,[52] including *Pseudomonas putida*, known to be the workhorse of bioremediation.[53] While these are named as individuals, they are part of a community of dozens of other microorganisms inhabiting the sediment. The bioremediation functions observed are more so the community's collective activity, rather than those of an individual.

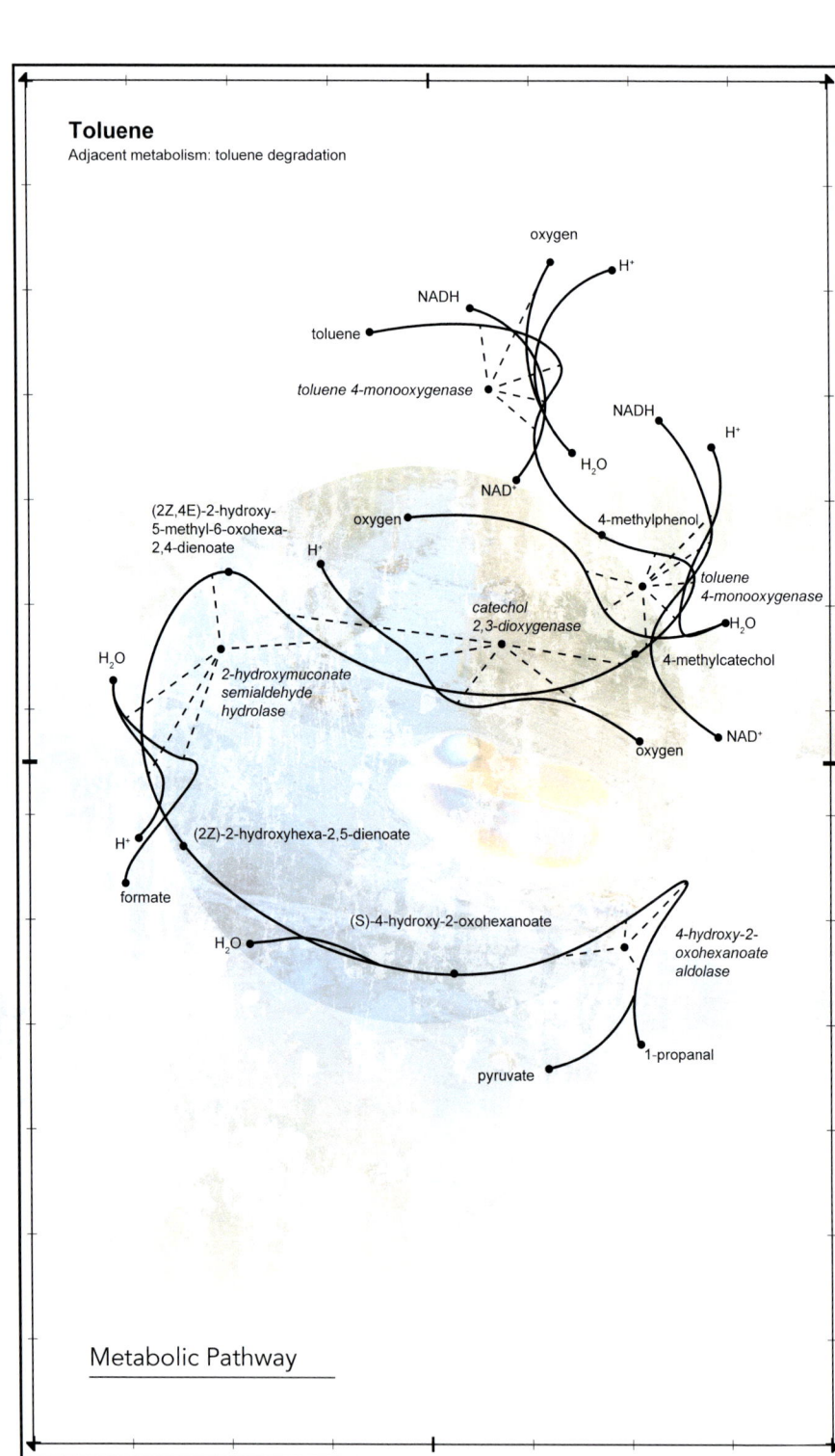

Toluene
Adjacent metabolism: toluene degradation

oxygen

H⁺

NADH

toluene

toluene 4-monooxygenase

H₂O

NAD⁺

NADH

H⁺

4-methylphenol

(2Z,4E)-2-hydroxy-
5-methyl-6-oxohexa-
2,4-dienoate

oxygen

H⁺

*catechol
2,3-dioxygenase*

*toluene
4-monooxygenase*

H₂O

4-methylcatechol

H₂O

*2-hydroxymuconate
semialdehyde
hydrolase*

oxygen

NAD⁺

H⁺

(2Z)-2-hydroxyhexa-2,5-dienoate

formate

(S)-4-hydroxy-2-oxohexanoate

H₂O

*4-hydroxy-2-
oxohexanoate
aldolase*

1-propanal

pyruvate

Metabolic Pathway

These pathways correspond to waste from various paint-making industries over the course of Gowanus' industrial history, including companies such as the Reliance Paint Co and W Devoe & CT Reynolds Paint Works that date to 1915.[54] The story of paint in Gowanus does not stop there, as the NYC Housing Authority (NYCHA) campuses still face ongoing issues with lead paint. Just a year after our sampling expedition described in the beginning of this chapter, a toddler in the Gowanus Houses was poisoned by lead paint chips.[55] The ground for the Gowanus Houses was broken in 1948 to house returning World War II veterans,[56] along with the Wyckoff Gardens and Warren Street NYCHA campuses.

One of the uses of lead as a paint additive was its biocidal properties. Lead paints are water resistant and therefore less prone to mold and mildew and also are used for their biocidal oligodynamic effects,[57] like a silver spoon or copper cup. While lead paint has been banned since 1978, we should not be surprised to find the newly developed condominiums touting waterfront views on the canal also be equipped with antimicrobial paint, surely deemed "safe" for families. If microbes are taken into consideration in architectural design, it is surely to eliminate them. In particular, one of the concerns of the rezoning plans have to do with gut microbes—specifically, those that will end up in the canal from the increased residential load on the neighborhood's already taxed sewer system, which regularly overflows in the canal with heavy rains.[58]

As gentrification of the neighborhood follows the Superfund declaration in 2013, many NYCHA residents are concerned that the neighborhood infrastructure will not sustain an influx of additional residents. Increased property values also displace longtime residents, who will not benefit from the remediated canal despite the fact that they and their families suffered from its exposure.

The human storytelling tendency is to want to find the actor, the hero,[59] but here what matters is the assemblage. The poetic relationship of historic record-keeping and remediation between microbes and the built environment in the canal was an inadvertent one, but we can find clues at these extremes that indicate the possibility for cross-species collaborations. The current plans for remediation of the site is to cap the canal with concrete, effectively eliminating the microbes which, given the time, could clean up the contaminants.

CONCLUSION

These metabolites, of the Gowanus Canal and the other environments outlined in this book, give us a framework to define material agency in human systems as mediated by microbial metabolisms. This microbial continuum fills the porous spaces between material and human, nature and culture, at telescoping scales.

The pathways that lead to a radiation resistant metabolite mediate our relationship to the cosmos and hold agency in promising the possibility of long-term space travel. Scaling down from the cosmic to the internal universe, neurotransmitter metabolites mediate the relationship between the inner and the outside world. Our bodies occupy the structures built with the metabolites of calcium carbonate, holding together soil and mud to define habitable environments. On the planetary scale, the commons of the beaches and their connection to deep time and deep sea reservoirs are fed by the metabolisms of the coral holobiont. Our daily bread and wine, and the rituals and trade that surround them are mediated by fermentation metabolisms, whether you pick them up at Whole Foods or from the Roman market. On a cultural scale, the environmental contamination related to fossil fuel extraction and usage is mediated by microbial metabolisms, both in remediation and as a historical record. These are a few examples, but I imagine we can find metabolites in the interstices of any material we choose to observe. The metabolic framework for materialism follows Donna Haraway's material animism: "It matters which worlds world worlds. It matters who eats whom and how. It remains a material question for cosmopolitical critters in the Communities of Compost."[60] Given the "task … to make kin in lines of inventive connection as a practice of learning to live and die well with each other in a thick present," the Communities of Compost is an imagination of a community in which each human individual is tasked with the stewardship of another species, and exists as a symbiont with that species. The symbiosis manifests in such a way that their actions must nurture the "ongoingness" of that species for the next five human generations. Here metabolic material agency suggests a way to frame and nurture ongoingness with microbial worlds and their material agency.

All of these examples discussed illustrate the emergent agencies of materials in relationship with microbes—they arose, evolved, adapted, without human input. But there are many intentional interfaces that people have designed to enact and harness these material agencies. Synthetic biologists dissect cellular processes and excise molecular mechanisms to use them as engineering tools and materials. Large-scale bioreactors are built to harness, like beasts of burden, metabolisms for fermentation or biosynthesis of materials. These apparatuses should actually not be an aspiration, but a cautionary tale.

Though biological, these practices are as extractive as material mining operations. It is not for naught that microbes are cast as "factories" in the biotechnological discourse, yet offered as "green" or "natural" approaches to our need for material resources. The caution here is not the issue of scale, but the implicitly finite nature of this extractive process. Here Donna Haraway's *ongoingness* is a meaningful meter to assess these relationships—can they be stewarded for five generations of Communities of Compost? So, if we acknowledge that material agency in human ecologies, societies, and economies is mediated by microbial metabolisms, how can we embrace this agency of metabolites?

We can think about this in the context of design. We owe meaningful aspects of our growth, development, health, and well-being to the denizens of microorganisms with which we interact. [61] Some are symbiotic—the human microbiome of the skin, gut, mouth. Some are environmental—the microbes that inhabit the indoor and outdoor spaces we live and work in. The distinction between self and the space around us is porous. Humans are entangled in a microbial continuum between organism and environment. This notion of the multispecies self—the holobiont—was developed early in the nineteenth century[62] and formalized in the late twentieth century in context of genetics by Lynn Margulis.[63]

How is this meaningful from a material perspective? Design for humans should be design for the more-than-human.[64] Relational inclusion, as opposed to eradication, of microbes should impact our choice of design materials, our clothing, our layout of spaces, our climate control systems—the interface technologies that mediate our corporeal presence in ecosystems. Architects and urban planners choose the materials with which we interact in our cities. These design material choices, in turn, dictate the kinds of microbes with whom we cohabitate. Designers map metrics of temperature, light, energy consumption, traffic, sound, and air quality to make these decisions and apply these abiotic measurements toward the health and comfort of human inhabitants. Microscopic life has not traditionally been considered in these metrics unless it threatens human health as a pathogen.

My longest running close observation of microbial relationships with the built environment is through our work at the Gowanus Canal. This contamination is not due to pathogenic microbes, in fact, as we have seen, the living and evolving microbiome is actively remediating the toxic waste of our past century's anthropocentric industrial design. The current plans for remediation of the site is to cap the canal with concrete, effectively eliminating the microbes which, given the time, could digest the contaminants.

Willing relationality of microbes and humans is a challenge of scale. Indeed, materials look and operate very differently at the human scale and the micron scale. What does designing for an organism mean when the organism is invisible and instruments are therefore necessary to mediate the relationship? The growing field of bioart is feeding our collective subconscious, and bio-designers are proposing interfaces and interactions that offer explicitly collaborative relationships. These tools should help us get a "feeling for the organism," to use a term favored by Barbara McClintock.[65] Here *feeling* is about empathy and establishing a relationship, but also about developing intuitive understanding through continued and careful observation.

What ongoingness and kin-making await us that employ sensitivities to the oft-overlooked actors and agents of our host environments—the microscopic, the nonhuman, the nonliving?

ENDNOTES

1 Ian Quate and Matthew Seibert, landscape designers, author and editor of this volume, respectively. Ellen Jorgensen, biologist and entrepreneur, co-founded Genspace, New York's first community biology lab, which hosted this sampling expedition.

2 Scott Enman, "Toxic Gowanus Canal to Get More Signs Reminding You Not to Fish or Swim in It," *Brooklyn Eagle*, December 6, 2020, https://brooklyneagle.com/articles/2019/12/06/toxic-gowanus-canal-to-get-more-signs-reminding-you-not-to-fish-or-swim-in-it/.

3 Scape Studio, "The Gowanus Lowlands," accessed October 18, 2020, https://www.scapestudio.com/projects/the-gowanus-lowlands/.

4 "The Gowanus Canal Conservancy," accessed October 18, 2020, https://gowanuscanalconservancy.org/.

5 Graciously lent to us by the Gowanus Dredgers, a local non-profit organization that offers human-powered water sports on the canal. In the summertime, don't miss their movie nights on the water.

6 Weill Cornell Medical College, "The Mason Lab, Integrative Functional Genomics," accessed October 18, 2020, http://www.masonlab.net/.

7 "Native American Archaeology," New York State Museum, accessed October 18, 2020, http://www.nysm.nysed.gov/research-collections/archaeology/native-american-archaeology/collections/west-athens-hill-site.

8 Joseph Alexiou, *Gowanus: Brooklyn's Curious Canal*. (S.l.: New York University Press, 2020).

9 The Environmental Protection Agency's Superfund program designates certain contaminated sites as priority for remediation, in part based on their threat to human health. US EPA, "Superfund," US EPA, accessed October 19, 2020, https://www.epa.gov/superfund.

10 Richard Plunz and Patricia Culligan, "Eco-Gowanus. Urban Remediation by Design," *New Urbanisms / Columbia University Urban Design Program* 8 (n.d.).

11 See the *Crude: The Bakken Fossil Fuel Frontier* chapter for an onto-cartographic exploration of this material.

12 The class, taught by Elaine Gan at NYU, focused on clocks and the walking tour was a discussion on how time is metered differently for different organisms.

13 A turning basin is a small offshoot of the canal, at a right angle. This basin allows large vessels to turn around as the narrow width of the canal otherwise does not afford it.

14 See the *Mud: and its Meaning in a Port Town* chapter for an onto-cartographic exploration of this material.

15 See the *Clay: Spies in the Making* chapter for more on this vessel.

16 See the *Sand: 825 Miles* chapter for more on the engineering of coastal commons.

17 "Community-Based Urban Design," Gowanus by Design, accessed October 18, 2020, https://www.gowanusbydesign.org/.

18 See the *Uranium: Big Bangs, Metal as Metaphor* chapter for the effects of radiation on humans.

19 Remediation of heavy metals is not degradation as they are elemental, but organisms can assimilate them and thus prevent them from interacting in other harmful ways.

20 A term coined by the poet and geneticist Christopher Mason, www.masonlab.net.

21 For a discussion of missing datasets and biases in data collection and dissemination, see Catherine D'Ignazio and Lauren F. Klein, *Data Feminism*, Strong Ideas Series (Cambridge, MA: The MIT Press, 2020).

22 "The Microbes in Your Home Could Save Your Life | Popular Science," accessed October 19, 2020, https://www.popsci.com/bugged/.

23 Ebrahim Afshinnekoo et al., "Geospatial Resolution of Human and Bacterial Diversity with City-Scale Metagenomics," *Cell Systems*, February 2015, https://doi.org/10.1016/j.cels.2015.01.001.

24 "XMP – Profiling Novel and Extreme Environments Using Advanced Genomic and Microbiological Technologies," accessed October 19, 2020, http://extrememicrobiome.org/.

25 Microorganisms consist in the second-largest category of biomass on Earth, after plants. Yinon M. Bar-On, Rob Phillips, and Ron Milo, "The Biomass Distribution on Earth," *Proceedings of the National Academy of Sciences* 115, no. 25 (June 19, 2018): 6506–11, https://doi.org/10.1073/pnas.1711842115.

26 "Homo Sapiens Ethanol Degradation II," accessed October 20, 2020, https://biocyc.org/HUMAN/NEW-IMAGE?type=PATHWAY&object=PWY66-21.

27 "ToxicSites," accessed October 20, 2020, http://www.toxicsites.us/.

28 Mark Smith et al., "Natural Bacterial Communities Serve as Quantitative Geochemical Biosensors," *MBio* 6, no. 3 (2015): 1–13, https://doi.org/10.1128/mBio.00326-15.Editor.

29 As did Madonna, who coined the popular use of the term in her 1984 track "Material Girl." Just as Madonna insists that "If they don't give me proper credit / I just walk away" we here try to give proper credit to the microbes in our material world.

30 Thomas Cavalier-Smith, Martin Brasier, and T. Martin Embley, "Introduction: How and When Did Microbes Change the World?," *Philosophical Transactions of the Royal Society B: Biological Sciences* 361, no. 1470 (June 29, 2006): 845–50, https://doi.org/10.1098/rstb.2006.1847.

31 "[...] a proposed genre of onto-cartographic stories [...], whose core protagonists are often not human, often not even living, but rather mineral, chemical, elemental. " Introduction, p 12.

32 Ekaterina Dadachova and Arturo Casadevall, "Ionizing Radiation: How Fungi Cope, Adapt, and Exploit with the Help of Melanin," *Current Opinion in Microbiology* 11, no. 6 (December 2008): 525–31, https://doi.org/10.1016/j.mib.2008.09.013.

33 Paridhi Gupta et al., "MDP: A Deinococcus Mn2+-Decapeptide Complex Protects Mice from Ionizing Radiation," *PLOS ONE* 11, no. 8 (August 8, 2016): 2, https://doi.org/10.1371/journal.pone.0160575.

34 Franco Cortese et al., "Vive La Radiorésistance!: Converging Research in Radiobiology and Biogerontology to Enhance Human

Radioresistance for Deep Space Exploration and Colonization," *Oncotarget* 9, no. 18 (February 12, 2018): 14692–722, https://doi.org/10.18632/oncotarget.24461.

35 "Notification to NASA," *Maitriyana Buddhist Community* (blog), accessed October 20, 2020, https://maitriyana.com/2020/05/29/notification-to-nasa/.

36 Stephanie A. Flowers, Kristen M. Ward, and Crystal T. Clark, "The Gut Microbiome in Bipolar Disorder and Pharmacotherapy Management," *Neuropsychobiology* 79, nos. 1–2 (2020): 43–49, https://doi.org/10.1159/000504496.

37 John F. Cryan and Timothy G. Dinan, "Mind-Altering Microorganisms: The Impact of the Gut Microbiota on Brain and Behaviour," *Nature Reviews. Neuroscience* 13, no. 10 (October 2012): 701–12, https://doi.org/10.1038/nrn3346.

38 Michael G. Gomez et al., "Biogeochemical Changes during Bio-Cementation Mediated by Stimulated and Augmented Ureolytic Microorganisms," *Scientific Reports* 9, no. 1 (August 8, 2019): 11517, https://doi.org/10.1038/s41598-019-47973-0.

39 "'Oyster Artist' Creates High-Tech Artificial Shells – Chesapeake Bay Magazine," accessed October 20, 2020, https://chesapeakebaymagazine.com/oyster-artist-creates-high-tech-artificial-shells/.

40 The term "holobiont" encompasses the notion of host and symbiont, and broadens the definition of identity to include both an individual and their microbiome.

41 National Oceanic and Atmospheric Administration, "How Does Sand Form?," accessed October 20, 2020, https://oceanservice.noaa.gov/facts/sand.html.

42 Janelle R Thompson et al., "Microbes in the Coral Holobiont: Partners through Evolution, Development, and Ecological Interactions," *Frontiers in Cellular and Infection Microbiology* 4 (January 7, 2014): 176, https://doi.org/10.3389/fcimb.2014.00176.

43 *A philsophical Enquiry into some of the most considerable phenomena's of Nature ... conformable to the doctrine of Fermentation* (London, 1715).

44 Cal BP denotes calibrated carbon dating, Before Present.

45 Adam Boethius, "Something Rotten in Scandinavia: The World's Earliest Evidence of Fermentation," *Journal of Archaeological Science* 66 (February 2016): 169–80, https://doi.org/10.1016/j.jas.2016.01.008.

46 Duccio Cavalieri et al., "Evidence for S. Cerevisiae Fermentation in Ancient Wine," *Journal of Molecular Evolution* 57 (August 1, 2003): S226–32, https://doi.org/10.1007/s00239-003-0031-2.

47 Eric A Dubinsky et al., "Succession of Hydrocarbon-Degrading Bacteria in the Aftermath of the Deepwater Horizon Oil Spill in the Gulf of Mexico," *Environmental Science & Technology* 47, no. 19 (October 1, 2013): 10860–67, https://doi.org/10.1021/es401676y.

48 Gene Zine, Elizabeth Henaff, NYC Tech Zine Fair, 2018.

49 "Field Pilot Test of Novel Biological EOR Process for Extracting Trapped Oil from Unconventional Reservoirs | SBIR.Gov," accessed October 20, 2020, https://www.sbir.gov/sbirsearch/detail/1505327.

50 Erick Rhoan, "The Rightful Position: The BP Oil Spill and the Gulf Coast Tribes," *San Joaquin Agricultural Law Review* 20, no. 173 (2010): 173–92.

51 Remediation of heavy metals is not degradation as they are elemental, but organisms can assimilate them and thus prevent them from interacting in other harmful ways.

52 "MetaCyc Proteobacteria," accessed October 20, 2020, https://biocyc.org/META/NEW-IMAGE?type=ORGANISM&object=TAX-1224.

53 Eduardo Díaz, "Bacterial Degradation of Aromatic Pollutants: A Paradigm of Metabolic Versatility," *International Microbiology: The Official Journal of the Spanish Society for Microbiology* 7, no. 3 (September 2004): 173–80.

54 Plunz and Culligan, "Eco-Gowanus. Urban Remediation by Design."

55 J. David Goodman, Al Baker, and James Glanz, "Tests Showed Children were Exposed to Lead. The Official Response: Challenge the Tests (Published 2018)," *The New York Times*, November 18, 2018, sec. New York, https://www.nytimes.com/2018/11/18/nyregion/nycha-lead-paint.html.

56 "Officials Start Work on Gowanus Houses | The Brooklyn Eagle," Brooklyn Public Library, accessed October 20, 2020, http://bklyn.newspapers.com/image/52811685/.

57 "Oligodynamic Effect," Wikipedia, accessed October 20, 2020, https://en.wikipedia.org/w/index.php?title=Oligodynamic_effect&oldid=964182067.

58 "Flooding and the Urban Microbiome," accessed October 20, 2020, https://wp.nyu.edu/floodingmicrobiomes/.

59 Ursula K. Le Guin, *Carrier Bag Theory of Fiction* (IGNOTA Books, 2020).

60 Donna J. Haraway, *Staying with the Trouble: Making Kin in the Chthulucene* (Duke University Press, 2016), https://doi.org/10.1215/9780822373780.

61 Pirkka V. Kirjavainen et al., "Farm-Like Indoor Microbiota in Non-Farm Homes Protects Children from Asthma Development," *Nature Medicine* 25, no. 7 (July 2019): 1089–95, https://doi.org/10.1038/s41591-019-0469-4.

62 Jan Baedke, Alejandro Fábregas-Tejeda, and Abigail Nieves Delgado, "The Holobiont Concept before Margulis," *Journal of Experimental Zoology Part B: Molecular and Developmental Evolution* 334, no. 3 (2020): 149–55, https://doi.org/10.1002/jez.b.22931.

63 Lynn Margulis and René Fester, eds., *Symbiosis as a Source of Evolutionary Innovation: Speciation and Morphogenesis* (Cambridge, Mass: MIT Press, 1991).

64 Stacy Alaimo, *Bodily Natures: Science, Environment, and the Material Self* (Bloomington: Indiana University Press, 2010).

65 Elizabeth Henry, "Toward a 'Feeling for the Organism,'" *NWSA Journal* 9, no. 3, (1997): 156–62.

BIBLIOGRAPHY

A philosophical Enquiry into some of the most considerable phenomena's of Nature ... conformable to the doctrine of Fermentation. London, 1715.

Afshinnekoo, Ebrahim, Cem Meydan, Shanin Chowdhury, Dyala Jaroudi, Collin Boyer, Nick Bernstein, Julia M. Maritz, et al. "Geospatial Resolution of Human and Bacterial Diversity with City-Scale Metagenomics." *Cell Systems*, February 2015. https://doi.org/10.1016/j.cels.2015.01.001.

Alaimo, Stacy. *Bodily Natures: Science, Environment, and the Material Self.* Bloomington: Indiana University Press, 2010.

Alexiou, Joseph. *Gowanus: Brooklyn's Curious Canal.* New York: New York University Press, 2020.

Baedke, Jan, Alejandro Fábregas-Tejeda, and Abigail Nieves Delgado. "The Holobiont Concept before Margulis." *Journal of Experimental Zoology Part B: Molecular and Developmental Evolution* 334, no. 3 (2020): 149–55. https://doi.org/10.1002/jez.b.22931.

Bar-On, Yinon M., Rob Phillips, and Ron Milo. "The Biomass Distribution on Earth." *Proceedings of the National Academy of Sciences* 115, no. 25 (June 19, 2018): 6506–11. https://doi.org/10.1073/pnas.1711842115.

Boethius, Adam. "Something Rotten in Scandinavia: The World's Earliest Evidence of Fermentation." *Journal of Archaeological Science* 66 (February 2016): 169–80. https://doi.org/10.1016/j.jas.2016.01.008.

Brooklyn Public Library. "Officials Start Work on Gowanus Houses | The Brooklyn Eagle." Accessed October 20, 2020. http://bklyn.newspapers.com/image/52811685/.

Cavalieri, Duccio, Patrick E. McGovern, Daniel L. Hartl, Robert Mortimer, and Mario Polsinelli. "Evidence for S. Cerevisiae Fermentation in Ancient Wine." *Journal of Molecular Evolution* 57 (August 1, 2003): S226–32. https://doi.org/10.1007/s00239-003-0031-2.

Cavalier-Smith, Thomas, Martin Brasier, and T. Martin Embley. "Introduction: How and When Did Microbes Change the World?" *Philosophical Transactions of the Royal Society B: Biological Sciences* 361, no. 1470 (June 29, 2006): 845–50. https://doi.org/10.1098/rstb.2006.1847.

Cortese, Franco, Dmitry Klokov, Andreyan Osipov, Jakub Stefaniak, Alexey Moskalev, Jane Schastnaya, Charles Cantor, et al. "Vive La Radiorésistance!: Converging Research in Radiobiology and Biogerontology to Enhance Human Radioresistance for Deep Space Exploration and Colonization." *Oncotarget* 9, no. 18 (February 12, 2018): 14692–722. https://doi.org/10.18632/oncotarget.24461.

Cryan, John F, and Timothy G Dinan. "Mind-Altering Microorganisms: The Impact of the Gut Microbiota on Brain and Behaviour." *Nature Reviews. Neuroscience* 13, no. 10 (October 2012): 701–12. https://doi.org/10.1038/nrn3346.

Dadachova, Ekaterina, and Arturo Casadevall. "Ionizing Radiation: How Fungi Cope, Adapt, and Exploit with the Help of Melanin." *Current Opinion in Microbiology* 11, no. 6 (December 2008): 525–31. https://doi.org/10.1016/j.mib.2008.09.013.

Díaz, Eduardo. "Bacterial Degradation of Aromatic Pollutants: A Paradigm of Metabolic Versatility." *International Microbiology: The Official Journal of the Spanish Society for Microbiology* 7, no. 3 (September 2004): 173–80.

D'Ignazio, Catherine, and Lauren F. Klein. *Data Feminism.* Strong Ideas Series. Cambridge, MA: The MIT Press, 2020.

Dubinsky, Eric A., Mark E. Conrad, Romy Chakraborty, Markus Bill, Sharon E. Borglin, James T. Hollibaugh, Olivia U. Mason et al. "Succession of Hydrocarbon-Degrading Bacteria in the Aftermath of the Deepwater Horizon Oil Spill in the Gulf of Mexico." *Environmental Science & Technology* 47, no. 19 (October 1, 2013): 10860–67. https://doi.org/10.1021/es401676y.

Enman, Scott. "Toxic Gowanus Canal to Get More Signs Reminding You Not to Fish or Swim in it." *Brooklyn Eagle*, December 6, 2020. https://brooklyneagle.com/articles/2019/12/06/toxic-gowanus-canal-to-get-more-signs-reminding-you-not-to-fish-or-swim-in-it/.

"Field Pilot Test of Novel Biological EOR Process for Extracting Trapped Oil from Unconventional Reservoirs | SBIR.Gov." Accessed October 20, 2020. https://www.sbir.gov/sbirsearch/detail/1505327.

"Flooding and the Urban Microbiome." Accessed October 20, 2020. https://wp.nyu.edu/floodingmicrobiomes/.

Flowers, Stephanie A., Kristen M. Ward, and Crystal T. Clark. "The Gut Microbiome in Bipolar Disorder and Pharmacotherapy Management." *Neuropsychobiology* 79, no. 1–2 (2020): 43–49. https://doi.org/10.1159/000504496.

Gomez, Michael G., Charles M. R. Graddy, Jason T. DeJong, and Douglas C. Nelson. "Biogeochemical Changes during Bio-Cementation Mediated by Stimulated and Augmented Ureolytic Microorganisms." *Scientific Reports* 9, no. 1 (August 8, 2019): 11517. https://doi.org/10.1038/s41598-019-47973-0.

Goodman, J. David, Al Baker, and James Glanz. "Tests Showed Children were Exposed to Lead. The Official Response: Challenge the Tests (Published 2018)." *The New York Times*, November 18, 2018, sec. New York. https://www.nytimes.com/2018/11/18/nyregion/nycha-lead-paint.html.

Gowanus by Design. "Community-Based Urban Design." Accessed October 18, 2020. https://www.gowanusbydesign.org/.

Gupta, Paridhi, Manoshi Gayen, Joan T. Smith, Elena K. Gaidamakova, Vera Y. Matrosova, Olga Grichenko, Barbara Knollmann-Ritschel, Michael J. Daly, Juliann G. Kiang, and Radha K. Maheshwari. "MDP: A Deinococcus Mn2+-Decapeptide Complex Protects Mice from Ionizing Radiation." *PLOS ONE* 11, no. 8 (August 8, 2016): e0160575. https://doi.org/10.1371/journal.pone.0160575.

Haraway, Donna J. *Staying with the Trouble: Making Kin in the Chthulucene*. Duke University Press, 2016. https://doi.org/10.1215/9780822373780.

Henry, Elizabeth. "Toward a 'Feeling for the Organism.'" *NWSA Journal* 9, no. 3, (1997): 156–62.

"Homo Sapiens Ethanol Degradation II." Accessed October 20, 2020. https://biocyc.org/HUMAN/NEW-IMAGE?type=PATHWAY&object=P-WY66-21.

Kirjavainen, Pirkka V., Anne M. Karvonen, Rachel I. Adams, Martin Täubel, Marjut Roponen, Pauli Tuoresmäki, Georg Loss, et al. "Farm-Like Indoor Microbiota in Non-Farm Homes Protects Children from Asthma Development." *Nature Medicine* 25, no. 7 (July 2019): 1089–95. https://doi.org/10.1038/s41591-019-0469-4.

Le Guin, Ursula K. *Carrier Bag Theory of Fiction*. IGNOTA Books, 2020.

Maitriyana Buddhist Community. "Notification to NASA." Accessed October 20, 2020. https://maitriyana.com/2020/05/29/notification-to-nasa/.

Margulis, Lynn, and René Fester, eds. *Symbiosis as a Source of Evolutionary Innovation: Speciation and Morphogenesis*. Cambridge, Mass: MIT Press, 1991.

"MetaCyc Proteobacteria." Accessed October 20, 2020. https://biocyc. org/META/NEW-IMAGE?type=ORGANISM&object=TAX-1224.

National Oceanic and Atmospheric Administration. "How Does Sand Form?" Accessed October 20, 2020. https://oceanservice.noaa.gov/facts/ sand.html.

New York State Museum. "Native American Archaeology." Accessed October 18, 2020. http://www.nysm.nysed.gov/research-collections/archae-ology/native-american-archaeology/collections/west-athens-hill-site.

"'Oyster Artist' Creates High-Tech Artificial Shells – Chesapeake Bay Magazine." Accessed October 20, 2020. https://chesapeakebaymagazine. com/oyster-artist-creates-high-tech-artificial-shells/.

Plunz, Richard, and Patricia Culligan. "Eco-Gowanus. Urban Remediation by Design." *New Urbanisms / Columbia University Urban Design Program* 8 (n.d.).

Rhoan, Erick. "The Rightful Position: The BP Oil Spill and the Gulf Coast Tribes." *San Joaquin Agricultural Law Review* 20, no. 173 (2010): 173–92.

Scape Studio. "The Gowanus Lowlands." Accessed October 18, 2020. https://www.scapestudio.com/projects/the-gowanus-lowlands/.

Smith, Mark, Andrea M. Rocha, Chris S. Smillie, Scott W. Olesen, Charles Paradis, Liyou Wu, James H. Campbell et al. "Natural Bacterial Communities Serve as Quantitative Geochemical Biosensors." *MBio* 6, no. 3 (2015): 1–13. https://doi.org/10.1128/mBio.00326-15.Editor.

"The Gowanus Canal Conservancy." Accessed October 18, 2020. https:// gowanuscanalconservancy.org/.

"The Microbes in Your Home Could Save Your Life | Popular Science." Accessed October 19, 2020. https://www.popsci.com/bugged/.

Thompson, Janelle R., Hanny E. Rivera, Collin J. Closek, and Mónica Medina. "Microbes in the Coral Holobiont: Partners through Evolution, Development, and Ecological Interactions." *Frontiers in Cellular and Infection Microbiology* 4 (January 7, 2014): 176. https://doi.org/10.3389/ fcimb.2014.00176.

"ToxicSites." Accessed October 20, 2020. http://www.toxicsites.us/.

US EPA. "Superfund." US EPA. Accessed October 19, 2020. https://www. epa.gov/superfund.

Weill Cornell Medical College. "The Mason Lab, Integrative Functional Genomics." Accessed October 18, 2020. http://www.masonlab.net/.

Wikipedia. "Oligodynamic Effect." Accessed October 20, 2020. https://en.wikipedia.org/w/index.php?title=Oligodynamic_effect&ol-did=964182067.

"XMP – Profiling Novel and Extreme Environments Using Advanced Genomic and Microbiological Technologies." Accessed October 19, 2020. http://extrememicrobiome.org/.

IMAGE CITATIONS + CREDITS

Listed by Page Number
281   Photo: Elizabeth Hénaff
283   Photo: Elizabeth Hénaff

285   Doering, Ulrich. Lake Natron, Tanzania, the Red Pigment in the Cyanobacteria Produce the Deep Reds of the Open Water - Image ID: A4K59H . n.d. Photograph. https://www.alamy.com/stock-photo-lake-natron-tanzania-the-red-pigment-in-the-cyanobacteria-produce-11068732.html; Blood Falls in Antarctica. n.d. Photograph. https://www.pikrepo.com/fpeqc/blood-falls-in-antarctica; Jackson, Martina. Pitch Lake, La Brea, Trinidad. June 19, 2011. Photograph. https://commons.wikimedia.org/wiki/File:Pitch_Lake.JPG; Marcos, Luis Bartolomé. Río Tinto, Cauce 34. October 28, 2014. Photograph. https://commons.wikimedia.org/wiki/File:R%C3%ADo_Tinto,_cauce_34.jpg; Salar-De-Uyuni-4634087__340. n.d. Photograph. https://cdn.pixabay.com/photo/2019/11/18/07/03/salar-de-uyuni-4634087__340.jpg; St. John, James. Spinifex Metakomatiite (Upper Komatiitic Unit, Kidd-Munro Assemblage, Neoarchean, 2.711-2.719 Ga; Pyke Hill, near the Potter Mine, East of Timmins, Ontario, Canada) 4. July 15, 2012. Photograph. https://www.flickr.com/photos/jsjgeology/46911126565; Submarine Ring of Fire 2006 Exploration. expl1497, 2006. NOAA Vents Program. Retrieved Retrieved September 7, 2020;   References: "Acid Mine Drainage." Wikipedia. Wikimedia Foundation, October 19, 2020. https://en.wikipedia.org/wiki/Acid_mine_drainage; Allred, Ashlee, and Bonnie K. Baxter. "Hypersaline Environments." Hypersaline, January 6, 2020. https://serc.carleton.edu/microbelife/extreme/hypersaline/index.html; Arora, Naveen Kumar, and Hovik Panosyan. "Extremophiles: applications and roles in environmental sustainability." (2019): 1-2; Bordenstein, Sarah. "Alkaline Environments." Alkaline, June 17, 2020. https://serc.carleton.edu/microbelife/extreme/alkaline/index.html; Bruckner, Monica. "Microbes Living within Rocks." Endoliths, October 15, 2020. https://serc.carleton.edu/microbelife/extreme/endoliths/index.html; Deiss, Julian. "Environments Without Water." Without Water, November 13, 2019. https://serc.carleton.edu/microbelife/extreme/withoutwater/index.html; Dominguez, Dallas. "Halophiles." microbewiki. Kenyon College, May 6, 2016. https://microbewiki.kenyon.edu/index.php/Halophiles; Google (n.d.). [Google Maps coordinate locations for identified extreme sites]. Retrieved September 7, 2020; Long, Kat. "New Species of Bacteria Eats Plastic." The Wall Street Journal. Dow Jones & Company, March 10, 2016. https://www.wsj.com/articles/new-species-of-bacteria-eats-plastic-1457634401; Merino, Nancy, Heidi S. Aronson, Diana P. Bojanova, Jayme Feyhl-Buska, Michael L. Wong, Shu Zhang, and Donato Giovannelli. "Living at the Extremes: Extremophiles and the Limits of Life in a Planetary Context." Frontiers in microbiology 10 (2019): 780; "PROJECTS." XMP. Accessed October 25, 2020. http://extrememicrobiome.org/projects/; Richlen, Mindy. "Acidic Environments." Acidic, October 4, 2018. https://serc.carleton.edu/microbelife/extreme/acidic/index.html; "Soda Lake." Wikipedia. Wikimedia Foundation, October 22, 2020. https://en.wikipedia.org/wiki/Soda_lake; Vedler, Eve. "Megaplasmids and the degradation of aromatic compounds by soil bacteria." In Microbial megaplasmids (2009), Springer, Berlin, Heidelberg, pp. 33-53, figure 1; Yoshida, Shosuke, Kazumi Hiraga, Toshihiko Takehana, Ikuo Taniguchi, Hironao Yamaji, Yasuhito Maeda, Kiyotsuna Toyohara, Kenji Miyamoto, Yoshiharu Kimura, and Kohei Oda. "A bacterium that degrades and assimilates poly (ethylene terephthalate)." Science 351, no. 6278 (2016): 1196-1199.

Research Assistant: Theodore Teichman

289   NYC Department of City Planning, NYC Department of Parks and Recreation, NYC Economic Development Corporation. Publicly Owned Waterfront Areas (December 20, 2019), digital map; NYC Open Data. Tidally Coordinated Shoreline (December 12, 2018); NYS GIS Program Office. Street Segment GDB - National Geospatial Data Asset (NGDA) (October 15, 2020), Telephone: (518) 242-5029 Email: its.sm.SAM.Maintenance@its.ny.gov; New York State Department of Environmental Conservation. Stormwater Regulated MS4 Areas - New York State (NYSDEC) (April 15, 2010), Albany, NY, New York State Department of Environmental Conservation

Research Assistant: Theodore Teichman

291   Basemap: Esri "Topographic World Map" Sources: Esri, HERE, Garmin, Intermap, INCREMENT P, GEBCO, USGS, FAO, NPS, NRCAN, GeoBase, IGN, Kadaster NL, Ordnance Survey, Esri Japan, METI, Esri China (Hong Kong), © OpenStreetMap contributors, GIS User Community; Data: Geospatial information for all publicly available FRS facilities that have latitude/longitude data. Facility Registry Service, (July 24, 2020), Environmental Protection Agency, retrieved from the Geospatial Data Download Service

Research Assistant: Theodore Teichman

297   Data: Elizabeth Hénaff. Basemap sources: Esri, DigitalGlobe, GeoEye, i-cubed, USDA FSA, USGS, AEX, Getmapping, Aerogrid, IGN, IGP, swisstopo, and the GIS User Community.

Research Assistant: Theodore Teichman

301   Metabolic Pathway: Caspi, Ron, Richard Billington, Ingrid M Keseler, Anamika Kothari, Markus Krummenacker, Peter E Midford, Wai Kit Ong, Suzanne Paley, Pallavi Subhraveti, and Peter D Karp. "The MetaCyc Database of Metabolic Pathways and Enzymes - a 2019 Update." Nucleic Acids Research 48, no. D1 (2019): D445–D453. https://doi.org/10.1093/nar/gkz862.

Research Assistant: Theodore Teichman

304   Ibid.

Research Assistant: Theodore Teichman

306   Ibid.

Research Assistant: Theodore Teichman

309   Ibid.
        Research Assistant: Theodore Teichman
311   Ibid.
        Research Assistant: Theodore Teichman

What can it mean to be human in this time when the human is something that has become sedimented in the geology of the planet? What forms of ethics and politics arise from the sense of being embedded in, exposed to, and even composed of the very stuff of a rapidly transforming material world?

- Stacy Alaimo[1]

Concl

by Matthew Seibert

unison.

We are in between stories. The old story, the account of how the world came to be and how we fit into it, is no longer effective. Yet we have not learned the new story.

Thomas Berry[2]

Davis Pond Diversion

CONTEXT MAP

Davis Pond Diversion Inlet

Standing on the structure, looking north toward the Mississippi River as 8,000 cubic feet / second get swallowed below.

oupy, SOFT, AND SUPPORTING WORLDS UPON WORLDS

As the gates slowly open, the first thing I notice are the full trees being tugged against the guard bars as if a hole opened in the earth, pulling weighty volumes of muddy water and river detritus down into a tempestuous belly. The second thing is a burgeoning vibration growing beneath my feet, carrying up and into my bones. And then the barges about 200 feet away, larger than tractor trailers, begin to awaken and casually—ominously—drift in our direction. I turn to our guide, a civil engineer with the US War Department's Army Corps of Engineers.[3] He shrugs his shoulders, "They're not supposed to be parked there." This, I admit, fails to assuage a growing eddy of unease, as we are standing on a barrier crafted of concrete and steel but anchored in soil and mud, meant to hold back the enormous weight of America's largest river, a river that has at times been referred to with the moniker of a Hollywood blockbuster, the Glistening Executioner.[4] The barrier was built not just to hold back a wall of water, but also mountains of muddy sediment—or to be more accurate: approximately 436,000 tons[5] of rolling sand dunes, billowing plumes of silt, and curling clouds of clay on any given day—waiting to lurch toward the quickest way to the ocean. Standing on the only thing between it and its downriver destination, vibrations still tremoring the body, lunch makes its presence known anew.

We were standing on the Davis Pond Diversion, a concrete and steel control structure 15 miles west of New Orleans, enabling the controlled release of water and—more importantly, though not originally designed for it—the sediment it carries, into the marshes and swamps of coastal Louisiana's Barataria Basin. This was one of several strategic perforations in the thousands of miles of the Mississippi River levee system, one of several mythic structures in the crusade to employ Old Man River's directed flooding and deposition of sediment to restore wetland health, build land, and confront sea level rise. The Mississippi River, and the material it hauls, embodies a deep history of a life-giving and life-taking force. This force, this propensity or power, this capacity for influence asserts an *agency* of nonliving materials.

Organic and inorganic materials mix, gradually washing out of the Rocky and Appalachian Mountains, from the Great Plains and fields of Middle America, nearly a half of the continental US, and then mobilize within the Mississippi River's water column. This muddy mix is not merely inanimate sand, silt, and clay. Rather, it is an elemental material that gives life to a whole range of vibrant cultures, economies, and ecologies in the Delta region. Sediment also contributes to cyclical gradients of death and decay, what geomorphologists call *the delta cycle*: following floods, newly deposited sediment builds lands, life and its ecologies follow, but the buildup causes the river to find a shorter, quicker way to the ocean, unloading its sediment elsewhere as the first landscape begins slowly settling and ultimately deteriorating back into open saltwater. A new delta lobe, as it is called, emerges every thousand years or so as the river avulses, switching channels, and restarting the cycle. This is, of course, complicated by human action, mostly isolated to the last hundred years or so, not to mention the increasing frequency and intensity of storms and hurricanes as the ocean warms. In fact, it is the vast expanses of these swamps and marshes, built upon deposited sediment, that form the first lines of defense against such growing storms, buffering the remote lifelines of oil and gas and shipping infrastructure upon which much of this country depends. They also buffer the cities and communities that call this imperiled landscape home.

The mud of the Mississippi does not *take* action with premeditated intent, no. But it does put action *into motion* in new and often unexpected ways. In the case of the Mississippi Delta region, it puts a whole human culture into motion, a rich culture of food, music, and parades. It puts one of the nation's richest ecologies of biodiversity into motion, with its roseate spoonbills, American alligators, and bald cypresses. It puts a national economy of oil and gas, fishing, and shipping into motion. The Mississippi's mud acts to give life. And take it away.[6] A material not inert, without power, but acting, full of liveliness, co-creating an entangled world of distributed agencies.

AN ETHICS, AN ETHOS, A PRACTICE

And so again, we return to the question posed in the introduction: Why does this matter? What does a recognition of an agency of nonliving material mean for the reader? The designer? As the previous chapters illustrate in text and image, the binaries of life and material, of nature and culture, of humans as supreme to their environments, are not as clear cut as we may think. In fact, as stories, open to alternative readings, they show an inherent and necessary messiness where conventional and embedded dualities crack and crumble under pressure, to reveal a flat ontology of things, beings, and events. From the Chilean lithium powering our smartphones and balancing our psyches, to the shifting sands and boundaries of Florida's beach-based commons, to the crude oil fueling not only the "man camps" of North Dakota but Man's enterprises across the globe, the preceding chapters facilitate the reader in seeing the world anew, from material up. This amounts to a new power. A new materialist sensibility can catalyze action to transpose a world in crisis—both of climate and social justice—to one compatible with the planet's carrying capacity; a sensibility that exhibits widespread equality across the human, nonhuman, and nonliving; and rich with entangled socioecological relationships. As discussed below, this sensibility is best cultivated through practice, resulting in a collective ethos, to confront and transform the dominant models of being in the world (i.e. the endless production and consumerism of neoliberal capitalism), and offer points and methods of intervention for both the professional designer and the nonprofessional designer, the everyday citizen.

Distributing agency to not only the nonhuman but the nonliving "need not be the denial of our own," as anthropologist Lucy Suchman reminds us.[7] In fact, it calls on us to exhibit an acutely disciplined agency in the cultivation of a new materialist practice of being. A practice of living *with* and *of* all things (biological, mineral, temporal). A type of speculative ethics that "provok[es] political and ethical imagination in the present."[8] A personal-collective ethos of everyday engagement.[9] It is no easy task as it is a battle against centuries of anthropocentrism, patriarchy, rationalism, industrialism, capitalism, among the many other names of the dominant models structuring modern life. A new materialist practice calls for new ontological stances about what our world is, what and how we fit into it, and how we come to know this shared world. This in turn informs our being, doing, and knowing.[10] "Worlds are enacted by practices," writes Colombian-American anthropologist Arturo Escobar.[11] In other words, we must co-create our worlds, and worldviews, anew, through shared and distributed agencies where we are one among many equal actors pursuing a daily, rich, and necessarily messy story of being.

Cultivating a new worldview requires practice and cultivating a new practice requires frequent questioning of the taken-for-granted, of one's assumptions: What invisible externalities am I paying for when purchasing a new phone and the rare metals it is made from? What cosmologies exist behind the salt in my shaker? Where did the sand of this beach come from and where will it go? How many prehistoric lives went into the combustive power of a piston firing in my car? How many contemporary lives for it simply to reach my gas station pump? What was life like when the river was visible, not hidden behind a levee, with frequent but mostly minor flooding?

Embracing interdependence and relationality is at the heart of a new materialist practice and can—should!—materialize at all scales. From alternative economic systems reorganizing geopolitics, to national governance, down to individual behaviors and choices, scales of intervention illustrate how seeing the world with a flat ontology can create concrete change in being within and relating to our environments. As seen both in the above questions and the below precedents, so much of our lives are determined by the economic systems behind them. Transforming how we assign value and worth to other people, other life forms, and other material landscapes, and its systems of exchange are critical for both a new materialist practice and a new worldview for an ecological, just, and vibrantly convivial future. Fortunately, tastes of the transition have already begun.

CONTEMPORARY TRANSITION DISCOURSES AND SCALES OF INTERVENTION

New materialism isn't the first discourse to critique the sociocultural modes and models that have come to dictate our lives over the past few centuries. Economists, activists, public servants, and humanities scholars have all offered criticisms and alternatives to the systems that have led to the rampant social and environmental crises of today. These crises have frequently, and diversely, been traced back to our dominant ways of being in the world in what can be called transition discourses.[12] And, in these same discussions, various alternatives have been proposed across a number of scales.

Among these criticisms and alternatives, some of the most promising have come from a growing contingency of twenty-first century economists. By proposing degrowth and post-development as models to shrink the global population's demands within the carrying capacity of the planet, they directly challenge neoliberal capitalism's dominion. By increasingly recognizing the rights of nature, nations are challenging the legal definition of personhood. By working toward regenerative, distributive, and embedded design of their economies, cities and regions are challenging nature and culture divides. In promoting change of individual behavior and perception, scholars and activists are challenging consumerism, traditional identity, and rationalism. The common thread across these scales and methods of intervention—both illustrative *of* and allied *with* a new materialist practice—is the "need to step outside existing institutional and epistemic boundaries if we truly want to strive for worlds and practices capable of bringing about the significant transformations seen as needed."[13] Let us take a closer look.

Transition discourses continue to expand the conversations around *extractivist*[14] models of neoliberal capitalism and

Barataria Basin

Land-building and wetland growth in Barataria Basin, south of the Davis Pond Diversion and north of Lake Cataouatche.

0   375   750   1125   1500
Feet

its replacement by alternative economic systems. Though aligned in intent, the Global North and Global South differ in their vocabulary and conceptualization of transitions and alternative economics. Standard proposals in the North revolve around the idea of degrowth, often focusing attention on the concepts of *commoning* and the commons, while in the South, the conversation orbits around ideas of post-development, often surrounding *Buen Vivir*,[15] the *rights of nature*, and post-extractivism. Both geographies' conceptualizations are important in articulating visions for a better and feasible tomorrow. Both respond to traditional economists' "physically impossible narratives of continuing growth that guide decisions made by national governments and companies," to use environmental scientist Vaclav Smil's words.[16] Or as teen activist and *Time* Person of the Year, Greta Thunberg, puts it, "We are in the beginning of a mass extinction, and all you can talk about is money and fairy tales of eternal economic growth. How dare you!" Degrowth calls for a shifting of values and systems, from eternal growth (usually measured in GDP) and mass-market production to low carbon, low resource, local services (or cosmopolitan localism[17]). Degrowth is further understood as a new imaginary, disentangling human identity from economic representations.[18] As the winners of the 2019 Nobel Prize in Economics, Abhijit Banerjee and Esther Duflo, argue, "nothing in either our theory or the data proves the highest GDP per capita is generally desirable."[19] Rather than chasing higher and higher growth, as we have done exponentially since the Reagan-Thatcher neoliberal revolution of market-led globalization in the 1980s, governments and institutions should focus on policies and initiatives with proven benefits, such as direct assistance to the poor[20] and growing the practice of *commoning*, or collaboratively managing natural or cultural resources as a commons for the greater good.[21]

In the Global South, in response to Western institutions' (the International Monetary Fund, the World Bank, etc) classification of countries in relation to development (underdeveloped, developing, developed), transition discourses are themselves positioned as *post*-development. Emerging in a refute of foreign worldviews based on economics are social movements like the Quechua and Aymara's *Buen Vivir* in South America and the Zapatista's *pluriverse* in Mexico. This emergence was largely a renewed call for acknowledgment of preexisting Indigenous practices and beliefs in opposition to globalization's homogenizing tidal wave across the world.[22] *Buen Vivir*'s regional, place-based practices—a "living well together" of individuals, communities, and their environments (living and nonliving), under constant construction and reproduction—counter previously unknown modern and Western conceptions of development, wealth, and poverty based on the amassing or scarcity of material goods.[23] Grounded in the belief of multiple ontologies and epistemologies—worldviews and ways of knowing—the

> *most substantive versions of* Buen Vivir *in the Andes reject the linear idea of progress, displace the centrality of Western knowledge by privileging the diversity of knowledges, recognize the intrinsic value of nonhumans (biocentrism), and adopt a relational conception of all life.*[24]

One way in which these Indigenous practices are, in turn, being increasingly realized both home and abroad is the growing awareness of the *rights of nature* and the formalization of those rights in policy and law. Like fundamental human rights and their authority born from the mere fact of human existence, the rights of nature are born from the existence of the natural world, with its species and ecosystems. In critiquing the view of natural worlds as "resource," as property to be owned and exploited, the *rights of nature* concept aspires to mobilize from a place driven by economic motives to that of systems ecology where nature and culture are inextricably entangled and interdependent. And within this web, honoring the *rights of nature* means promoting human survival and well-being. Following Latin America's pink tide,[25] *Buen Vivir* and the *rights of nature* have been written into the Ecuadorian and Bolivian constitutions, in 2008 and 2009 respectively. Article 71 of Ecuador's constitution states that "Nature or Pachamama [Mother Earth], where life is reproduced and exists, has the right to exist, persist, maintain itself and regenerate its own vital cycles, structure, functions and its evolutionary processes."[26] In 2012, Bolivia passed the *Law of Mother Earth and Integral Development for Living Well,* which states, "violation of the rights of Mother Earth, as part of comprehensive development for Living Well, is a violation of public law and the collective and individual rights."[27] The *rights of nature*, so formalized, give representation to nonhumans and thus enable people to voice nonhuman interests in legal proceedings.

Forged in the fires of Indigenous activism in Latin America, the *rights of nature* have since spread. Similar rights of personhood were codified in places like Colombia's Amazon Forest, New Zealand's Whanganui River, and India's Ganges River. A measure in Toledo, Ohio which grants legal rights to the Great Lakes passed in February 2019, setting up the ability to sue on behalf of the threatened ecosystem. National actions have reverberated in the halls of international governance. The United Nations has considered adopting a *Universal Declaration of the Rights of Nature* similar to its lauded *Universal Declaration of Human Rights;*[28] the International Union for Conservation of Nature issued a *World Declaration of the Environmental Rule of Law;*[29] and a number of transnational knowledge networks have emerged, such as *The International Rights of Nature Tribunal.*[30] An more inclusive jurisprudence is gaining ground.

Policy and legal formulation by governing powers are critical—an ultimate goal even. But in a world staring down the barrel of climate catastrophe, a more comprehensive, elemental, and ultimately personal kinship must be cultivated in parallel, not just with the charismatic beings like pandas and pangolins, but with the nonliving and forgotten too: our material substrates, devoid of the fuzziness of fur and usually clothed in uninspiring shades of brown and beige. But so much of our personal, daily lives are predicated around the economic models that structure these substrates. As radical economist Kate Raworth writes, "Reversing consumerism's financial and cultural dominance in public and private life is set to be one of the twenty-first century's most gripping psychological dramas."[31] And as quoted in the introduction of this atlas, we return to Jane Bennet's new materialist presage, "there will be no greening of the economy, no redistribution of wealth, no enforcement or extension of rights without human dispositions, moods, and cultural ensembles hospitable to these effects."[32] Degrowth, *Buen Vivir*, and the *rights of nature* are critical allies in this projection of new dispositions, moods, and cultural ensembles.

But in many ways (with the exception of *Buen Vivir* perhaps), these concepts are indicative of the larger effects we want to see in the systems structuring contemporary life. New materialism is certainly embodied in these efforts, but, as a philosophy, is perhaps best invoked as a personal practice which in turn informs larger and larger collectives as a shared ethos. When we begin to see ourselves and our nonhuman peers as kin, we lay the foundation for the paradigmatic shift needed to upturn consumer culture, reconfigure capitalist economies, and thrive in more-than-human assemblages of life and matter. Donna Haraway calls this messy assemblage the *compost* of the *Chthulucene*: a rich and heterogeneous mix of all things, fundamentally of the earth, that conjures a new era of complex and rewarding relationships beyond the capital and hubris of the Anthropocene.[33] When we begin to see ourselves made of the same carbon as the forests we burn, the same lithium as the salars we siphon, our livelihoods and joy based on the same sand, mud, and clay we manipulate for recreation, industry, and sustenance. When we begin to see ourselves as more than human, linked to microscopic worlds invisibly operating for both our benefit and detriment, tying us to material worlds animated through metabolic processes. We can begin to rethink the economy, the distribution of wealth, and the application of intrinsic rights. Many of us already are. But most of us are not, unconditioned to see that alternative worlds are possible and our place within it malleable.

A new materialist practice can break down the hegemonic dominance of one worldview and open it to a pluriverse of many worlds, grounded in material foundations. It starts here at the individual. One day at a time.

DESIGNING WITH MATERIAL AGENCY

The scalar suite of interventions discussed above instigates transitions away from an extractivist carbon-based capitalism for the few, and informs the many ways in which a quotidian new materialist practice can begin to reverberate up and out. It also articulates a new way of conceptualizing design's role in the world for the designer. As illustrated, in working through this conceptualization, it's important to keep in mind the scales at which professional design happens. Trained in landscape architecture, I find myself frequently communicating to those who ask about the nature of the discipline that landscape designers often engage with everything from the park bench to the park system, from the community engagement meeting to the mega-regional master plan. Designers of the built environment often design the bollard and the means of communication, the physical and the semiotic, the object and, to a degree, the meaning. More often than not this is in service to a capital or private project, because that is how firms keep the lights on, how they support their staff with reliable salaries, and in turn support families and their loved ones. Because that is how our world is designed—the dominant economic model underpinning contemporary civilization. But, little by little, designers are recognizing the common results of working within traditional modes of operations. They are recognizing their historical role in spatializing inequality and defining racialized topographies, erasing cultural histories and damaging ecosystems.[34] The election in the US of a racist and xenophobic administration in 2016 and the maturation of the Black Lives Matter movement in mid-2020 has certainly accelerated the acknowledgment of designers' role in facilitating systemic racism and inequitable public spaces. But clearly there is a lot more work to do in unlearning racist, colonial, and plutocratic practices.

Many firms are beginning to formulate or reground their missions in activist language, prioritizing community engagement and local histories.[35] While these measures are notable and important, a new materialist practice in design would likely take them further. Just as new materialism itself calls for a complete reorganization of one's ontology, leveling life (human and nonhuman) and matter along equal fields of agency, a new materialist design practice would require an ontological reorientation and adoption of an epistemic humility. This reorientation and new humility frequently overlap as they are mutually informative. How one sees the world and their place in it is directly informed by how one acquires knowledge and how they decide what knowledge is true, or untrue. New materialist design practice requires not only a recognition of design's historical role in racist and destructive histories—seeing particular communities of humans, nonhumans, and inanimate resources as undesirable, replaceable, or exploitable— but also an active structuring of projects around multiple and diverse agencies, cultures, economies, geographies, and politics. The designer is no longer the sole genius, or starchitect, but rather the *facilitator* of local life, matter, and the vital relationships between the two. Rather than top-down execution of the designer's brilliant vision, it's much more of a lifting of local voices and empowering of the many to catalyze inspiration and change. This epistemic humility[36]—or humble valuing of different beliefs, ways of knowing, and systems of ethics—brings the designer down to a flat hierarchy with nondesigners,[37] nonhumans, and the nonliving alike.

In more specific terms, designing with material agency might look like the exploratory fieldwork of the Swiss landscape architecture firm Vogt Landschaftsarchitekten whose "work begins not with a hypothesis to prove, but with a search for relevant questions."[38] It might look like the multidisciplinary Public Sediment collaboration, bringing together professional landscape architects, design academics, hydrologists, and ecologists together with community groups to rethink "sediment as a core building block of resilience in San Francisco Bay."[39] It might look like the Sand Engine's designed long-term management of a dynamic coastline;[40] the continuous planned maintenance of landscapes and ecologies like the former landfill Freshkills Park;[41] the annual controlled burns of Nelson Byrd Woltz Landscape Architects' conservation agriculture.[42]

Or, in the Global South, often forgotten in the design discourse of the wealthy, it might look like what Arturo Escobar calls autonomous design. "As a design praxis with communities that has the goal of contributing to their realization as the kinds of entities they are," autonomous design is based on the recognition that every person, and as a collective, every community, practices the design of themselves.[43] In this framing, design must be predicated upon the "presupposition that people are practitioners of their own knowledge," designing social organizations, cultural practices, and relations to the surrounding environment based on the understanding of their realities.[44] Escobar cites his work in Southwestern Colombia's Cauca Valley in collaboration with the Cauca Regional Autonomous Development Corporation (similar to the US's Tennessee Valley Authority) and the Black Communities Process (or PCN, a national group bringing together the majority of Black community organizations in Colombia) as illustrative of autonomous design practices. Key conclusions from this collaboration of governmental officials, academics, activists, and community members include:

> the Colombian-Ecuadorian Pacific as a "region-territory of ethnic groups," the conceptualization of the territory as the space for the "life projects of the communities," a framework for the conservation

> *of biodiversity based on the defense of territory and culture (very different from the established frameworks designed by conservation biologists and economists), and a set of guiding principles for the region's own vision of development and perspective on the future.*[45]

Much like new materialism's recognition of agency and self-organization of matter, or the mineral evolution of organic and inorganic compounds as discussed in the introduction, autonomous design is structured around the idea of *autopoiesis*, or a system's ability to reproduce and maintain itself.[46] With such framing, designing with material agency necessarily requires affording the system the freedom and power to express itself. This is relatively simple if the system doesn't contain that particularly problematic species, Homo sapiens, but few to none do. Thus, design as practiced with new materialist leanings or with Escobar's autonomy involves the facilitation, balancing, and, often, equalizing of actors and stakeholders over time and in consideration of uneven power, voice, and traditional understandings of agency.

NEW MATERIALIST PRACTICE AS TRANSITIONAL NEW STORY

The stories held within this book's chapters are thus a series of first attempts to offer alternative narratives about how we relate to our surrounding contexts, from the physical environments of our towns and cities to the logistical networks wedding us to invisible foreign lands. As the science increasingly reveals, we are not singular bodies operating within an environment. We ourselves are environments continually becoming, evolving, and co-constituting amid, against, and with other environments, sharing matter, information, and livelihoods. The agency of nonliving materials is, and has always been, a defining but unacknowledged feature of the universe. It's time we acknowledge this and embrace its fundamental necessity to our own lives and well-being. A new and evolving value regime for a new and evolving climate regime. A bold and vibrant future is open to our making. Intimately entangled with our material kin.

Evoking Bruno Latour's 1999 query concerning the omission of nonhumans in conceiving the "good life,"[47] used to conclude the Introduction, poet Susan Barba writes in 2020:

> Is it by virtue of this immense life-
>
> giving labor
>
> that the river is not a rights-holder
>
> but a natural object,
>
> meant for profit,
>
> like slaves like women
>
> an order apart
>
> like the roe and the deer?[48]

As global crises intensify, are we, at long last, ready to live the good life together, with and of our more-than-human kin, as rights-holders, as actors of agency? In response to Thomas Berry's opening orientation, let us move from between stories and learn the new. The stories comprising this book are just the beginning. In calling for a new materialist practice, it is incumbent upon us all to take a fresh look at the material substrates running through our daily lives. My ambition as editor of this collection is that the chapters provide precedents, blueprints, and maps to chart the way for these new stories. Might you use them in tandem with onto-cartographic principles in identifying key drivers, relationships, energy sources, and metabolites to see and act in the world from material up?[49]

What new story can you tell?

What new world can you materialize?

There is no time to waste in cultivating a new materialist practice.

ENDNOTES

1   Stacy Alaimo, *Exposed: Environmental Politics and Pleasures in Post-human Times* (Minneapolis: University of Minnesota Press, 2016), 1. Copyright 2016 by the Regents of the University of Minnesota.

2   Thomas Berry, *The Dream of the Earth* (San Francisco: Sierra Club Books, 1988), 123. Reprinted by permission of Counterpoint Press. Copyright © 1988 by Thomas Berry, from *The Dream of the Earth*. Reprinted by permission of Counterpoint Press.

3   This trip was organized by the Dredge Research Collaborative as part of *DredgeFest Louisiana* held in January of 2014 in Baton Rouge and New Orleans, LA.

4   Hugh Auchincloss Brown, correspondence to the Mississippi River Commission, 27 June 1952, New Orleans District Archives, US Army Corps of Engineers.

5   In fact, it used to be more than three times this back in 1951. *See* "The Mississippi River Delta Basin," The Louisiana Coastal Wetlands Planning Protection and Restoration Act Program, accessed May, 27, 2020, https://lacoast.gov/new/About/Basin_data/mr/#gsc.tab=0.

6   Stephen Ambrose, "Man vs. Nature: The Great Mississippi Flood of 1927," *National Geographic*, May 1, 2001, https://www.nationalgeographic.com/culture/2001/05/mississippi-river-flood-culture/; James A. Smith and Mary Lynn Baeck, "'Prophetic vision, vivid imagination': The 1927 Mississippi River flood," *Water Resources Research* 51, no. 12 (November 30, 2015), https://doi.org/10.1002/2015WR017927. The Great Mississippi Flood of 1927 was the most destructive river flood in the history of the US, killing approximately 500 people, leaving another 700,000 homeless, and resulting in around $1 trillion dollars in damage if measured in 2007.

7   Lucy Suchman, *Human-Machine Reconfigurations: Plans and Situated Actions* (Cambridge: Cambridge University Press, 2006), 285.

8   María Puig de la Bellacasa, *Matters of Care: Speculative Ethics in More than Human Worlds* (Minneapolis: University of Minnesota Press, 2017), 7.

9   Ibid., 22.

10   Arturo Escobar, *Designs for the Pluriverse: Radical Interdependence, Autonomy, and the Making of Worlds* (Durham: Duke University Press, 2018).

11   Ibid., 92.

12   Ibid., 137-164. Initiatives around transition discourses include the UK's Transition Town Initiative, the Tellus Institute's Great Transition Initiative, Joanna Macy and Chris Johnstone's the Great Turning, Thomas Berry's the Great Work, Charles Eisentein's transition from the Age of Separation to an Age of Reunion, Tony Fry's from Enlightenment to Sustainment, Andreas Weber's Enlivenment, and Greene's ecological-cultural civilization.

13   Ibid., 139.

14   Extractivism, from Latin America's *extractivismo*, is an economic model defined by exploitation and environmental damage in the extraction of natural resources. See *Lithium: Tracing the Green Energy Paradox across Battery, Body, Landscape, and Cosmos* chapter of this collection and Macarena Gómez-Barris, *The Extractive Zone: Social Ecologies and Decolonial Perspectives* (Durham: Duke University Press, 2017).

15 *Good-Living* in English, or collective well-being. Or in its original form, *Sumac Kawsay* in Quechua, the Andean Indigenous originators of the worldview and subsequent socioecological movement.

16 Vaclav Smil, *Growth: From Microorganisms to Megacities* (Cambridge: MIT Press, 2020), 507.

17 Arturo Escobar, *Designs for the Pluriverse*; Ezio Manzini, *Design, When Everybody Designs* (Cambridge: The MIT Press, 2015). Cosmopolitan localism aspires to bring the means of production and consumption closer together, informed and supported by distributed systems that link the local and the global, with the potential to cultivate a new sense of place.

18 Frederico Demaria, Francois Schneider, Filka Sekulova, Joan Martinez-Alier, "What is degrowth? From an activist slogan to a social movement," *Environmental Values*, 22 no. 2 (2013): 191–215, https://dx.doi.org/10.2307/23460978.

19 Abhijit Banerjee and Esther Duflo, *Good Economics for Hard Times: Better Answers to Our Biggest Problems* (New York: PublicAffairs, 2019), 226.

20 John Cassidy, "Can We Have Prosperity without Growth?" *The New Yorker*, February 10, 2020, https://www.newyorker.com/magazine/2020/02/10/can-we-have-prosperity-without-growth.

21 As another winner of the Nobel Prize in Economics (2009), Elinor Ostrom, has shown. See Elinor Ostrom, *Governing the Commons: The Evolution of Institutions for Collective Action* (Cambridge: Cambridge University Press, 1990); Basudeb Guha-Khasnobis, Ravi Kanbur, and Elinor Ostrom, eds., *Linking the Formal and Informal Economy: Concepts and Policies* (Oxford: Oxford University Press, 2006).

22 Escobar, *Designs for the Pluriverse*.

23 "Degrowth Info," Degrowth, accessed August 30, 2020, https://www.degrowth.info/en/dim/degrowth-in-movements/buen-vivir/

24 Escobar, *Designs for the Pluriverse*, 148.

25 The pink tide was a progressive wave of left-leaning democracies coming to power in Latin America, solidifying in the early 2000s but often cited as beginning with Hugo Chavez's election in 1998. Generally speaking, these left-wing and center-left governments represented a shift away from the reigning neoliberal economic model in engaging a legacy of widespread inequality and exploitation of natural resources by the Global North.

26 Republic of Ecuador, Constitution of 2008, title 2, ch.7, https://pdba.georgetown.edu/Constitutions/Ecuador/english08.html.

27 Plurinational State of Bolivia, *Law of Mother Earth and Integral Development for Living Well, Act No. 300*, introduced October 15, 2012, https://www.lexivox.org/norms/BO-L-N300.xhtml?dchttps://www.lexivox.org/norms/BO-L-N300.xhtml?dcmi_identifier=BO-L-N300&format=xhtmlmi_identifier=BO-L-N300&format=xhtml.

28 United Nations General Assembly, Seventy-first session, "Harmony with Nature: Note by the Secretary-General," A/71/266, August 1, 2016, https://www.un.org/ga/search/view_doc.asp?symbol=A/71/266.

29 "World Commission on Environmental Law," International Union for Conservation of Nature, accessed August 16, 2020, https://www.iucn.org/commissions/world-commission-environmental-law; World Commission on Environmental Law, *IUCN World Declaration on the*

Environmental Rule of Law (Rio de Janeiro: IUCN 1ˢᵗ World Congress on Environmental Law, 2016).

30 "What is an International Rights of Nature Tribunal?" Global Alliance for the Rights of Nature, accessed August 4, 2020, https://therightsofnature.org/rights-of-nature-tribunal/.

31 Kate Raworth, *Doughnut Economics: Seven Ways to Think Like a 21st Century Economist* (White River Junction, VT: Chelsea Green Publishing, 2017): 281.

32 Jane Bennett, *Vibrant Matter: A Political Ecology of Things* (Durham: Duke University Press, 2010), xii.

33 Donna Haraway, "Anthropocene, Capitalocene, Plantationocene, Chthulucene: Making Kin," *Environmental Humanities* 6 (2015): 159-165, https://environmentalhumanities.org/arch/vol6/6.7.pdf. In response to the Anthropocene, as driven by humans' accumulated waste, and the Capitalocene, as driven by biological exterminism at the hands of the global wealthy, Chthulucene is offered as denoting a past and future intimately of the earth. It is derived from *chthon*, a term meaning "earth" in Greek and associated with that that dwells in or under the earth.

34 Sarah Schindler, "Architectural Exclusion: Discrimination and Segregation Through Physical Design of the Built Environment," *The Yale Law Journal* 124, no. 6 (April 2015): 1836–2201, https://www.yalelawjournal.org/article/architectural-exclusion.

35 Lauren Ro, "Meet SCAPE, the Architects Designing Urban Landscapes to Stand the Test of Time," *Curbed*, November 15, 2017, https://www.curbed.com/2017/11/15/16644270/scape-landscape-architecture-kate-orff-groundbreakers-2017; "What Does it Mean to Engage in Activism through Design? To engage in design through activism?" The McHarg Center, accessed September 2, 2020, https://mcharg.upenn.edu/conversations/what-does-it-mean-engage-activism-through-design-engage-design-through-activism. "In the aftermath of Donald Trump's ascension to the Presidency, activism is once again en vogue for designers. Everywhere we look, projects and firms are rebranding themselves as 'activist practices,' replacing or augmenting their 'sustainability' and 'resilience' in the pursuit of new clients. Like so many other new and rediscovered trends in design, we often see activism being treated as a means to an end— an instrument for virtue signaling and business development, not for the kind of community and electoral organizing that others might define as activism."

36 Rob Holmes, "The Problem with Solutions," *Places Journal* (July 2020): https://doi.org/10.22269/200714. I credit Rob Holmes, author within this collection, with the timely reminder of this term, a term with a robust history of discussion in the philosophy of science. Additionally, his examples of fieldwork, synthetic cartography, exploratory scenarios, landscape modeling, collaborative processes, and designed maintenance were particularly informative in pointing to precedents of potential new materialist practice in the Global North as can be seen in the following paragraph.

37 Nondesigners being those not professionally trained in design. Like Ezio Manzini, I believe everyone is necessarily a designer in how they go about their daily life, a designer of their life project so to speak.

38 Alice Foxley and Günther Vogt, *Distance and Engagement: Walking, Thinking, and Making Landscape* (Zurich: Lars Müller Publishers, 2010), 22.

39   "Public Sediment," Multiplier/Bay Area Resilient by Design, accessed September 18, 2020, http://www.resilientbayarea.org/public-sediment#:~:text=PUBLIC%20SEDIMENT%20is%20a%20multidisciplinary,resilience%20in%20San%20Francisco%20Bay.&text=Yet%20the%20Bay%20Area's%20ecological,coastal%20edges%2D%20are%20at%20risk.

40   "Sand Motor – building with nature solution to improve coastal protection along Delfland coast (the Netherlands)," Climate Adapt, last modified February 15, 2019, accessed September 3, 2020, https://climate-adapt.eea.europa.eu/metadata/case-studies/sand-motor-2013-building-with-nature-solution-to-improve-coastal-protection-along-delfland-coast-the-netherlands.

41   Robert Sullivan, "How the World's Largest Gargabe Dump Evolved into a Green Oasis," *The New York Times*, August 14, 2020, https://www.nytimes.com/2020/08/14/nyregion/freshkills-garbage-dump-nyc.html; Irina Vinnitskaya, "Landfill Reclamation: Fresh Kills Park Develops as a Natural Coastal Buffer and Parkland for Staten Island," *ArchDaily*, March 3, 2013, https://www.archdaily.com/339133/landfill-reclamation-fresh-kills-park-develops-as-a-natural-coastal-buffer-and-parkland-for-staten-island.

42   Andrew Wright, "Beyond Mow and Blow: New Approaches to Park Maintenance," *The Dirt*, October 20, 2018, https://dirt.asla.org/2018/10/20/moving-public-park-maintenance-beyond-mow-and-blow/; "Native Meadow,"Nelson Byrd Woltz, accessed October 14, 2020, https://www.nbwla.com/projects/garden/native-meadow.

43   Escobar, *Designs for the Pluriverse*, 184.

44   Ibid., 184.

45   Escobar, *Designs for the Pluriverse*, 186; Arturo Escobar, *Territories of Difference: Place, Movements, Life, Redes* (Durham, NC: Duke University Press, 2008).

46   The term was first coined by Chilean biologist-philosophers Humberto Maturana and Francisco Varela in 1972 referring to the biochemistry of living cells. Humberto R. Maturana, *Biology of Cognition: Biological Computer Laboratory Research Report BCL 9.0* (Urbana, IL: University of Illinois, 1970), Reprint, *Autopoiesis and Cognition* (Dordecht: Reidel Publishing Co., 1980); Humberto R. Maturana, and Francisco J. Varela, *The Tree of Knowledge: The Biological Roots of Human Understanding* (Boston: Shambhala Publications, 1998).

47   Bruno Latour, *Pandora's Hope* (Cambridge: Harvard University Press, 1999), 297.

48   Susan Barba, *Geode* (Boston: Black Sparrow Press, 2020), 51.

49   As discussed in the Introduction, I build on philosopher Levi Bryant's geo-philosophy and onto-cartography. *See* Levi Bryant, *Onto-Cartography: An Ontology of Machines and Media*, Speculative Realism (Edinburgh: Edinburgh University Press, 2014).

BIBLIOGRAPHY

Alaimo, Stacy. *Exposed: Environmental Politics and Pleasures in Posthuman Times*. Minneapolis: University of Minnesota Press, 1993.

Ambrose, Stephen. "Man vs. Nature: The Great Mississippi Flood of 1927." *National Geographic*. May 1, 2001. https://www.nationalgeographic.com/culture/2001/05/mississippi-river-flood-culture/.

Banerjee, Abhijit and Esther Duflo. *Good Economics for Hard Times: Better Answers to Our Biggest Problems*. New York: PublicAffairs, 2019.

Barba, Susan. *Geode*. Boston: Black Sparrow Press, 2020.

Bennett, Jane. *Vibrant Matter: A Political Ecology of Things*. Durham: Duke University Press, 2010.

Berry, Thomas. *The Dream of the Earth*. San Francisco: Sierra Club Books, 1990.

Brown, Hugh Auchincloss. Correspondence to the Mississippi River Commission. June 27, 1952. New Orleans District Archives, US Army Corps of Engineers.

Bryant, Levi. *Onto-Cartography: An Ontology of Machines and Media*. Edinburgh: Edinburgh University Press, 2014.

De la Bellacasa, María Puig. *Matters of Care: Speculative Ethics in More than Human Worlds*. Minneapolis: University of Minnesota Press, 2017.

"Degrowth Info." Degrowth. Accessed August 30, 2020. https://www.degrowth.info/en/dim/degrowth-in-movements/buen-vivir/.

Demaria, Frederico, Francois Schneider, Filka Sekulova, and Joan Martinez-Alier. "What is degrowth? From an activist slogan to a social movement." *Environmental Values*, 22 no. 2 (2013): 191-215. https://dx.doi.org/10.2307/23460978.

Escobar, Arturo. *Designs for the Pluriverse: Radical Interdependence, Autonomy, and the Making of Worlds*. Durham: Duke University Press, 2018.

Escobar, Arturo. *Territories of Difference: Place, Movements, Life, Redes*. Durham, NC: Duke University Press, 2008.

Foxley, Alice and Günther Vogt. *Distance and Engagement: Walking, Thinking, and Making Landscape*. Zurich: Lars Müller Publishers, 2010.

Gómez-Barris, Macarena. *The Extractive Zone: Social Ecologies and Decolonial Perspectives*. Durham: Duke University Press, 2017.

Guha-Khasnobis, Basudeb, Ravi Kanbur, and Elinor Ostrom, eds. *Linking the Formal and Informal Economy: Concepts and Policies*. Oxford: Oxford University Press, 2006.

Haraway, Donna. "Anthropocene, Capitalocene, Plantationocene, Chthulucene: Making Kin." *Environmental Humanities* 6 (2015): 159–165. https://environmentalhumanities.org/arch/vol6/6.7.pdf .

Holmes, Rob. "The Problem with Solutions." *Places Journal* (July 2020): https://doi.org/10.22269/200714.

Latour, Bruno. *Pandora's Hope*. Cambridge: Harvard University Press, 1999.

Manzini, Ezio. *Design, When Everybody Designs*. Cambridge: The MIT Press, 2015.

Maturana, Humberto R. *Biology of Cognition: Biological Computer Laboratory Research Report BCL 9.0*. Urbana, IL: University of Illinois, 1970. Reprint. *Autopoiesis and Cognition*. Dordecht: Reidel Publishing Co., 1980.

Maturana, Humberto R., and Francisco J. Varela. *The Tree of Knowledge: The Biological Roots of Human Understanding*. Boston: Shambhala Publications, 1998.

"The Mississippi River Delta Basin." The Louisiana Coastal Wetlands Planning Protection and Restoration Act Program. Accessed May, 27, 2020. https://lacoast.gov/new/About/Basin_data/mr/#gsc.tab=0.

"Native Meadow." Nelson Byrd Woltz. Accessed October 14, 2020. https://www.nbwla.com/projects/garden/native-meadow.

Ostrom, Elinor. *Governing the Commons: The Evolution of Institutions for Collective Action.* Cambridge: Cambridge University Press, 1990.

Plurinational State of Bolivia. *Law of Mother Earth and Integral Development for Living Well, Act No. 300.* Introduced October 15, 2012. https://www.lexivox.org/norms/BO-L-N300.xhtml?dchttps://www.lexivox.org/norms/BO-L-N300.xhtml?dcmi_identifier=BO-L-N300&format=xhtml-mi_identifier=BO-L-N300&format=xhtml.

"Public Sediment." Multiplier/Bay Area Resilient by Design. Accessed September 18, 2020. http://www.resilientbayarea.org/public-sediment#:~:text=PUBLIC%20SEDIMENT%20is%20a%20multidisciplinary,resilience%20in%20San%20Francisco%20Bay.&text=Yet%20the%20Bay%20Area's%20ecological,coastal%20edges%2D%20are%20at%20risk.

Raworth, Kate. *Doughnut Economics: Seven Ways to Think Like a 21st Century Economist.* White River Junction, VT: Chelsea Green Publishing, 2017.

Republic of Ecuador, Constitution of 2008, title 2, ch.7. https://pdba.georgetown.edu/Constitutions/Ecuador/english08.html.

Ro, Lauren. "Meet SCAPE, the Architects Designing Urban Landscapes to Stand the Test of Time." *Curbed.* November 15, 2017. https://www.curbed.com/2017/11/15/16644270/scape-landscape-architecture-kate-orff-groundbreakers-2017.

"Sand Motor – Building with Nature Solution to Improve Coastal Protection along Delfland Coast (the Netherlands)." Climate Adapt. Last modified February 15, 2019. Accessed September 3, 2020. https://climate-adapt.eea.europa.eu/metadata/case-studies/sand-motor-2013-building-with-nature-solution-to-improve-coastal-protection-along-delfland-coast-the-netherlands.

Schindler, Sarah. "Architectural Exclusion: Discrimination and Segregation through Physical Design of the Built Environment." *The Yale Law Journal* 124, no. 6 (April 2015): 1836 – 2201. https://www.yalelawjournal.org/article/architectural-exclusion.

Smil, Vaclav. *Growth: From Microorganisms to Megacities.* Cambridge: MIT Press, 2020.

Smith, James A. and Mary Lynn Baeck. "'Prophetic vision, vivid imagination': The 1927 Mississippi River flood." *Water Resources Research* 51, no. 12 (November 30, 2015). https://doi.org/10.1002/2015WR017927.

Suchman, Lucy. *Human-Machine Reconfigurations: Plans and Situated Actions.* Cambridge: Cambridge University Press, 2006.

Sullivan, Robert. "How the World's Largest Gargabe Dump Evolved into a Green Oasis." *The New York Times.* August 14, 2020. https://www.nytimes.com/2020/08/14/nyregion/freshkills-garbage-dump-nyc.html.

United Nations General Assembly, Seventy-first session. "Harmony with Nature: Note by the Secretary-General." A/71/266. August 1, 2016. https://www.un.org/ga/search/view_doc.asp?symbol=A/71/266.

Vinnitskaya, Irina. "Landfill Reclamation: Fresh Kills Park Develops as a natural Coastal Buffer and Parkland for Staten Island." *ArchDaily.* March 3, 2013. https://www.archdaily.com/339133/landfill-reclamation-fresh-kills-park-develops-as-a-natural-coastal-buffer-and-parkland-for-staten-island.

"What Does it Mean to Engage in Activism through Design? To Engage in Design through Activism?" The McHarg Center. Accessed September 2, 2020. https://mcharg.upenn.edu/conversations/what-does-it-mean-engage-activism-through-design-engage-design-through-activism.

"What is an International Rights of Nature Tribunal?" Global Alliance for the Rights of Nature. Accessed August 4, 2020. https://therightsofnature.org/rights-of-nature-tribunal/.

"World Commission on Environmental Law." International Union for Conservation of Nature. Accessed August 16, 2020. https://www.iucn.org/commissions/world-commission-environmental-law.

World Commission on Environmental Law. *IUCN World Declaration on the Environmental Rule of Law*. Rio de Janeiro: IUCN 1st World Congress on Environmental Law, 2016.

Wright, Andrew. "Beyond Mow and Blow: New Approaches to Park Maintenance." *The Dirt*. October 20, 2018. https://dirt.asla.org/2018/10/20/moving-public-park-maintenance-beyond-mow-and-blow/.

IMAGE CITATIONS + CREDITS

Listed by Page Number

325   Photo: Matthew Seibert

327   Basemap sources: Esri, DigitalGlobe, GeoEye, i-cubed, USDA FSA, USGS, AEX, Getmapping, Aerogrid, IGN, IGP, swisstopo, and the GIS User Community; Copyright: © 2014 National Geographic Society, i-cubed

       Research Assistant: Aleksander De Mott

329   Photo: Matthew Seibert

335   Basemap sources: Esri, DigitalGlobe, GeoEye, i-cubed, USDA FSA, USGS, AEX, Getmapping, Aerogrid, IGN, IGP, swisstopo, and the GIS User Community

by Thomas Woltz

My hope in contributing an afterword to this book is to offer reflections on how the tools of onto-cartography and the new materialist sensibility described in *Atlas of Material Worlds* might be embedded from design theory, through design practice, and into the built landscapes that surround us. When the editor invited me to contribute some closing words, it was requested that I take the voice of one who designs and oversees construction of landscapes at a range of scales and remains involved decades after they are set in motion; in other words, a maker. The preceding essays of common and uncommon substances have opened my eyes to the power of material agency and offered a new understanding of the rich continuum of the living and non-living suggested by Seibert's "familial world."

We have read how, if we commit to it, the shared ethos of a new materialism, "seeing our non-living partners as our material kin," could deflate individualist ego, improve global environmental balance and dramatically alter political policy. This reorientation of our human sense of self in designing the material world could lead to a deep humility in reimagining the designer's role in making the world around us. Nothing short of a design revolution. So now the task is to imagine committing to this ethos firmly enough in the design and making of landscapes that it shapes the construction process and endures in the built environment.

I have introduced myself as a "maker," as mentioned, at the request of the editor, but that bears some clarification, given that as a design practice we are not literally constructing these places. So much of the credit for realizing these landscapes is due to the earthmovers, horticulturists, artisans, masons, and maintenance personnel. I am trained as a landscape architect and have, alongside my colleagues in the firm, led the design and performed construction administration for a broad range of landscapes over the past two decades. We have had commissions in 12 countries and 25 of the 50 States and have been fortunate to see a high percentage of these projects come to fruition. Thanks in great part to the influence of Warren Byrd, founding partner of the firm, a design process that continues to evolve is grounded in deep research, both ecological and cultural, that shapes the essence of every landscape we design.

With full-time positions at Nelson Byrd Woltz Landscape Architects for a cultural landscape historian and a conservation biologist, the design teams begin research alongside these specialists, to uncover the deep narratives that have shaped a particular site. Beginning with the pre-historic, geologic activity that brought the site into being and evaluating its current ecological state, we then seek the cultural responses that in turn reshaped that ecosystem. An ecological phenomenon attracts a human intervention that, in turn, reshapes the very thing that attracted them, and on it goes. This dance whirls back and forth over time blurring into a fascinating continuum of enmeshed culture and ecology. This empathic, research-based design process for imagining contemporary landscapes is one that grows from respect and reverence for the interconnectedness we observe in the world around us and seemed like a fairly thorough investigative process... until now.

The call for an increased "relationality" laid out in *Atlas of Material Worlds* has opened my eyes to a new level of interconnectedness of material, place, and culture that alters the very lens through which I see my personal practice and my firm's design process. Reading each chapter, as a designer already deeply committed to interrogating history and building narrative, I felt my anthropocentric preconceptions challenged and destabilized in healthy ways that expanded my thinking toward new pathways of research and understanding of the material world. In our firm's cultural landscape research, agendas of race, Indigeneity, injustice, abuse, and political structures that have been intentionally hidden, are revealed and examined leading to a design response. In our firm's biological research, the invisibility and legacy of damaged ecosystems is revealed, inspiring design strategies for healing and rebuilding fragile and compromised landscapes. Now a third vein of deep research into material agency and interrelatedness takes on a new urgency for me. Committing to understanding, for example, the formation and origin of sands we specify, the externalities of lithium extraction for "green" technologies, the life force of crude embedded in every product we assemble can inculcate a new habit of interrogation beyond the materials outlined in this book. In order for it to truly embed in our practice as makers of landscapes, we have to find the point at which we fuse the perception of ourselves to these materials and discover the shared continuum… the "familial world."

This potential for the designer, in my case human, to participate in the material continuum was brought home personally to me in Elizabeth Hénaff's essay, *Metabolite*. "This microbial continuum fills the porous spaces between material and human, nature and culture, at telescoping scales." You see, I was one of the intrepid landscape designers working with her, clad in a Tyvek suit, rubber gloves, boots, goggles and mask using the unwieldy PVC straw, sucking up and bottling samples of Gowanus Mayonnaise for her to take to the laboratory for analysis. Would there be life forms metabolizing in this toxic stew that would be lost once the life there is sealed under a concrete floor? Despite my participation in and support of the project, I thought of the extremophiles revealed in the Gowanus stew unempathetically as "out there" doing their thing. Her call for "relational inclusion" of microbes into the design process and her coda recapping the role of microbes in each of the materials described in this book, shook me into a new awareness of kinship. With kinship comes stewardship, with awareness comes care.

This work will be hard. But this work is already hard. Unwavering diligence is required to steward the theoretical underpinning of a designed landscape through zoning approval, building code compliance, conflicting public opinion, emerging donor opinion, then into cost estimating, construction budgeting, inevitable value engineering, quality control during construction, and finally assuring sufficient care and maintenance of the new landscape. Each step is fraught with potential dilution or even total loss of the essential ideas of the design developed in the research process. But I take heart that a commitment to a new materialist practice could offer powerful antidotes to the threats outlined above. Over the years I have found that the narratives of culture and ecology embedded in our design work have been the very spark that draws community support, construction funding, and a commitment to maintenance. I now see that revealing the deep relatedness between materials and humans, the continuum of living and non-living, would further elevate the public awareness of the landscape as an *integrated commons of belonging.*

It is admittedly less convenient when inanimate objects have agency, when externalities of materials and their extraction are deeply considered and when familiar tropes of consumer and consumed are brought into a respectful, interdependent partnership of shared resources. If we design the world with these sensibilities, it becomes harder to plunder, to hate, to damage, to steal, to destroy, because we ultimately are performing these actions upon ourselves. We know too much, or rather, we finally know enough. When there is no distance, no empty space between the minerals, molecules, and microbes that shape us and our world, we enter a new relationship of accountability and stewardship. A new humility and ethos emerge and we take a substantial step closer to Bruno Latour's "good life."

Stacy Alaimo is Professor of English and Core Faculty Member in Environmental Studies at the University of Oregon. Her books include *Undomesticated Ground: Recasting Nature as Feminist Space* (2000); *Material Feminisms* (co-edited, 2008); *Bodily Natures: Science, Environment, and the Material Self* (2010); and *Exposed: Environmental Politics and Pleasures in Posthuman Times* (2016).

Denise Hoffman Brandt is a professor of landscape architecture at the City College of New York, and principal of Hoffman Brandt Projects in New York City.

Kristi Cheramie, FAAR, is Associate Professor and Head of Landscape Architecture at the Ohio State University's Knowlton School. Her research explores the ways we respond to and cope with environmental fluctuation. Her first book, *Through Time and the City: Notes on Rome*, examines historical notions of environmentalism and perceptions of flooding, climate exigencies, and debris.

Brian Davis is Associate Professor of Landscape Architecture at the University of Virginia. He is a professional landscape architect, a member of the Dredge Research Collaborative, a Fellow of the American Academy in Rome, and the co-founder of Proof Projects. He works on muddy places as cultural landscapes.

Elizabeth Hénaff is a computational biologist with an art practice. She holds an Assistant Professor position at the NYU Tandon School of Engineering, where she teaches biodesign. She leads the Laboratory for Living Interfaces, studying the interaction of organisms and their environment through scientific and design enquiries.

Rob Holmes teaches landscape architecture at Auburn University. He is also co-founder of the Dredge Research Collaborative, an independent nonprofit organization that seeks to improve sediment systems through design research, building public knowledge, and facilitating transdisciplinary conversation.

Ian Quate and **Colleen Tuite** co-direct Other Fields, an activist-directed, experimental landscape design studio based in Asheville, NC and New York City.

Matthew Seibert is an Assistant Professor of Landscape Architecture at the University of Virginia and co-founder of Landscape Metrics, a visualization studio specializing in science communication. Beyond his present studies in the agency of nonliving materials, his work employs representation as interrogative and speculative tools within historical trajectories, crafting rich parafictions as both critique and potential future.

Thomas Woltz is the owner and principal of Nelson Byrd Woltz Landscape Architects, a 50-person firm with offices in New York City; Charlottesville, VA and Houston, TX. Woltz holds masters degrees in architecture and landscape architecture and was named the Design Innovator of the Year by the Wall Street Journal Magazine in 2013.